Physics History from AAPT Journals

Edited by
Melba Newell Phillips

published by
American Association of Physics Teachers

Physics History from AAPT Journals

ISBN # 0-917853-14-8

For order information, write:

American Association of Physics Teachers
Publications Department
5110 Roanoke Place, Suite 101
College Park, MD 20740, U.S.A.

Contents

III. Historical Aspects of Physics Education

Physics History from AAPT Journals

Introduction

A book of reprints of historical articles from its journals was considered by the Committee on the Fiftieth Anniversary of the American Association of Physics Teachers (AAPT) during 1979 but not given priority at that time. Instead, the Association published *On Teaching Physics*, with authors whose names appear in the history of AAPT as officers or as recipients of honors. The ideas expressed were usually those that seemed particularly important for the furtherance of physics teaching at various times, and so were related to the history of the organization itself. Even with fairly strict principles of omission it was not possible to publish more than a sample of articles that met the criteria.

Although several pieces of physics history got into *On Teaching Physics*, e.g., that on Thomas Young by Robert W. Pohl, reminiscences by E.H. Hall about teaching physics at Harvard a century ago now, and recollections of Paul Ehrenfest by George Uhlenbeck, there is again an embarrassment of riches along with space limitations. To get an idea of the abundance from which the present choice has been made one has only to look at the cumulative indexes for the *American Journal of Physics* and that for *The Physics Teacher* under the heading "History and Biography." This book consists of only twenty-seven articles—twenty from *AJP* and seven from *TPT*—a ratio roughly the same as the age ratio of the two journals in 1983 when the selection was made, although the coincidence is completely fortuitous. Their order is determined by the subject matter, and the evident differences in format should not prove disturbing. It has been the policy of *TPT* to give more detailed information on the authors than does *AJP*, but no differentiation is made in the brief author identifications appended at the end of this volume.

It must be emphasized that although the selection includes articles by several historians—Lawrence Badash, Marjorie Malley, Albert Moyer, Sidney Rosen, Barbara Swartz—this book, like the journals themselves, is intended for the use of physics teachers, not professional historians of science. The physics teachers for whom the articles were written are interested in history and the people who made it, but in the main they have neither access to the original papers and voluminous libraries on which professional historians depend nor time to read volumes instead of short articles. Moreover, whether the teacher is explicitly interested or not, history is implicitly woven into the fabric of physics and its pedagogy. It is well known that myths are often perpetuated unwittingly. This was pointed out long ago by Florian Cajori in "The Pedagogic Value of the History of Physics" (School Review, May 1899); he includes warning against the misuse of history, as do contributors to a symposium on "Use of Historical Material in Elementary and Advanced Instruction" (Am. J. Phys. 18, 332, 1950), and "Errors in the History of Science," a letter to the editor also in Volume 18 of the journal. The same message has been given in two international conferences on the role of history of physics in physics education: See *Proceedings of the International Working Seminar on the Role of History of Physics Education*, S. Brush and A. King, eds. (University Press of New England, Hanover NH, 1972), and *Proceedings of the International Conference on Using History in Innovatory Physics Education*, F. Bevilacqua and P.J. Kennedy, eds., (Centro Studi per Didattica, Della Facolta di Scienze Mathematiche, Fische e Naturali-Universita di Pavia, Pavia, Italy 1983). The primary purpose of most historical articles is, of course, to supply information, not to correct misinformation, but many function in both ways.

Some of the omissions for the present collection require special comment. One is the series of four papers that were presented in a symposium on "History of the Atom" at the annual meeting of the APS and AAPT in January 1980. (Am. J. Phys., 49, 205, 1981). They are sufficiently recent to be readily accessible, rather lengthy as a whole, and somewhat more formal than the other articles. We must be content to recommend but not reprint them. Important papers from outside sources have been published in our journals from time to time. In general these are not included here, but there is an exception, a lecture delivered by Heinrich Hertz in 1899, which appears as part of an article by E.C. Watson. It is impossible to include several excellent articles celebrating anniversaries of important discoveries, e.g., "The Birth and Early Infancy of X-rays" (Am. J. Phys. 13, 382, 1945), and "The Early Years of Radioactivity" (Am. J. Phys. 14, 226, 1946). "The Jubilee of the Electron" by E.C. Watson (Am. J. Phys. 15, 458, 1947), which incorporates a lecture by J.J. Thompson, is also omitted. Special mention should be made of Van Vleck's Richtmyer lecture, "Landmarks in the Theory of Magnetism" (Am. J. Phys. 18, 495, 1950). But the list of omissions could be extended almost indefinitely.

For this book we start with a short collection on classical physics, chosen in part because most of the pieces involve or touch upon early contributions of American physicists. An exception is the article on Oersted, who has been adopted in a special way by AAPT. By his own account we find that Oersted's greatest discovery was not accidental, contrary to the familiar story recounted by a former student. A lecture was indeed involved in his first observation of electromagnetism, but "the experiment made no strong impression on the audience." It was some three months later that he found it convenient to resume his researches, then to formulate the law that "the magnetical effect of the electrical circuit has a circular motion around it." Most of Oersted's attention was devoted to teaching, lecturing, and textbook writing. We shall not attempt notes on other individual articles in this section, but should point out that although Count Rumford was only one of Sandy Brown's many interests it was an abiding one, and his full-length biography of Rumford appeared in 1979.

For the next section we are very fortunate in having available so many reminiscences from people who participated in the great advances that took place in the 1920's—Chandrasekhar, A. H. Compton, Condon, Dennison, Heisenberg, and G. P. Thomson. Historians remind us that memories are not entirely reliable, but they do convey the "feel" of an era in a way that secondary accounts cannot, however accurate they may be. We are reminded, for instance, that when Condon generalized in his 1926 PhD thesis what has become known as the Franck-Condon Principle, there was no uncertainty principle. Good results were obtained nevertheless because the masses of nuclei involved during an electronic transition in a diatomic molecule are relatively large. (The quantum mechanical treatment, also due to Condon, came later.) Also in 1926 came Dirac's derivation of what came to be called Fermi-Dirac statistics, and Chandrasekhar notes that its first application to a new field was the resolution (by R.H. Fowler) of Eddington's paradox that "The stars will need energy in order to cool." Even earlier, in 1926, Heisenberg recalled that "...I did not know what a matrix was and did not know the rules of matrix multiplication. But one could learn these operations from physics, and later it turned out that it was matrix multiplication, well known to the mathematicians." As to the discovery of the positron, even the encyclopedias tell us its existence was prediced by Dirac and confirmed by Anderson, but Carl Anderson himself assures us the discovery was wholly accidental. The reader is apt to find other surprises in the articles of this section.

Our last section is devoted to the history of physics education, a subject much neglected by both physicists and historians. The piece on Oersted would have been equally appropriate here, even though he is much better known for his discovery of electromagnetism than as a pedagog. Joseph Henry and Robert A. Millikan are also remembered primarily for scientific discoveries but the two articles from *TPT* emphasize their contributions to physics education. In this section, as in the others, the articles are arranged in roughly chronological order of their subject matter, although it is impossible to set up a rigorous time sequence for the development of student laboratories in high schools and colleges. College-level laboratory work for students (in distinction to demonstrations) was actually introduced a little earlier than that at the secondary level—the role of MIT in this innovation is sometimes overlooked—but did not become the usual practice at established colleges and universities for another twenty years or so. There is considerable overlap between the two pieces on the introduction and development of high school laboratories, but both have unique features. Especially interesting are the critical comments on what tended to become overemphasis on laboratory at the expense of general principles. As for more advanced work, it is well known that graduate schools and the research training of graduate students for all sciences arose in Germany, but the pioneering work in physics by Magnus in Berlin is not so familiar. The graduate school tradition is related to the professionalization of physics, and was introduced into the United States with the founding of Johns Hopkins University in 1876. As a final article in this section we include an illuminating commentary on the teaching of science and of history.

All articles in this volume contain references for further reading, but for convenience and more complete historical treatment of the subject matter some suggestions are in order. Those interested in Oersted's original publication, for instance, would no doubt find the appendix to *Oersted and the Discovery of Electromagnetism* by Bern Dibner (Blaisdell Publishing Co., 1962) more accessible than the *Annals of Philosophy* where the English version first appeared. On other subjects an even more accessible book is *History of Physics, Readings from Physics Today Number Two,* published by the American Institute of Physics in 1985. Many subjects covered in our book are much the same as those in that selection, and the articles, like ours, are designed to appeal to a diverse audience. A short but carefully selected bibliography in the *Physics Today* book will be found useful. Two of the AAPT articles, "A Study of the Discovery of Fission" and "The Two Maps," have already been reprinted in the teachers' guide for *Moments of Discovery, Unit 1: The Discovery of Fission.* This is an audio-visual package prepared by Arthur Eisenkraft and the Center for History of Physics (principally Joan Warnow) for use in the schools. It is marketed by the American Institute of Physics.

Further references are needed for articles that appeared in AAPT journals some time ago. There is now, for example, *Radioactivity in America* by Lawrence Badash (Johns Hopkins Press, 1979), as well as *The Compton Effect* by Roger H. Stuewer (Science History Publications, 1975). Carl Anderson prepared an extension of his article in our book for *The Birth of Particle Physics* edited by Laurie Brown and Lillian Hoddeson (Cambridge University Press, 1983). Some of the other articles in the Brown-Hoddeson collection will also be of interest. More diverse sources include the journal *Historical Studies in the Physical Sciences,* in which a number of related articles have appeared for the first time. In Volume 12 (1981), for instance, there is one on the transistor by Lillian Hoddeson and one on the discovery of electron diffraction by Arturo Russo. Most encyclopedias have articles on scientists of historical importance, but the most reliable source is the *Dictionary of Scientific Biography,* a series of volumes prepared by experts. For more extensive study the best thorough bibliography on modern physics is surely *Literature on the History of Physics in the 20th Century* by John L. Heilbron and Bruce R. Wheaton (University of California, Berkeley, 1981). Teachers will be particularly interested in *Resources for the History of Physics,* Stephen G. Brush, editor (Hanover, New Hampshire: University Press of New England, 1972). This is a valuable bibliography of books, audio-visual materials, and primary historical works, including translations, of particular importance for teaching. Stephen Brush is currently preparing a Resource Letter with this same title, scheduled to appear in 1986, which will bring this Resource up to date and thus make it even more valuable.

BENJAMIN FRANKLIN:

"Let the Experiment be made."

Albert E. Moyer

Benjamin Franklin helped kindle two revolutions, one political and the other scientific. One revolution, in the period around 1776, molded thirteen British colonies into a consolidated and self-governing nation. The other revolution, in the decades around 1750, transformed a jumble of inconclusive electrical studies into a unified and established science. The political revolution was fueled by the Declaration of Independence, which Franklin helped draft. The electrical revolution was quickened by a widely-read collection of Franklin's scientific letters titled *Experiments and Observations on Electricity* — a collection he is shown holding in the engraving of Fig. 1.

Although in his later years Franklin deemed politics and public service to be duties that took precedence over the personal joys of science, he always retained a passion for physical science, or as it was known, "natural philosophy." Thus, soon after the War of Independence, Franklin wrote the following note to the president of the leading English scientific organization, the Royal Society of London: "Be assured that I long earnestly for a Return of those peaceful Times, when I could sit down in sweet Society with my English philosophic Friends, communicating to each other new Discoveries, and proposing Improvements of old ones. . .Much more happy should I be thus employ'd in your most desirable Company, than in that of all the Grandees of the Earth. . . ."

Franklin's political achievements while in the company of the "Grandees of the Earth" in London, Paris, and Philadelphia are well-known. But what specifically were the "new Discoveries" that he nostalgically recalled communicating to his "English philosophic Friends"? In other words, for what specific reasons does Franklin merit the designation of revolutionary in science as well as in politics?

Enjoying a great Happiness, Leisure to read, study, make Experiments

Franklin's introduction to electricity was abrupt. In the mid-1740s, around age 40, he attended a public program of "electric experiments" given in Boston and then Philadelphia by a visiting lecturer "lately arrived from Scotland." Franklin afterward recalled that the lecturer's demonstrations "were imperfectly perform'd, as he was not very expert; but being on a Subject quite new to me, they equally surpriz'd and pleas'd me." Enthralled by sparks and shocks,

Albert E. Moyer *is presently attending the University of Wisconsin-Madison, where he is writing a doctoral dissertation on the history of American science. He has degrees in physics from Oberlin College and the University of Colorado. He also taught physics for three years at a junior college in eastern Kentucky. (Department of the History of Science, University of Wisconsin, Madison, Wisconsin 53706)*

Fig. 1. Benjamin Franklin was born in 1706 in Boston, where as a boy he assisted in his father's soap and candle business and served as an apprentice in his brother's printing shop. In 1723 he moved to Philadelphia, eventually distinguishing himself there as a printer, public servant, and scientist. Beginning in the late 1750s he represented the Colonies in London, returning to Philadelphia in 1775 to participate in the Second Continental Congress. Finally, he helped in Paris to negotiate the peace with Britain and then helped in Philadelphia to draft the Federal Constitution in 1787, three years before his death at age 84.

This 1761 engraving of Franklin by James McArdell was based on a prior oil painting by Benjamin Wilson. Wilson was one of England's leading electrical experimenters as well as a popular portrait painter. The book that Franklin holds is marked "Electric^{l} /Exp^{ts}"; on his right is the glass globe of a static electricity machine. (Engraving courtesy of the Franklin Collection, Yale University Library.)

Franklin immediately bought the lecturer's apparatus. Within a year or two he was carrying out his own electrical experiments, some of which by then involved a long glass tube that could be charged by rubbing with buckskin; he had received this new "Present of a Glass Tube, with some Account of the Use of it in making such experiments" from Peter Collinson, the British purchasing agent of the public library that Franklin had founded in Philadelphia. During these early years, Franklin may have rounded out his initiation into electricity by reading one of the increasingly common surveys of this popular subject, such as that appearing in 1745 in the *Gentleman's Magazine*, a London-based monthly with a broad circulation.

What might such a survey have included? Or to phrase the question more broadly, what was known about electricity in the mid-1740s when Franklin was beginning his researches?

Generally speaking, as a newcomer to electricity, Franklin would have encountered many curious phenomena and devices that had been discovered or invented since about 1700. Englishman Francis Hauksbee had helped initiate this 18th-century surge of empirical knowledge when his pioneering studies of spark phenomena led him to design a frictional electric generator (Fig. 2). Building on Hauksbee's successes, Stephen Gray went on both to discover electrical conduction and to describe electrostatic induction; he also distinguished between conductors and insulators. One of Gray's colleagues, Frenchman Charles Dufay, then identified two types of electrification, "vitreous" and "resinous," and demonstrated that bodies having like kinds repel while those having unlike kinds attract. Finally, in 1746, on the eve of Franklin's researches, Pieter Van Musschenbroek, of the University of Leiden, added to the building wave of knowledge with his discovery of the electrical condenser or "Leyden jar."

Franklin was therefore fortunate to begin his researches at a time when "electricians," as these scientists called themselves, were no longer constrained to study only miniscule and transient electrical effects produced by either rubbing amber, as the ancient Greeks had done, or rubbing glass tubes. Rather, electricians were now able to examine large-scale and sustained effects thanks to refined

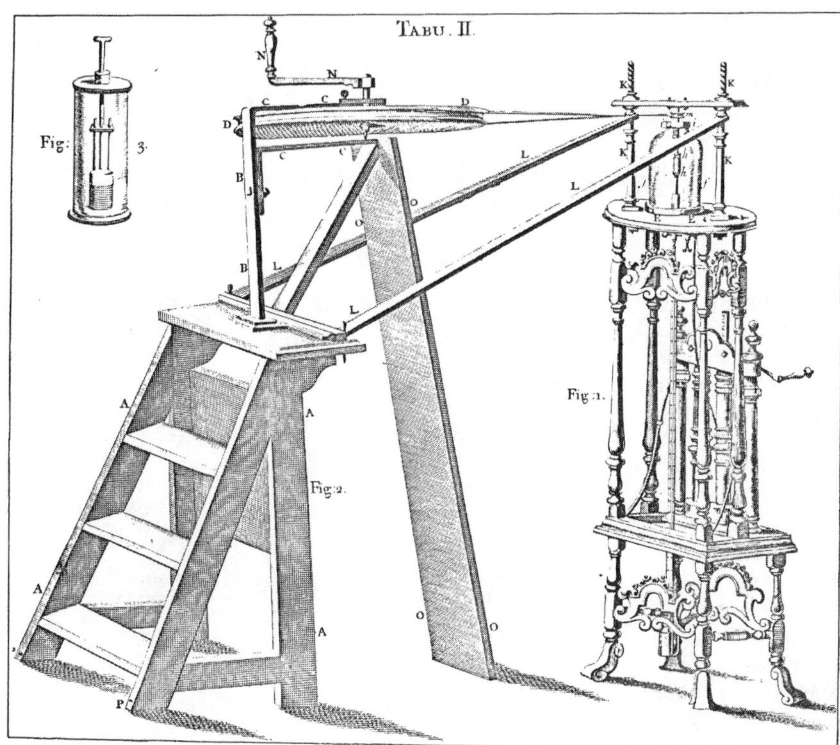

Fig. 2. In the early stages of his electrical research, Francis Hauksbee used hand-operated "machines" such as this to study the light given off when certain materials are rubbed together in a vacuum. In the machine shown, an air pump is on the right, topped off with an evacuated jar containing a spindle that could be rotated by the wheel and belt on the left. Flashes of light emanated from within the jar when, for example, Hauksbee placed amber on the spindle and caused it to rub at high speeds against fixed woolen pads. Such machines eventually led him to construct a simple frictional electric generator. (From Hauksbee, *Physico-Mechanical Experiments on Various Subjects*, 2nd ed., London, 1719.)

electrostatic generators and Leyden jars. Franklin himself soon supplemented the original glass tube, which Collinson had sent him, with both "M. Musschenbroek's wonderful bottle" and a frictional "machine" (Fig. 3).

Franklin was also fortunate to be setting about his researches in a stimulating atmosphere of conceptual ferment. Through the 1740s electricians were often perplexed and usually in disagreement about the theory behind the many new electrical phenomena and devices. Hauksbee, for example, had explained the electrical effects of the frictional generator as an emission of an invisible, material "effluvium." In comparison, Gray was more noncommittal about the cause of conduction, referring vaguely to an "electric virtue." Meanwhile, Dufay's two types of electrification were conducive to a theory that involved two interacting electrical "fluids."

Finally, Franklin had the additional advantage of commencing his researches with a relatively open and unprejudiced mind. Although he soon became aware of and drew on the prevailing European, and particularly British, electrical outlooks, he was unencumbered and unrestrained by a prior commitment to any one of these outlooks. In this era of conceptual uncertainty and divergence, there were advantages to entering the field as a middle-aged, well-read gentleman with only a general, but firm, knowledge of post-Newtonian natural philosophy. And there were even more advantages in being a provincial American colonist living far from the scientific cliques of Paris, London, and Berlin. In the late 1750s, when

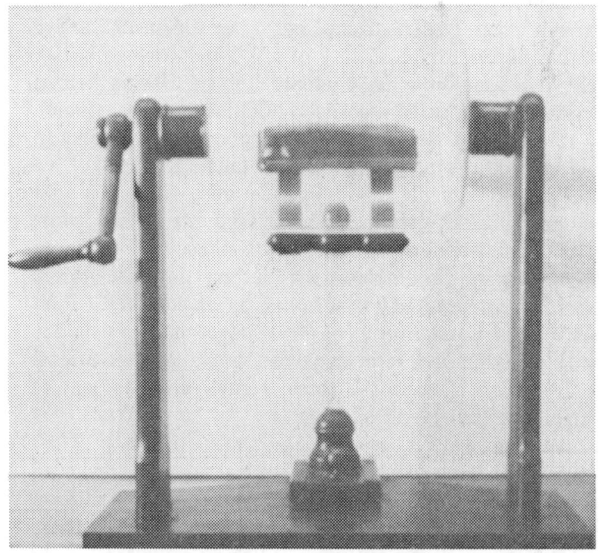

Fig. 3. Electrostatic generator presented to Yale College by Franklin soon after he received his honorary M.A. degree in 1753. While discussing such a generator in 1749, Franklin had explained that "the whirling glass globe, during its friction against the cushion, draws [electrical] fire from the cushion, the cushion is supplied from the frame of the machine, that from the floor on which it stands. Cut off the communication by thick glass or wax, placed under the cushion, and no fire can be *produced*, because it cannot be *collected.*" (Photo courtesy of the Franklin Collection, Yale University Library.)

Franklin, by then a respected scientific innovator, wrote to Musschenbroek in Leiden requesting current European literature on electricity, Musschenbroek perceptively replied: "I should wish . . .that you would go on making experiments entirely on your own initiative and thereby pursue a path entirely different from that of the Europeans, for then you shall certainly find many other things which have been hidden to natural philosophers throughout the space of centuries."

By 1747 Franklin had become deeply immersed in his electrical researches. "I never was before engaged in any study," he wrote to Peter Collinson in London, "that so totally engrossed my attention and my time as this has lately done. . . ." Luckily, this forty-year-old Philadelphian was currently free of both financial cares and public duties; thus he confided to an American scientific colleague: "I . . .am taking the proper Measures for obtaining Leisure to enjoy Life and my Friends more than heretofore, having put my Printing house under the care of my Partner David Hall, absolutely left off Bookselling, and remov'd to a more quiet Part of the Town, where I am settling my old Accounts and hope soon to be quite a Master of my own Time, and no longer (as the Song has it) *at every one's Call but my own* . . . .Thus you see I am in a fair Way of having no other Tasks than such as I shall like to give my self, and of enjoying what I look upon as a great Happiness, Leisure to read, study, make Experiments, and converse at large with such ingenious and worthy Men as are pleas'd to honour me with their Friendship or Acquaintance. . . ."

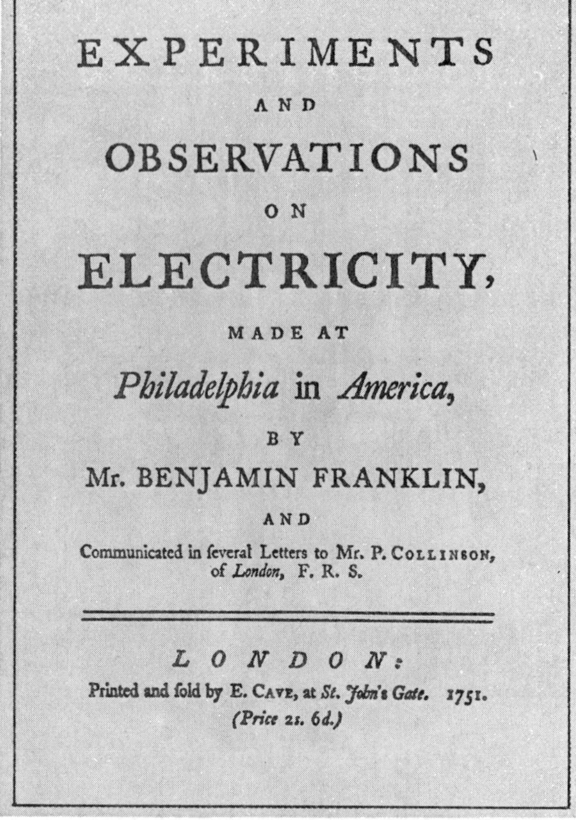

EXPERIMENTS

AND

OBSERVATIONS

ON

ELECTRICITY,

MADE AT

Philadelphia in *America*,

BY

Mr. BENJAMIN FRANKLIN,

AND

Communicated in ſeveral Letters to Mr. P. COLLINSON, of *London*, F. R. S.

LONDON:

Printed and ſold by E. CAVE, at *St. John's Gate*. 1751.

(Price 2s. 6d.)

Fig. 4. Title page of the first English edition of Franklin's letters on electricity.

With these Experiments, how many pretty systems do we build

One of the ingenious and worthy men honoring Franklin with his friendship was Peter Collinson, active member of the Royal Society of London as well as British agent for Franklin's Philadelphia library. Having helped introduce Franklin to electricity, Collinson soon assumed the role of his main scientific correspondent. Beginning in 1747, Franklin sent to Collinson detailed accounts of his ongoing experiments and speculations. Not only did Collinson present certain of these accounts to his fellow natural philosophers at the Royal Society, but more important, he assembled the letters into a book, *Experiments and Observations on Electricity.* First published in London in 1751, this collection of Franklin's scientific letters was later expanded, going through a total of five English editions, three French editions, and a German and Italian edition (Fig. 4).

William Watson, one of Franklin's overseas mentors and perhaps England's leading electrical scientist, reviewed the first edition of the book in 1751 and concluded, "Upon the whole, Mr. Franklin appears in the work before us to be a very able and ingenious man. . . .I think scarce any body is better acquainted with the subject of electricity than himself." Such enthusiasm for the book was to grow, particularly in England and France; hence by 1770, a London reviewer of the fourth edition of the letters even suggested a parallel with Newton's *Principia:* "The philosophical world would have been too long acquainted with the merit of these justly celebrated publications [the letters] to require, at this time, any character of them from us. The light thrown by them on a new and extensive branch of physical science has already diffused itself throughout Europe; where the experiments and observations of Dr. Franklin constitute the *principia* of electricity, and form the basis of a system equally simple and profound."

In his letters Franklin presented a single-fluid theory of electricity. That is, he explained electrical phenomena in terms of minute electrical corpuscles that together constituted an invisible "electrical fluid" — a fluid he also called "electrical fire" or "electrical matter." Such corpuscular explanations, by the way, were common in 18th-century studies of phenomena such as heat and light; indeed, Franklin and many of his scientific colleagues were working within a general tradition of nonquantitative experimentation and corpuscular speculation that had its most immediate roots in Isaac Newton's widely-read *Opticks* (1704). In Franklin's theory, the particles of the electrical fluid were attracted to and normally diffused among the elementary corpuscles of regular matter. All macroscopic bodies therefore contained a certain natural amount of usually unnoticed electrical fluid. An excess or deficiency of this fluid resulted in the neutral body becoming, to use Franklin's new terminology, "electrised *positively*" or else "*negatively.*" And although the fluid could be collected, transferred, or circulated, Franklin further emphasized, it was always conserved. The "plus" charge added to a body must always be exactly balanced by

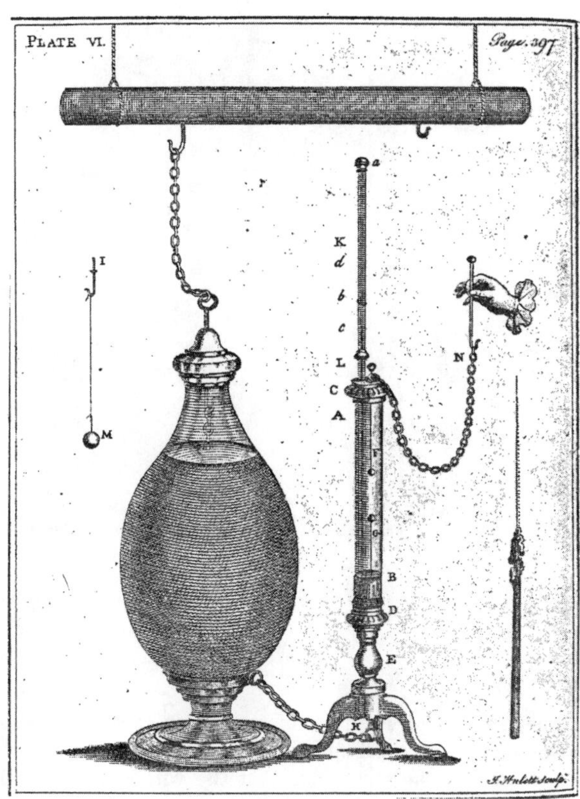

Fig. 5. Later editions of Franklin's book also included letters from Ebenezer Kinnersley, Franklin's main American collaborator. Shown here is Kinnersley's "electrical air thermometer" with which he demonstrated in 1761 that "the electric fire, though it has no sensible heat when in a state of rest, will, by its violent motion, and the resistance it meets with, produce heat in other bodies when passing through them. . . ." The apparatus consisted primarily of a Leyden jar (left) and a glass tube (right) containing both a thermometer and the particular conducting body to be tested.

a "minus" charge subtracted from a donor body. In other words, implicit in the theory was what we today designate the principle of conservation of charge.

Besides outlining this innovative concept of positive and negative electrification in connection with the conservation of charge, Franklin went on in his letters to present an even more detailed picture of the atomistic mechanisms behind electrical phenomena. "Electrical matter differs from common matter in this," Franklin elaborated, "that the parts of the latter mutually attract, those of the former mutually repel each other. . . .But though the particles of electrical matter do repel each other, they are strongly attracted by all other matter. . . ." Being attracted to common matter, therefore, the particles of electrical matter permeate all normal bodies; but being self-repulsive they can permeate the bodies to only a certain maximum level of concentration — to the level where the internal attractive and repulsive forces balance. If additional electrical matter is added to such a normal, balanced body, then, according to Franklin, "It lies without upon the surface, and forms what we call an electrical atmosphere; and then the body is said to be electrified." With this detailed hypothesis, Franklin could account for a variety of effects. Two positively charged bodies repel each other, for example, due to the repulsive force of the atmospheres created by excessive electrical matter. Similarly, Franklin could explain the attraction of a positively charged body for a negatively charged one; the atmosphere produced by the excessive electrical matter of the positive body is drawn to the exposed common matter of the electrically deficient body.

Although, as we shall see, this hypothesis also proved fruitful both in explaining the Leyden jar and in suggesting the electrical character of lightning, it did have a fundamental weakness: it could not account for the mutual repulsion of two negatively charged bodies. Franklin was aware of this problem, considering such repulsion to be "a phaenomenon difficult to be explained." One of Franklin's followers, Franz Aepinus, working in Germany and Russia, finally rectified this defect in the late 1750s by modifying one of Franklin's basic postulates. Aepinus asserted that unadorned common matter is not self-attractive but rather self-repulsive. Thus, he not only brought the repulsion of two negatively charged bodies within the domain of Franklin's theory, but he also introduced a satisfying symmetry — although attracted to one another, electrical matter and common matter were now each self-repulsive.

Any brief summary of Franklin's one-fluid theory, such as in the last few paragraphs, perhaps leaves the impression that Franklin was equally terse and detached in his letters to Collinson. This is not so. His letters were actually chatty and personal; eighteenth-century readers must have felt a sense of participation as they progressed through these diary-like musings in which Franklin gradually articulated his ideas. In later letters, Franklin would forthrightly acknowledge and then clarify earlier misconceptions. Once he even personally confided to Collinson that he was now "asham'd" about a prior, rash hypothesis; with typical humility he added: "In going on with these Experiments, how many pretty systems do we build, which we soon find ourselves oblig'd to destroy! If there is no other Use discover'd of Electricity, this, however, is something considerable, that it may *help to make a vain man humble.*" Incidentally, Franklin was not alone in Philadelphia building "pretty systems." As he readily conceded, he was assisted in his researches by various able Americans, including primarily his "ingenious Neighbor," Ebenezer Kinnersley (Fig. 5).

Fig. 6. A set of Franklin's original Leyden jars joined together to form what he called an "electrical-battery." Franklin requested in his will that this box of jars, each 15 in. high, be given to the American Philosophical Society in Philadelphia; this organization still possesses the historic apparatus. Franklin perhaps had this "battery" in mind when, in 1751, he wrote: "There are no bounds (but what expence and labour give) to the force man may raise and use in the electrical way: For bottle may be added to bottle *in infinitum*, and all united and discharged together as one, the force and effect proportioned to the number and size." (Photo courtesy of the American Philosophical Society.)

The Leyden jar — "this miraculous bottle!"

To merit the designation of "revolutionary," Franklin's theory did more than account for the two states of electrification and explain electrostatic attraction and repulsion — although these were significant accomplishments in this period of experimental turmoil and theoretical uncertainty. The theory also explained certain well-known but more complex phenomena, such as the Leyden jar, and predicted various new phenomena, most important, the electrification of clouds and lightning. By presenting detailed explications of phenomena such as the Leyden jar and lightning, Franklin enriched the flavoring of his already appetizing research program — a program made extra savory to European electricians, we should add, by Franklin's being from the exotic frontiers of North America.

Franklin had experimented much with the Leyden jar or, as he once termed it, "this miraculous bottle!" (Fig. 6). Not all of this Philadelphian's experiments, however, were successful. Once while investigating whether the shock from two large jars would kill a turkey, he was knocked unconscious when he inadvertently discharged the jars through his own body. In relating this embarassing incident to this son, Franklin implored him not to publicize "so notorious a blunder; a match for that of the Irishman . . . who, to divert his wife, poured the bottle of gunpowder on the live coal; or of that other, who, being about to steal powder, made a hole in the cask with a hot iron." Fortunately, most of Franklin's experiments with Leyden jars went more smoothly, leading him swiftly to a clear and unprecedented understanding of their operation.

One type of glass jar with which Franklin frequently experimented had its outer, bottom portions encased in a grounded metal foil; the inside was filled with water that could be electrified through a wire penetrating a cork at the top. Drawing on his single-fluid theory, he explained that such a Leyden jar functioned by the accumulation on its inner surface of electrical matter which repelled an equal amount of electrical matter from the grounded exterior surface. "At the same time that the wire and top of the bottle, &c. is electrised *positively* or *plus*," he wrote to Collinson, "the bottom [outside] of the bottle is electrised *negatively* or *minus*, in exact proportion: *i.e.* whatever quantity of electrical fire is thrown in at the top, an equal quantity goes out of the bottom." To demonstrate that the inside and outside surfaces were equally but oppositely charged, Franklin suspended a small cork on a silk thread between the main wire coming out of the bottle and a parallel wire attached to the metal foil that encased the bottom of the bottle (see the second Leyden jar in the top row of Fig. 7). He reported that with such an arrangement of two wires, the hanging cork "will play incessantly from one to the other, 'til the bottle is no longer electrised; that is, it fetches and carries fire from the top [inside] to the bottom [outside] of the bottle, 'till the equilibrium is restored."

Later, Franklin further explained that it was impossible to "force the electrical fluid through glass." On the other hand, "the particles of the electric fluid have a mutual repellency" that does act through glass. Consequently, the electrical fluid that is added to the inner surface of a Leyden jar "acts by its repelling force on the particles of the electrical fluid contained in the other surface"; that is, this repulsive force acts through the glass to drive away the fluid of the grounded outer surface. The result is a "charged" jar. Implicit in this imaginative and pioneering explanation, by the way, was also the explanation of electrostatic induction, a phenomenon Franklin later focused on explicitly. In a subsequent letter he described how, for example, the electrical atmosphere of a positively charged glass tube drives the electrical matter of a nearby insulated conductor to the far side of the conductor. Being able to explain how charges were induced in such general situations was perhaps as important for the acceptance of Franklin's theory as being able to account for the Leyden jar itself.

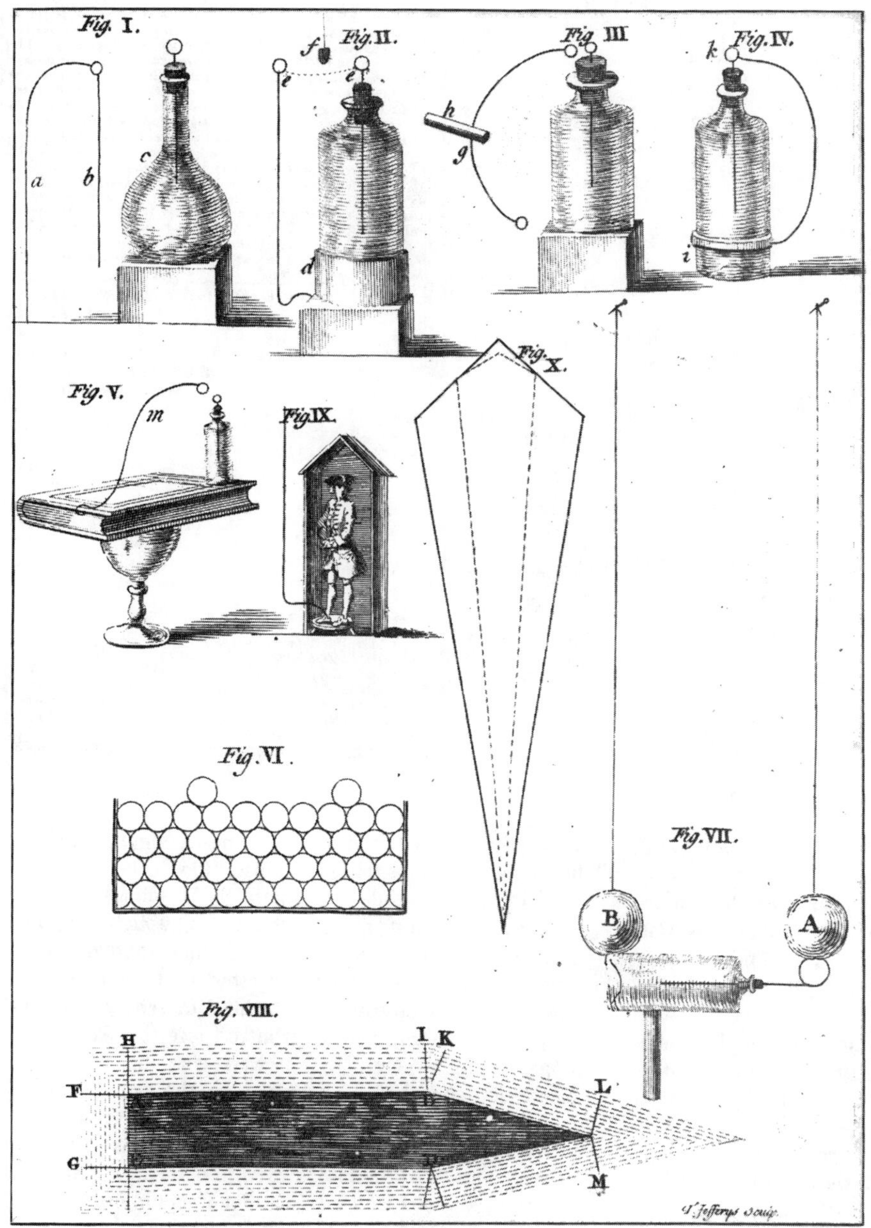

Fig. 7. Main plate of illustrations from Franklin's *Experiments and Observations on Electricity* (1751). Included are those figures that pertain to Franklin's research on Leyden jars and pointed metal conductors as well as his "sentry-box" experiment on lightning. (Photo courtesy of the Rare Book Department, University of Wisconsin Library.)

Drawing electric fire from the clouds

Not only did Franklin's research program clarify certain familiar electrical phenomena, it also helped establish the electrical identity of previously obscure phenomena, most important, lightning. Various natural philosophers had earlier speculated on the electrical nature of lightning. Franklin, however, was the first to devise a means of verifying the identity. Furthermore, he coupled this scientific breakthrough with a practical invention: the lightning rod.

Personal research notes from November 1749 reveal Franklin's thoughts that led to his prediction that a pointed metal rod would draw lightning, thus enabling experimenters to "ascertain its sameness with the electric fluid." In these informal but portentous notes, he wrote:

> "Electrical fluid agrees with lightning in these particulars: 1. Giving light. 2. Colour of the light. 3. Crooked direction. 4. Swift motion. 5. Being conducted by metals. 6. Crack or noise in exploding. 7. Subsisting in water or ice. 8. Rending bodies it passes through. 9. Destroying animals. 10. Melting metals. 11. Firing inflammable substances. 12. Sulphureous smell. — The electric fluid is attracted by points. — We do not know whether this property is in lightning. — But since they agree in all the particulars wherein we can already compare them, is it not probable they agree likewise in this? Let the experiment be made."

Poor RICHARD improved:

BEING AN

ALMANACK

AND

EPHEMERIS

OF THE

MOTIONS of the SUN and MOON;

THE TRUE

PLACES and ASPECTS of the PLANETS;

THE

RISING and *SETTING* of the *SUN*;

AND THE

Riſing, Setting *and* Southing *of the* Moon,

FOR THE

YEAR of our LORD 1753:

Being the Firſt after LEAP-YEAR.

Containing alſo,

The Lunations, Conjunctions, Eclipſes, Judgment of the Weather, Riſing and Setting of the Planets, Length of Days and Nights, Fairs, Courts, Roads, *&c.* Together with uſeful Tables, chronological Obſervations, and entertaining Remarks.

Fitted to the Latitude of Forty Degrees, and a Meridian of near five Hours Weſt from *London*; but may, without ſenſible Error, ſerve all the NORTHERN COLONIES.

By *RICHARD SAUNDERS*, Philom.

PHILADELPHIA:

Printed and Sold by B. FRANKLIN, and D. HALL.

Mayor's Courts for the City

ARE held quarterly at *Annapolis*, viz The laſt tueſday in *January*, *April*, *July* and *October*.

How to ſecure Houſes, &c. *from* LIGHTNING.

IT has pleaſed God in his Goodneſs to Mankind, at length to diſcover to them the Means of ſecuring their Habitations and other Buildings from Miſchief by Thunder and Lightning. The Method is this: Provide a ſmall Iron Rod (it may be made of the Rod-iron uſed by the Nailers) but of ſuch a Length, that one End being three or four Feet in the moiſt Ground, the other may be ſix or eight Feet above the higheſt Part of the Building. To the upper End of the Rod faſten about a Foot of Braſs Wire, the Size of a common Knitting-needle, ſharpened to a fine Point; the Rod may be ſecured to the Houſe by a few ſmall Staples. If the Houſe or Barn be long, there may be a Rod and Point at each End, and a middling Wire along the Ridge from one to the other. A Houſe thus furniſhed will not be damaged by Lightning, it being attracted by the Points, and paſſing thro the Metal into the Ground without hurting any Thing. Veſſels alſo, having a ſharp pointed Rod fix'd on the Top of their Maſts, with a Wire from the Foot of the Rod reaching down, round one of the Shrouds, to the Water, will not be hurt by Lightning.

QUAKERS *General Meetings are kept*,

AT Philadelphia, the 3d Sunday in March. At Cheſter-River, the 2d Sunday in April At Duck-Creek, the 3d Sunday in April. At Salem, the 4th Sunday in April. At Weſt River on Whitſunday. At Little Egg-Harbour, the 3d Sunday in May. At Fluſhing, the laſt Sunday in May, and laſt in Nov. At Setacket, the 1ſt Sunday in June. At New-town, (Long-Iſland) the laſt Sunday in June. At Newport, the 2d Friday in June. At Weſtbury, the laſt Sunday in Auguſt, and laſt in February. At Philadelphia, the 3d Sunday in September. At Nottingham, the laſt Monday in September. At Cecil, the 1ſt Saturday in October. At Choptank the 2d Saturday in October. At Little-Creek, the 3d Sunday in October At Shrewſbury the 4th Sunday in October. At Matinicok the laſt Sunday in October.

FAIRS are kept,

At Noxonton April 29, and October 21. Cohanſie May 5, and October 27. Wilmington May 9, and November 4. Salem May 12, and October 31. Newcaſtle May 14, and Nov. 14. Cheſter May 16, and Oct. 16. Briſtol May 19, and Nov. 9. Burlington May 21, and Nov. 12. Philadelphia May 27, and November 27. Lancaſter June 12, and Nov. 12. Marcus-Hook Oct. 10. Annapolis May 12, and Oct. 10. Charleſtown May 3, and Oct. 19.

Fig. 8. Cover of *Poor Richard's Almanack* juxtaposed next to the page containing Franklin's public announcement of the lightning rod.

How had Franklin come to this last and experimentally fruitful insight? Actually, from as early as his first substantive letter to Collinson, he had been studying this "wonderful effect of pointed bodies" (see, for example, the two pointed bodies in Fig. 7). He never completely succeeded, however, in fitting this effect to his single-fluid theory. At best, he vaguely suggested that the unelectrified common matter in the small point of a metal rod would produce a concentrated attractive force on the electrical matter in a correspondingly small portion of a nearby electrified body. This focused interaction would allow the pointed rod to draw away, "particle by particle," the normally intractable electrical matter of the other body. Franklin himself warned his readers that this hypothesis was inadequate and factitious; he assured them, however, of the empirical reality of "the power of points." Speaking metaphorically, he emphasized that "it is of real use to know that china left in the air unsupported will fall and break; but *how* it comes to fall, and *why* it breaks, are matters of speculation. It is a pleasure indeed to know them, but we can preserve our china without it."

Accordingly, by 1750, prior to adequate theoretical explanation, let alone full-scale tests, Franklin in his letters urged people to protect buildings from lightning by erecting "upright rods of iron made sharp as a neddle. . . ." Such pointed rods when properly grounded would, Franklin believed, "draw the electrical fire silently out of a cloud before it came nigh enough to strike. . . ." If the rods did not thus completely prevent the usual sharp lightning stroke, Franklin later added, they would at least safely conduct the stroke past the building to the ground. In *Poor Richard's Almanack* for 1753, he recommended such lightning rods to his fellow colonists (Fig. 8).

In 1750, while first publicly speculating on the efficacy of lightning rods, Franklin also publicly proposed a possible scientific experiment for testing whether lightning was indeed electrical. He suggested that an observer be placed on a high tower in "a kind of sentry-box" from which rose a long, pointed iron rod (see the sketch in the upper center of Fig. 7). If it were true that "the fire of electricity and that of lightning be the same," then the observer should detect "the rod drawing fire to him from a cloud." Even though the program for this experiment appeared in 1751 in the first edition of *Experiments and Observations on Electricity*, neither Franklin nor his British scientific readers rushed to carry it out.

The French, however, were more immediately curious. They succeeded with the experiment soon after reading about it in 1752 in the French translation of Franklin's book. One French natural philosopher related that Louis XV's enthusiasm for Franklin's various other discoveries prompted this first test: "The Philadelphian experiments, that Mr. Collinson . . .was so kind to communicate to the public, having been universally admired in France, the King desired to see them performed . . . His Majesty saw them with great satisfaction, and greatly applauded Messieurs Franklin and Collinson. These applauses of his Majesty having excited in Messieurs de Buffon, D'Alibard, and De Lor a desire of verifying the conjectures of Mr. Franklin upon the analogy of thunder and electricity, they prepar'd themselves for making the experiments." On the 10th of May 1752, the assistant of Thomas-François D'Alibard

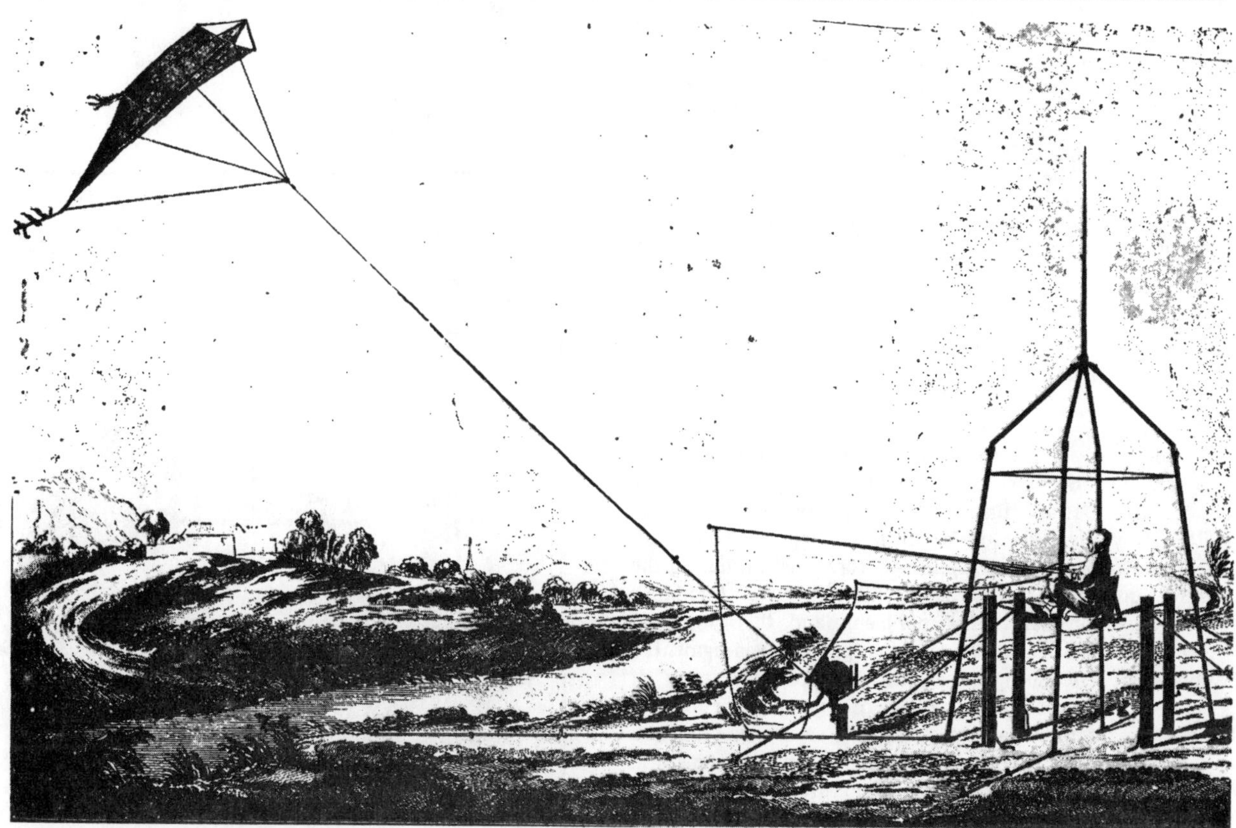

Duc de Chaulnes's Improvement of Dr. Franklin's Electrical Kite.

Presented by Dr Franklin June 20, 1788

Fig. 9. Not one of Rube Goldberg's fantastic contraptions, but — as the caption in Franklin's hand reads —"Duc de Chaulnes's Improvement of Dr. Franklin's Electrical Kite." This French print was "presented by Dr. Franklin, June 20, 1788" to the American Philosophical Society — an intercolonial group that Franklin helped found in Philadelphia, dedicated to "the promoting of useful Knowledge in general, and such branches thereof in particular, as may be more immediately servicable to the British Colonies. . . ." (Print courtesy of the American Philosophical Society.)

successfully drew sparks from the rod of a Franklin sentry-box apparatus.

One month later, apparently before learning of the French verification, Franklin also confirmed the electrical identity of lightning "in a different and more easy Manner." Supposedly during June of 1752, in a field outside of Philadelphia, he and his son coaxed "a very evident spark" from the famed "electrical kite." This kite, as all subsequent generations of American school children were to learn, was a large silk handkerchief stretched across two cedar sticks carrying a pointed wire, all held aloft by twine terminating in a dangling key and a short silk ribbon. In the meantime, back in Europe, word spread of D'Alibard's success with the sentry-box apparatus, and soon other French as well as English and German natural philosophers were eagerly duplicating the experiment. Perhaps they were too eager — a physicist in Russia, while attempting the experiment in 1753, was electrocuted. With safer results, however, enthusiastic European electricians also refined and repeated the Philadelphia kite experiment (Fig. 9).

These unprecedented experiments assured Franklin's already considerable fame among both natural philosophers and laymen. Following the initial lightning confirmations, Collinson wrote to Franklin from London: "Our papers are full of Electrical Experiments. Thou sees a Little Electrical Hint give[n] at Philadelphia has stimulated all Europe." By establishing that lightning was an electrical phenomenon, Franklin had dramatically extended the domain of electrical inquiry from the Leyden jars and frictional generators of man's meager laboratories to the clouds and thunder of Nature's grand landscapes. Electricity was no longer just an artificial contrivance or puny plaything of learned gentlemen. Moreover, natural philosophers, including Franklin, were pleased that seemingly impractical research had led to the invention of the lightning rod; this tangible benefit sustained their belief that esoteric science could be of "use to mankind." As we have seen, educated laymen such as those in the French court also admired Franklin's theoretical and experimental breakthroughs, electricity being one of the subjects in vogue during this Age of Enlightenment. Indeed, Franklin's popular renown in Europe as a scientist guaranteed in later years his cordial reception as a statesman.

Not everyone, however, was enthusiastic about Franklin's researches. His staunchest critic was the

Fig. 10. Typical diagram from one of the popular books of Abbé Nollet, Franklin's main French rival. Nollet wrote for the cultured aristocracy as well as for fellow natural philosophers; hence in this diagram experiments are being performed both by fashionable ladies and dapper gentlemen. The woman seated in the foreground, incidentally, is holding a cat on her insulating silk dress; after rubbing the cat's back with one hand, she uses a finger of her other hand to draw a spark from the cat's nose. Nollet advised his readers that this was a rare phenomenon, occurring only with "un chat très-electrisable!" (From Nollet, *Leçons de Physique Expérimentale,* 2nd ed., vol. VI, Paris, 1765.)

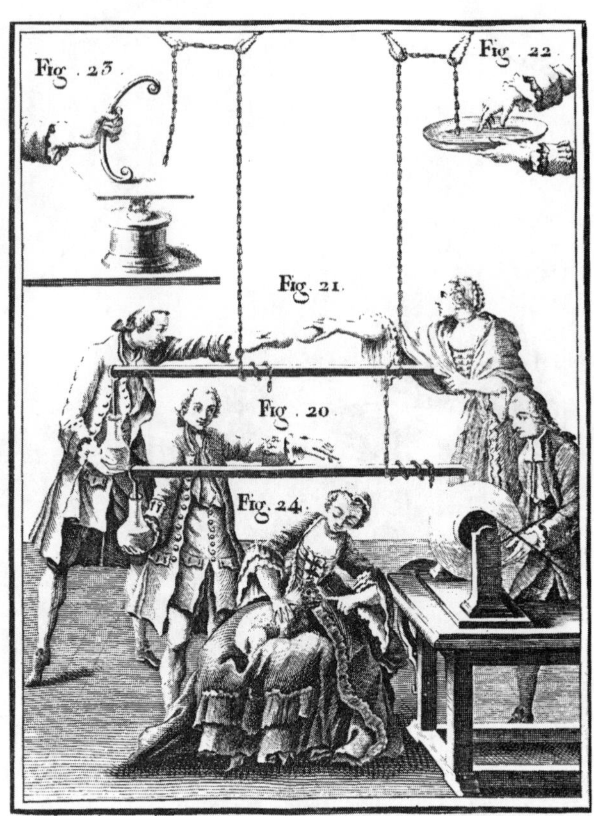

prominent French scientist Abbé Nollet, a prior student of Dufay and advocate of a two-fluid theory of electricity (Fig. 10). Certain British natural philosophers, such as Robert Symmer, also preferred the concept of two distinct and contrary kinds of electricity. Nevertheless, during the third quarter of the eighteenth century, Franklin's innovative program of experiment and theory gained widespread allegiance among the previously fragmented community of European electricians, providing many of them with a common model or example for further research. By thus helping to institute secure experimental and theoretical foundations, Franklin had helped transform the study of electricity into a mature and established science.

The high regard of natural philosophers and academic men for Franklin was reflected in the important honors they bestowed on him. The Royal Society awarded him the esteemed Copley Medal in 1753 and then three years later elected him a Fellow. During the same period, this man, who had only briefly attended grammar school, also received honorary degrees from Harvard, Yale, and William and Mary — Harvard acclaiming him for "great Improvements in Philosophical Learning, particularly wth. Respect to Electricity. . . ." Finally, in 1773 natural philosophers in Paris welcomed him as one of only eight foreign associates of the prestigious Royal Academy of Sciences.

The power of man

When Franklin, in anticipating the true identity of lightning, had written, "Let the Experiment be made," he had chosen a phrase that could also have been applied 30 years later to the "experiment" of American independence. In 1780, long after the electrical experiment but still in the tumult of the political one, the aged Franklin pondered the relative standing of his two lifelong pursuits: science and human affairs.

"The rapid progress true science now makes," he wrote, "occasions my regretting sometimes that I was born so soon. It is impossible to imagine the height to which may be carried, in a thousand years, the power of man over matter. . . .O that moral science were in as fair a way of improvement, that men would cease to be wolves to one another, and that human beings would at length learn what they now improperly call humanity."

Although this latter judgment was pessimistic, Franklin was generally hopeful about improving "moral science" as well as "true science." In both of these realms, Franklin was foremost a constructive and forward-looking revolutionary.

Bibliographical note

The leading authority on Franklin's physics is I. Bernard Cohen. Cohen has given a convenient overview of his historical findings in his article on Franklin in the *Dictionary of Scientific Biography,* edited by Charles C. Gillispie, vol. IV (Scribner's, New York, 1972), pp. 129-139; an illustrated version of this article is also available in book form from the same publisher under the title *Benjamin Franklin: Scientist and Statesman* (1975). With an eye toward historical revision, John L. Heilbron has recently written "Franklin's Physics," Phys. Today, 29, 32-37 (1976). Teachers wanting to introduce Franklin's electrical researches to their students might specifically use Duane Roller and Duane H.D. Roller's pamphlet "The Development of the Concept of Electrical Charge," Harvard Case Histories, Case 8 (Harvard University Press, Cambridge, 1954). Teachers might also consider Richard E. Orville's recent article "The Lightning Discharge," Phys. Teach., 14, 7-13 (1976).

To fully appreciate Franklin's science, one needs to read his writings in the original. Spencer R. Weart, the editor of the "Bicentennial Commemorative Volume" of the American Physical Society, has reprinted two key Franklin excerpts; see the first chapter of *Selected Papers of Great American Physicists* (American Institute of Physics, New York, 1976). I. Bernard Cohen has presented and described the bulk of Franklin's scientific papers in *Benjamin Franklin's Experiments: A New Edition of Franklin's Experiments and Observations on Electricity* (Harvard University Press, Cambridge, 1941). Finally, Raymond J. Seeger has assembled an anthology, with commentary, titled *Benjamin Franklin: New World Physicist* (Pergamon Press, Oxford, 1973).

Benjamin Thompson, Count Rumford

Sanborn C. Brown[1]

As the United States of America goes through its bicentennial celebrations it is clear that there is no one date on which all will agree as to when this country became itself. Towns, cities, and organizations all over the country are celebrating "their" bicentennial. The first defiant act, the first bloodshed, the first British soldier killed, the first shots from the Minutemen, the first city taken, the first declaration of independence, the first formation of a rebel government — all these and many more mark the gradual transition from loyal to violent opposition.

And so it was with the people. What determined who was an "American"? Was it where you were born and brought up? Were you an American if you were a Whig, and an Englishman if you were a Tory, even if you were born and brought up together? There is no answer, even though at the time it was vitally important. So let us celebrate Benjamin Thompson, later to become Count Rumford, as an American physicist, from birth to manhood a New Englander even though he had to flee for his life from the wrath of George Washington's troops and he later officered a regiment of the hated "red coats" on Long Island.

Benjamin Thompson was born in Woburn, Massachusetts on March 26, 1753 in a farmhouse which belonged to his grandfather and which, still standing, has become a museum and memorial to the fame that the future Count Rumford gained in later years.[2] His father died in November 1754 and in March of 1756 Benjamin's mother married Josiah Pierce, Jr. and they moved to a new home a short distance away. People remembered him as a bright child but his early biographers tell us that he was sullen and resentful when forced to take his share of the drudgery of the farm life of his family, and by the time he was 13 years old his family was pretty much at its wit's end to know what to do with him. Their decision to indenture him to a merchant, John Appleton, in Salem, Massachusetts suggests a remarkable understanding of his potentialities.

The Appleton family had been intellectual and social leaders for generations. John Appleton's father was a Doctor of Sacred Theology and the minister of the church in Cambridge. The younger Appleton graduated from Harvard College and at the time Benjamin came to live with him he was operating a retail variety store out of a shop in his house. Thompson was taken in as a member of the family, as was the custom of the time, so that he had the opportunity not only to learn a business, but also to absorb whatever education he could from a highly motivated and intellectually stimulating family. Many years later the then Count Rumford told his University of Geneva physicist friend Pictet[3] that he learned "algebra, geometry, astronomy, and even the higher mathematics. Before the age of fourteen, I had made sufficient progress in the class of studies to be able . . . to calculate and trace rightly the elements of a solar eclipse . . . my computation was correct

Sanborn C. Brown *is professor of physics emeritus at the Massachusetts Institute of Technology. For many years active in AAPT affairs, he was a recipient in 1962 of its Distinguished Service Citation. Professionally a plasma physicist, he has been studying and writing about Count Rumford for over thirty years. (Hemlock Corner, Henniker, New Hampshire 03242)*

Fig. 1. Count Rumford at the age of 55, painted by Rembrandt Peale, owned by the American Academy of Arts and Sciences, Boston.

within four seconds." Eventually one of Thompson's scientific experiments got the better of him and his apprenticeship came to an abrupt end. In the process of grinding and pounding the ingredients of some black powder he was investigating, an explosion resulted. Badly burned, he was sent home to recuperate.

The luckiest break that Benjamin Thompson had as a result of his mother's remarriage to Josiah Pierce was the location of their new home. It was directly across the road from the home of a young man named Loammi Baldwin who was to grow up to be one of the leading engineers in Massachusetts. He was nine years older than Benjamin but he developed a genuine concern for his young friend, did much to cultivate his interest in science and was ready and eager to be his mentor and advisor. They studied together much of the year it took Thompson to recover from his burns. Eventually Benjamin was well enough to go back to work and he located a job with Hopestill Capen who kept a drygoods shop in the heart of Boston in a building that now houses the Old Union Oyster House.

It seems doubtful that Benjamin went to work for Mr. Capen because he was interested in the drygoods business. To be stuck behind a counter when the excitement of revolution was all about him was hardly congenial to Benjamin Thompson's restless and ambitious nature, and Capen, writing to Benjamin's mother in the spring of 1770 that "he oftener found her son *under* the counter, with gimlets, knife, and saw, constructing some little machine, or looking over some book of science, than *behind* it, arranging the cloths or waiting upon customers"[4], fired his young apprentice.

Back in Woburn Thompson applied himself to looking for a profession that would get him out of his farmer-family sphere and the next thing he tried was to apprentice himself to Dr. Hay of Woburn and to set himself seriously to study medicine. He kept a diary[5] of his activities and for a brief period, beginning June 12, 1771, he recorded that he attended lectures on natural philosophy at Harvard by Professor John Winthrop, although there is no record at Harvard that he had any official attendance at the college.

After the initial excitement of coming to Dr. Hay, Thompson seemed to look on his study of medicine more as an educational opportunity than as a preparation for a career. An amusing tale of Thompson's less serious investigations tells of how Dr. Hay "returning home one day, was much surprised at hearing a hog squealing up chamber, ran up, in great haste. Behold, Thompson had got the brochia of a hog, blowing into them."[6] Thompson's first recorded scientific observation was made while he was under Dr. Hay's tutelage. He wrote a detailed description of a monstrous child born in Woburn on April 16, 1771 in his diary with a careful drawing of the anatomical anomolies and two years later he submitted these observations to the American Philosophical Society of Philadelphia. They were recorded as received, but no further notice was taken of them.[7]

Living with Dr. Hay was costing Benjamin money which his family could not afford and to meet his bills he went searching for a school teaching job.For a while he was "keeping school" at the neighboring town of Wilmington, and then he got a steadier teaching position in Bradford, a particularly fortunate circumstance. It put him under the influence of the minister of that town, the Reverend Samuel Williams. Williams was a remarkable scholar who was interested in science, history, and journalism as well as his ministry. Later he became professor of natural philosophy at Harvard. He took a real interest in young Thompson and convinced him that medicine was not his field and that he should become a schoolmaster. Following this advice, Thompson closed out his accounts with Dr. Hay in June 1772 and went to live with Rev. Williams. While he was with Williams, Thompson submitted another paper to the American Philosophical Society, this one "An Account of an Aurora Borealis Observed at Bradford in New England."[8] The paper received slightly more attention than his previous attempt since it was read to a meeting of the Society on April 2, 1773. However it was never published.

Through the help of Samuel Williams and his own family, Benjamin Thompson was offered a permanent position as a schoolmaster in Concord, New Hampshire, and there he went in the summer of 1772 to take up his duties. His period of education as it was commonly carried out in Colonial America had been completed.

Many of the leading citizens of Concord came from Woburn and first among these was the minister, Rev. Timothy Walker, under whose sponsorship Thompson arrived. Fully as important in the affairs of Concord through the years had been Col. Benjamin Rolfe. As the history of Concord records it:[9] "By inheritance, and by his own industry and prudent management, Col. Rolfe acquired a large property in lands. He lived a bachelor until he was about 60 years of age, when he married Sarah, the eldest daughter of Rev. Mr. Walker, whose age was 30 . . . . Col. Rolfe dies December 21, 1771, in the 62d year of age."

Although Benjamin Thompson's school was a big one, as he wrote to his mother,[10] "I have 106 Scholars at my School," teaching was for him only a means of getting ahead. Far more useful were the attentions he immediately found bestowed on him by his sponsor's recently widowed daughter Sarah. Even though she was over 30, a 19-year-old aspirant to wealth and social rank could hardly fail to take advantage of such an opportunity, and less than eleven months after Sarah Rolfe was left a widow she became Sarah Thompson and Benjamin became one of the wealthiest men in New Hampshire. Col. Rolfe had held a high military rank in the provincial militia, and it is not surprising that on their wedding day Sarah and Benjamin Thompson were honored guests at Governor Wentworth's table in the capital city of Portsmouth. At this point school teaching was forgotten and Thompson threw himself wholeheartedly into cultivating a friendship with the Royal Governor.

In 1773, with riots and violent disturbances cropping up all over the colonies, to anyone aspiring to fame and fortune among the wealthy and aristocratic, soldiering for the Royal Governor seemed an obvious step. Viewed from today's conditions, it would seem preposterous to give an officer's rank to a young man not yet 21 with absolutely no military experience, but in the 18th century, the militia, like one's tenants and one's servants, could be hired and fired at will by a governor who looked on the colony almost as a family possession. For Governor Wentworth, commissioning Benjamin Thompson as an officer in the Fifteenth Regiment of Militia was nothing he had to justify as long as it might prove to be useful, and so starting on January 20, 1774 Thompson became Major Benjamin Thompson, and most useful to the British cause did he become.

Being now a large landowner, Major Thompson was constantly looking for farmhands and he hired all that came his way. If he discovered that they were deserters from the British army he purposely worked them so hard and presented such a miserable picture of colonial working conditions that they were eager to quit, since even life in the army seemed greatly to be preferred. When they appeared ripe for the suggestion, they were visited by a secret agent who was a British soldier disguised as one of Thompson's laborers who offered them not only a release from their commitment to Thompson and free passage back to Boston, but a military pardon as well. For a while the scheme worked so well that Thompson was commended in several dispatches to London, and his reputation with the British rose by the month. But such an operation could not be kept secret forever, and eventually the citizens of Concord became suspicious of what was going on and charged Thompson with being a rebel to the state and demanded that he appear on December 12, 1774 before the Committee of Correspondence of Concord. The committee formally discharged him, but though the Committees of Correspondence tried to preserve their dignity as pseudo-legal bodies, the firebrand "patriots" often took matters into their own hands, to descend upon the object of their wrath to tar and feather their unfortunate victims and ride them on a rail out of town.

Thompson was not the man to be caught in such a situation. He hurriedly gathered his effects together and fled to Boston. That night the incensed mob converged upon his house. In a way typical of his whole life, he thought only of himself. He may not have even told his wife where he was going; he certainly did not tell his father-in-law to whose care he abandoned Mrs. Thompson. Also nowhere in his correspondence did he allude to the fact that in the middle of this tumultuous autumn he had become a father. A daughter, Sarah, was born in October to begin life in the shadow of her father's troubled existence, a shadow from which she never emerged throughout her whole long life.

At the very time that Benjamin Thompson arrived in Boston, General Thomas Gage, The Royal Governor of Massachusetts was actively trying to strengthen his espionage coverage of the area around Boston and he saw Thompson as an important and useful addition to his intelligence organization. Thompson accepted this assignment and again proved most useful to the British. At least once while he was active in the Woburn area he put his scientific knowledge to a useful purpose by employing secret ink in a letter which successfully eluded the Americans trying to seal off Boston and was delivered without detection.[11] Technically it was as good as many of the secret inks used in World War I. But again the native citizens got suspicious and in October Thompson fled behind the British lines.

While Thompson was in Boston he undertook to write a long report on the "state of the Rebel Army"[12] which covered a wide range of information on military and political affairs. It is a document that underscores the writer as a shrewd observer and a careful reporter. By March 1776 the position of the British in Boston became quite untenable and the army abandoned the town, sailing with over a thousand loyalist sympathizers to Halifax, Nova Scotia. The official dispatches of this reverse in the fortunes of the King's army were entrusted to a Judge William Brown to take to London, and in Judge Brown's company sailed the aggressive young Loyalist from Massachusetts, Major Benjamin Thompson.

We know that before Thompson went aboard the British frigate that took him to London he supplied himself with letters of introduction to important people and copies of all of his intelligence reports. He gained almost immediate access to the highest political circles in Britain at the end of April 1776. He made a tremendous impression on Lord George Germain, the Secretary of State for the Colonies, and with the young American at his elbow, Germain became an expert on the many details and affairs of Massachusetts Bay, to the great advantage of both men. Thompson was immediately added to Germain's entourage and he ingratiated himself so much to the politically

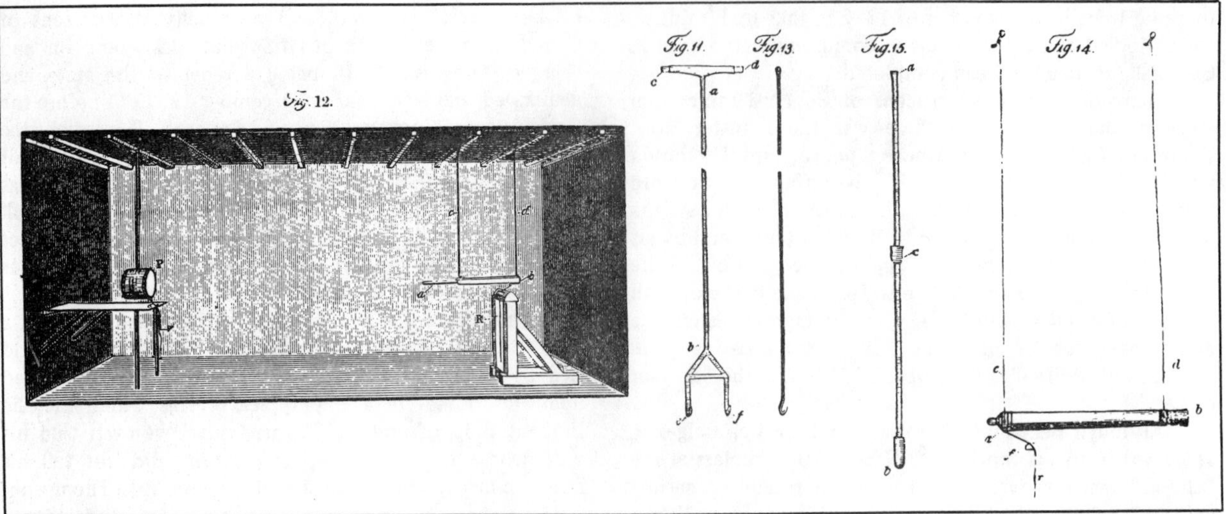

Fig. 2. Ballistic pendulums for testing the force of gunpowder.

powerful that in December 1776 King George appointed him Secretary of the Province of Georgia.

As Benjamin Thompson's fortune and prestige grew so did his leisure and he turned his attention back to one of his first interests, gunpowder, which resulted in his first published paper "New Experiments upon Gun-powder, with occasional Observations and practical Inference; to which are added, an Account of a new Method of determining the Velocity of all Kinds of Military Projectiles, and the Description of a very accurate Eprouvette for Gun-powder."[13]

The "new Method" was to fire a small cannon at a heavy pendulum and to determine the swing of both the target and the gun. The distance each swung was measured by tapes connected to them which ran through friction slots to hold them at their maximum extension. Reproducing his illustration, Fig. 2 clearly shows our now most familiar ballistic pendulum. Most of this very long (99-page) paper deals with trying to explain gunpowder explosion, the force of fired gunpowder, and how to design cannon most effectively. To physicists it is interesting to see in this paper that Thompson's ideas about the nature of heat were being developed. Quite by accident he measured the heating of the gunbarrel after firing blank charges and much to his surprise he found that the gun became hotter than when a bullet was fired. He tells us that he had assumed that the heat came entirely from the burning gunpowder and had expected that if the fire were enclosed for a longer time by the ball this would make the barrel hotter. In worrying about this he came to the conclusion that the heat must be due to the sudden expansion and contraction of the barrel itself and resulted from "the motion and friction of the internal parts of the metal among themselves, occasioned by the sudden and violent effort of the powder upon the inside of the bore." Thus we see that even in 1778 Thompson was thinking of heat as generated by mechanical action. This paper was widely read and acclaimed and it established Thompson as a serious scientist. It won him fellowship in the Royal Society of London and he could proudly sign himself "Benjamin Thompson Esq. F.R.S."

In September 1779 Thompson was appointed Deputy to the Inspector General of Provincial Forces. In this position he was responsible for all clothing and accoutrements sent from England to the officers and men serving in the armed forces in the colonies. The customary procedure was to speculate as aggressively as one could, buying material as cheaply as possible in England and selling to the government for shipment overseas at as high a price as would pass the various Secretaries of War, State, and Treasury. Besides worrying about saddles and swords, his interest had to include uniforms and cloth to be sent out to be made into garments in the colonies. Here again we get a glimpse of Thompson's scientific curiosity at work in the areas of his job. He turned his attention to the "Specific Gravity, Diameter, Strength, Cohesion etc. of Silk."[14] He was interested in a method of finding and testing the strength of silk fibers. To determine their diameter he measured their volume by loss of weight in water. He found it difficult to do this experiment because of the great variability of the weight of the "dry" skeins, and to get a true weight he had to dry the silk in front of a fire. This got him thinking about the implications of this to his job as Deputy Inspector. Carefully measuring the loss of weight as he dried out the skeins he concluded that "silk possesses a power of attracting and imbibing water from the air .... Hence it appears that a Merchant or Manufacturer who purchases 100 lbs. of raw silk, in a common state as to dryness actually pays for at least 8 lbs. of Water; and if it be Kept for a considerable time in a place the quantity of moisture imbibed will be still greater." With his newly discovered information Thompson could make a profit shipping silk even without a markup on the cost per pound if he bought his bolts of silk by weight in their driest possible state in London, and sold by weight on the docks of New York or Charleston after their long damp voyage across the Atlantic.

In September 1780, Benjamin Thompson was made Undersecretary of State for the Colonies. For the first time he wielded real power and the files of the Public Record Office in London are full of his letters and orders as he carried on the business assigned to him. Unfortunately it shows us the picture of an official not only willing to promise much and deliver relatively little, but also malicious and vindictive toward those from America whom he felt he could push around. He made himself thoroughly disliked by those who came in contact with him.

Undersecretary of State for the Colonies, Deputy Inspector of Provincial Forces, Secretary to the Colony of Georgia — Benjamin Thompson had come a long way in a few short years, but despite this impressive array of titles, his future looked dim because of one overriding fact. He was completely the product of Lord George Germain, and any unbiased observer could confidently predict that the days of Germain's political power were numbered. Taxes were high, the American war was unpopular and going badly for the British, and Germain himself was hated by most of Parliament. To unhook himself, late in 1780 Thompson started to raise a regiment to be "composed intirely of young Gentlemen of the first families and connections in America" to be known as the King's American Dragoons. In February 1781 he purchased the commission of Lieutenant Colonel and was appointed Commandant. In October he sailed for America to recruit men and officers for his new venture.

When Thompson landed in New York he was by no means starting from scratch in raising his regiment. He had done some recruiting in London before he left, and since June 1780 a major in his regiment had been working full time for Thompson in getting not only the manpower but in arranging the organizational details as well. Thompson, with plenty of money and excellent connections,completed his regiment on August 6, 1782 and to demonstrate his political power, he arranged that the formal presentation of his regimental colors be timed with the presence in New York of the King's third son, Prince William Henry, the future William IV, so that he could receive his colors from a royal hand. He was assigned to active duty on Long Island.

To the inhabitants of Huntington, Long Island, Benjamin Thompson was the devil incarnate.[15] He chose for the site of his fort two acres of land on which was built the town church. He ordered the building razed and its timbers used for constructing fortifications. All available men in the vicinity were pressed into hard labor for the construction. All the apple trees in the vicinity were ordered cut down and delivered to the regiment, as were all the chestnut rails from the townspeople's fences. The gravestones were not used for construction, but for fireplaces, tables and ovens. Many of the townspeople remembered being made sport of that winter by receiving from the soldiers loaves of bread with the reverse inscription from the tombstones of their loved-ones baked into the lower crust. When the regiment was ordered back to New York to join the defeated British army in fleeing for London in the middle of the winter of 1783, Thompson ordered his troops to burn up all the wood around the fort so that the populace would have neither fences nor firewood.

Back in London, Thompson made sure that if he did not get any further military assignment he would receive an army pension at half pay for the rest of his life. He also arranged to be promoted to a full colonel before he went on half pay. He then went seeking his fortune on the continent, ending up with a position of military aide to the reigning duke of Bavaria, Elector Carl Theodor,

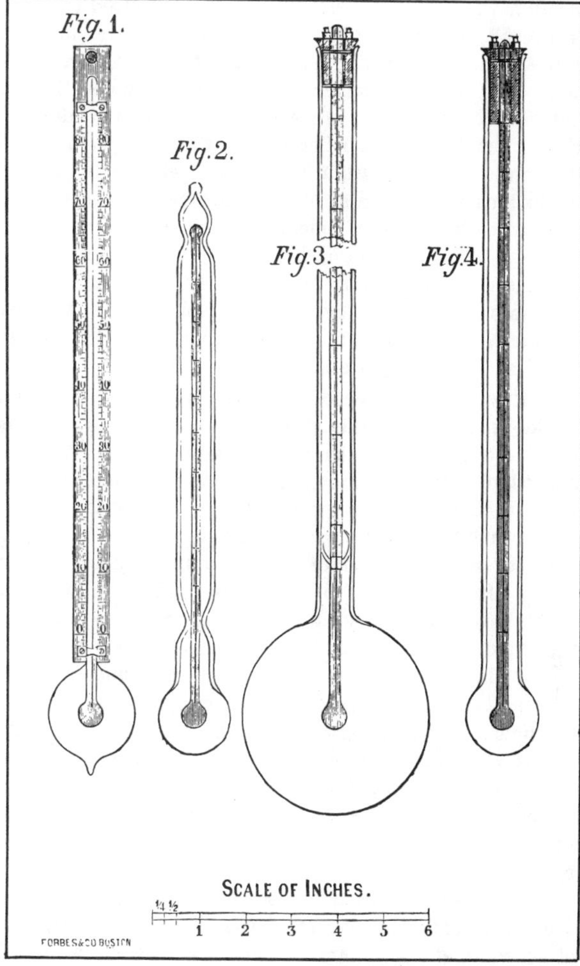

Fig. 3. Thermometers for measuring heat conduction through many substances.

headquartered in Munich. While negotiating the exact nature of his job, the elector suggested to Thompson that some mark of royal approbation from George III would considerably enhance Colonel Thompson's position in Bavaria. Thompson took one quick trip back to London to see what he could do and actually made out very well as he persuaded the King to knight him, and he returned to Munich as Sir Benjamin.

Besides being Carl Theodor's Adjut , General, Thompson was a Colonel Commandant of a Regiment of Cavalry. However he had no actual regiment and so he turned his attention to scientific experiments. Connected with the Mannheim Academy of Sciences there was an excellent instrument maker and Thompson took full advantage of him in designing a long series of experiments on the propagation of heat through many different kinds of substances.[16] We can see what Thompson did by referring to his own illustrations of his thermometers, reproduced in Fig. 3. First he studied the heat conduction of a "Torricelli" vacuum. The tubes surrounding the thermometers were sealed to the top of a mercury barometer and then evacuated by tipping over the barometer and sealing off the tube. He measured the response of the thermometers to change of temperature with the bulbs evacuated and full of air and found that his vacuum was a better insulator than air. He tested to see if this result was true for both increasing and decreasing heat and whether it depended on the rate at which the heat was applied or on the geometry of his thermometers. He also designed thermometers to measure other liquids and gases. He exploded the one labelled Fig. 3 while trying to measure the conductivity of carbon dioxide and with No. 4 he tried to do a careful job on saturated water vapor. Generalizing from these latter measurements he showed that dry air was a better insulator than moist air. He also measured the conductivity of water and mercury. Although these measurements were too crude to give real values for thermal conductivity he was successful in arranging all of his test samples in their proper order as heat insulators.

It was not because of Thompson's science that he got his job in Bavaria. Although the Elector was pleased by the international fame which Sir Benjamin had achieved as a natural philosopher, Thompson's primary responsibility as aide to Carl Theodor was military, and with a basic commitment to order and efficiency, he began to work on a plan for reorganization of the structure of the Bavarian army. After four years of careful planning[17] in February 1788, Thompson, by then promoted to a General, presented to His Electoral Highness his detailed plan. One of the first reforms that Sir Benjamin introduced was to reduce the indebtedness of the common soldier. All equipment and clothing was to be issued by the state to the ordinary soldier free of charge, and his pay was to be increased. Food was also to be available from the military establishment rather than the then current system of having the soldiers buy (or steal) their supplies from the common people.

Food was one of the most expensive commodities for the army, and wherever there was a garrison, Thompson attached to it a "Military Garden" and taught the soldiers to grow their own food. He made an estimate that the monetary value of the produce from the military garden at Mannheim alone amounted to 10% of the total cost of provisions and forage for the whole army. Sir Benjamin's reorganization reached further than the strictly military. Within the army he set up schools and workshops for the soldiers and also for their children, a fact that has led some historians to credit him with setting up the first state supported public schools. The very large military garden in Munich, besides vegetable gardens, fields for grain and hay, also embodied a school of agriculture, model dairies, a cattle breeding station and was open to the public for edification and amusement. Half of its area still exists today as the famous "English Garden" in Munich.

Many of Sir Benjamin's organizational changes were inspired by economy. He had assured the Elector that not only would the army be better equipped, larger, and with greatly improved morale, but also less expensive. Having instituted the military gardens to reduce the cost of food, he turned his attention to the cost of clothes, shoes, and other supplies, but in trying to do something about these, he promptly ran out of manpower.

In an underdeveloped and backward country like Bavaria the poor were not just depressed, they were destitute and starving. Those who did not live on the streets were crowded into hovels in the utmost squalor and misery. Whereas most writers of the period saw in the beggars an insuperable social problem, Thompson recognized in them a highly profitable source of cheap labor. On New Year's Day, 1790, he put the army into the streets of Munich to arrest every beggar they could find. After being picked up, each was assigned a number, released, and told to report the next day at a military workhouse to be taught a trade. As an inducement they were assured of shelter, warmth, and one good meal a day. Then, apparently with great patience, the beggars were taught to spin, to weave, and to sew, producing cloth and clothes, hats and shoes for the Bavarian army.

Since Thompson had collected the dregs of society as his workers, the problems of theft, fraud, and cheating were very real. To combat these, every detail was rigidly controlled and recorded by duplicate or triplicate printed forms which were constantly checked and cross-checked.[18] He is often credited with having introduced the use of multicopy forms as a bureaucratic technique in government operation.

Granted the economy of the poor people's workhouse concept and the fact that this manpower could be organized to make cloth and clothes, what was its purpose? It was, of course, to keep the army warm. But what kind of clothes would keep the soldiers the warmest (for the least money)? That was a question for which there was no answer, so that was where Thompson started to look for basic scientific facts. Where does the warmth of clothing come from? Until Thompson understood that, he could not be sure that his slaving workers were maximizing their efficiency. Using his thermometer No. 4 (Fig. 3), which he called his "passage thermometer," he analyzed the heat conductivity of various clothing materials with the same technique he had used for his liquid measurements. He studied silk, wool, cotton, fur, feathers. First he took equal weights of these substances and measured how fast his thermometers responded to temperature changes. Then he studied how the heat passed through the same material packed tightly or loosely around the thermometer bulb, and how the fineness, the fluffiness, and the density of the insulator affected the measurements. After an extensive study, he came to the remarkable conclusion that it was air, and how it was trapped in the cloth, that was the real key to the insulating properties.

Not only was this discovery important to his own affairs, but it was judged to be such a forward step in the contemporary understanding of heat insulation that the Royal Society of London awarded him its highest award, the Copley Medal, for his paper on this subject.[18]

Another tremendous technological problem that stared the economy-minded Thompson in the face came from the nature of the military workhouse in Munich. Built toward the middle of the 17th century, in modern terms it would be called monolithic prison architecture. For us brought up in today's society, it is difficult to appreciate how miserable the artificial lighting was in the 18th century, and under what dark and gloomy conditions factory workers customarily toiled through the many hours of winter darkness. Tallow candles or oil torches were the standards of the time, and these being a continuous expense, were rationed to a minimum. It was obvious to Sir Benjamin that improvements in this area might be both dramatic and money saving. He turned his full scientific energy to "Measuring the comparative Intensities of the Light emitted by luminous Bodies."[19]

The first thing Sir Benjamin had to do was devise a method for determining not only whether he was making

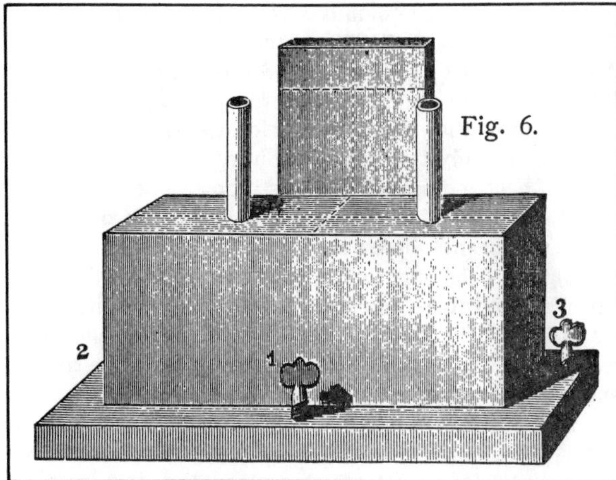

Fig. 4. Rumford's shadow photometer.

improvements, but how great they were. He turned to the relative blackness of shadows as a way of comparing various forms of light sources. He mounted a piece of white paper on the wall of a darkened room, and a little way in front of the wall he fastened a vertical dowel. Arranging his candles or lamps on tables so that the shadow of the dowel from each light source fell close together on the wall, he adjusted the distances of the lights so that the darkness of each shadow was identical. Knowing that the intensity of the light fell off inversely as the square of the distance between the source and the dowel, he could compare the relative strengths of each of his light sources. What he wanted to determine was what kind of lighting would be most efficient for his beggar-workers, so he tested all of the available lamps and candles.

His greatest problem in the beginning was some kind of standard. The light intensity of the candles not only flickered, but rapidly decreased in brightness as the candle burned. Oil lamps were steadier, but he could find no way to set them to give a repeatable brightness, since the light was low when the wick was short and rapidly became brilliant as the wick was turned up. His solution was destined to have a profound effect on the science of lighting for the next century or more. He found that he could get a reproducible light intensity from a candle if he properly snuffed it to a given wick length and burned it only for a short time so that the wick length was constant. He used this candle to adjust an oil lamp. Years later, when carrying out some much more sophisticated experiments, he defined the details of this candle so precisely that it was adopted as the international standard, and illumination to this day is measured in "candle power." He also designed a very simple shadow photometer. His illustration is shown in Fig. 4.

While Sir Benjamin was totally involved in his reorganization of every facet of the Bavarian Army, by a complicated chance of politics, his patron, Carl Theodor, became for four months during 1792, Vicar of the Holy Roman Empire. The Elector seized upon this opportunity to reward his friends and one of his acts was to raise his faithful aide to Count of the Holy Roman Empire and Thompson chose for his name, Rumford, the ancient name of Concord, New Hampshire where his first rise to fame and fortune occurred.

Having risen to the zenith of his power in Bavaria, the new Count Rumford left Munich for a leisurely 16-month tour of Italy, visiting the famous Alessandro Volta, who demonstrated his experiments on the electrical stimulation of frogs' legs which was just then involving him in a controversy with Galvani on the nature of electricity. Rumford spent two months in Florence in the laboratory of Gregorio Fontana, a world famous builder of scientific instruments, and the Count turned his attention to studies in both the fields of light and heat, and he undertook an extensive study of colored shadows which he concluded were optical illusions.[20]

During his travels the city of Verona asked him to design two large institutional kitchens in the hospital of La Pietà and La Misericordia. The Count was already well known for advanced construction of kitchen machinery in Munich, both in his military workhouses and in the military academy; advances stimulated by his desire to spend the least money in the process of cooking food for his beggar-workers as well as the army. The essence of his invention was the modern kitchen range. Basically what he did was enclose the fire in a box, and having experimented with heat insulators, he minimized the loss of heat by the proper choice of stove construction. He was well aware of the need to provide proper circulation of air and to regulate the amount of heat under the pots and pans. He accomplished this by building an adjustable register for the air intake below the fire and installing a damper in the output flue. He also found, by measuring the flue temperature, that more heat was lost up the chimney from big fires than from little fires. He therefore designed his kitchen fireplaces with many small fires and if large quantities needed to be cooked at once, like the food for the children in the La Pietà Hospital, such grates should be in series with other pots before being exhausted up the chimney.

Rumford spent a great deal of time building and rebuilding kitchen equipment and as he experimented with ways of enclosing stoves and cooking equipment, he designed a roaster for which he is justly famous. He studied the difference between roasting and baking and concluded that baking was accomplished in a sealed container at a constant temperature, while roasting required free flow of air and a fire that could be adjusted at will. So successfully did the Count solve these problems that hundreds of his roasters were installed all over Europe and America and in old New England houses they are still to be found unchanged in form from his original design.

In 1796 Rumford published an essay entitled "of Food; and particularly of Feeding the Poor."[21] This very long and detailed essay ends by giving a large number of recipes which constitute a cookbook of wholesome, filling foods at a minimum cost using such ingredients as Indian corn, macaroni, potatoes (which he is credited with introducing into Central Europe), barley, and rye. At the beginning, Rumford gives us his theory of nutrition. What he purports to have discovered was that the basic nutrient element in food was water, not water in a normal state, but water decomposed by the action of food. One of his establishments in the English Garden was a veterinary school; there not only did he breed cattle but experimented with ways to minimize the cost of cattle food while still maintaining the animals fat and healthy. As a result of a long series of tests he demonstrated that feeding hogs, bullocks, and milch cows on potatoes, bran, oatmeal, turnips, and grains cooked in large quantities of water, not only made the animals fatter, stronger, and healthier than if they ate uncooked food, but required less solids. This convinced him that the water was far from a passive agent in the nutritional process. He was quick to carry over these ideas to his House of Industry, and it was not long before the poor were being fed soup morning, noon and night.

Rumford was not a good linguist. He never felt at home in German and what letters we have written in French are not very good. As his duties in Munich became less and less time consuming he spent more of his time experimenting with all kinds of technological improvements which he wanted to write about. But writing and publishing in English was hard to do in Bavaria so he obtained the Elector's permission to spend about a year in England to write up a trunkload of notes and rough drafts of papers. Landing in London, where the universal method of home heating was wood or coal burned in open fireplaces, he was profoundly shocked by the waste and filth of the open fireplace and set to work to improve the situation. So successful was he that no major improvements have been made since. He introduced the smoke shelf, to separate the rising hot air up the front of the chimney from the falling colder air at the back to keep the chimney from smoking, and invented the damper to close off the chimney when not in use. He beveled the sides and back of the fireplace to throw the heat out into the room and his essay "of Chimney Fire-places, with Proposals for improving them to save Fuel; to render Dwelling-houses more Comfortable and Salubrious, and effectually to prevent Chimnies from Smoking"[22] is to this day a standard treatise on how to build an excellent fireplace.

All these technological inventions were making the Count a wealthy man and in July 1796 he used some of his fortune to guarantee that his name would be known forever in the scientific community. He endowed two Rumford Premiums, one with \$5000 to the American Academy of Arts and Sciences in Boston, and one with £1000 to the Royal Society of London. These were the largest prizes ever offered up to that time in the field of science and were to be awarded every two years for outstanding research in the fields of heat and light. They are still so awarded.

Back in Munich in 1796, Rumford made one of his most significant discoveries. He tells us that he was motivated by the urge to find out why he burned his mouth when he tried to eat too hot apple pie! Building directly on his work showing that heat insulation of clothing and stoves was due to the lack of conduction of air if it were trapped and not free to move, he tested to see if the same were true for water. Was it the apple fibers in the pie that kept the water from moving which let it retain its heat? Was stationary water a nonconductor of heat?

His clue came from a chance observation of the motion of the liquid inside one of his thermometers.[23] He discovered that dust inside his thermometer liquid columns demonstrated what we now call "convection" currents, although this name was not given to the phenomenon until many years later.[24] These were, of course, the same fluid currents that Rumford had made use of in his lamps and fireplaces, and in air they had been known for a long time, but the Count was the first to recognize their importance in the heat conduction in liquids.[25]

In September 1797 Rumford wrote[26], "I have been constantly at Munich, and much employed in casting Cannon and preparing them to be able to give our enimies a warm reception should they pay us another visit, which, considering present appearances I do not think at all impossible." He was about to undertake his most famous experiment (see Fig. 5).

The caloric theory of heat[27] visualized the caloric

Fig. 5. Model of the cannon-boring experiment.

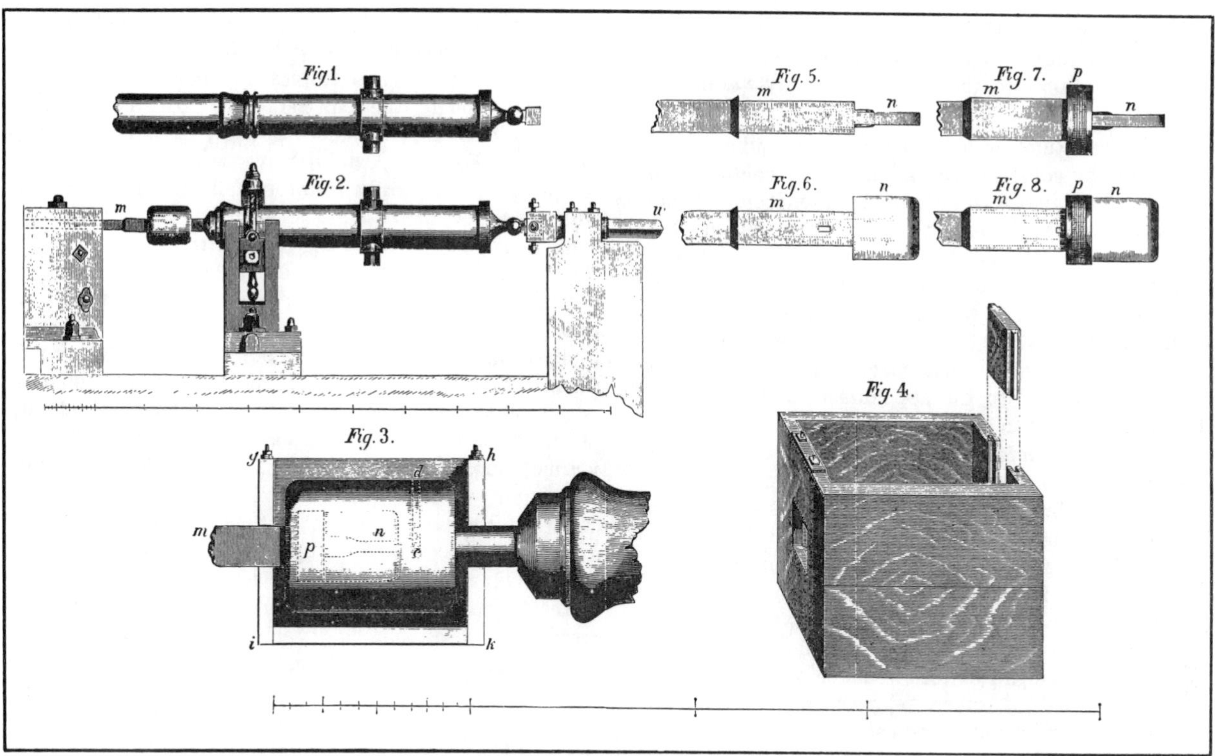

Fig. 6. Rumford's cannon-boring experiment to measure heat from friction.

fluid as squeezed out of a body under the action of mechanical work. Considering the atoms of a gas as essentially unhindered by the gravitational attraction of neighboring gas atoms, they were held apart by the mutual repulsion of their caloric atmospheres. By pushing the atoms closer together, one was merely overcoming the thermorepulsion of mechanical force, and in so doing, squeezing the extra caloric out of the gas, which appeared as heat. If the experiment were reversed, and the gas expanded, the temperature fell, since the atoms were farther apart than their caloric atmospheres would keep them. This would result in a lack of caloric surrounding their atoms, making them colder. Experimental study of the removal of heat by compression was by no means confined to gaseous bodies. For example a piece of iron heated by hammering could not be heated again in that way unless it was meanwhile introduced into the fire, a fact that was explained by supposing that the caloric which had been pressed out of the iron by the percussion was recovered in the fire although now we explain this same observation in terms of the work-hardening of the metal.

It was common knowledge that doing work against friction produced heat, and the caloricists pointed out that friction only came into play when surfaces were pushed together. The advocates of the material theory therefore argued that this heat from friction came from the caloric that was squeezed out from the surface by this pressure. Count Rumford's experiment is well illustrated in Fig. 6. Fig. 1 of his series of figures shows how the cannon came from his foundry. The cannon were cast vertically and the large metal header extended beyond where the barrel was to be bored. The Count had this extra metal formed into the shape illustrated in his Figs. 2 and 3. In the normal production of cannon, the bore was drilled out by mounting the casting horizontally, connecting the cannon part to a horse-driven mill through the bar *w* and the stationary drill *m* was advanced down the casting. For the purpose of this experiment the bit of the drill *n*, shown in greater detail in Fig. 3 through 8, was purposely dull, and the test cylinder of metal was enclosed in a wooden box which could be filled with water. The purpose of the experiment was to see if the quantity of the heat produced by the boring was always the same independent of how long the cannon was drilled. If heat were a material substance it should eventually be drained out but what Rumford showed was that no matter how long the horses were kept at driving the cannon, the length of time it took the dull drill to heat the water to boiling was always the same. A model of this experiment was exhibited for many years in the Deutsches Museum in Munich until destroyed by American bombs during World War II. A picture of this model is shown in Fig. 5.

Rumford carried out many other experiments trying to disprove the caloric theory of heat. Among these were his studies of the fact that water goes through a point of maximum density before it freezes. If caloric was a material substance, it should occupy some space. Removing it by cooling should always lead to contraction, and Rumford felt that this anomolous expansion cast doubt on the caloric theory. Material substances have weight and the Count attempted unsuccessfully to measure the weight of heat, again questioning the material theory of heat. Throughout his life he hammered away at the caloric theory, but in spite of all of his studies he was unable to find definitive proof that the material caloric did not exist. It was not until the development of thermodynamics, a half a century after Rumford, that the caloric theory was finally laid to rest.

For years Rumford had been trying to find a way to get free of the court of Bavaria, and as he continued to upset the *status quo* with his reorganizations and schemes fewer and fewer people wanted him to stay. Finally in 1798 the elector was pleased to name him as his Minister Plenipotentiary to the Court at London. This appointment was made, however, without consulting the British government, and when Rumford arrived in London he immediately discovered that George III had a most "decisive Objection" to the appointment and Rumford's political career came to an abrupt end.

For a while Rumford negotiated with the United States to use his military talents and he sent to Washington his entire military library and a collection of improved cannon which he had designed. He was actually offered the position as the first superintendent of the then forming military academy at West Point. During the discussion of this offer it became clear that President John Adams was in no mood to have an ex-Tory in his military establishment and it was arranged that although Rumford would be officially offered the job, because of other "engagements which great obligations have rendered sacred and inviolable," he would refuse the offer. What these obligations were soon emerged.

In February 1799 he wrote:[28] "We are considering of a scheme for forming in London a Public Institution for Diffusing the knowledge and facilitating the introduction of useful mechanical improvements, and for *teaching* by means of regular Courses of Philosophical Lectures and Experiments the application of Science to the common purposes of life." Rumford's institution, which was shortly to become the Royal Institution of Great Britain, is today one of the leading scientific institutions of the world, but as the Count envisioned it, it was much more like a museum of science than a research institution, although it was to have "a lecture-room . . . fitted up for philosophical lectures and experiments; and a complete LABORATORY AND PHILOSOPHICAL APPARATUS, with the necessary instruments . . . provided for making *chemical* and other *philosophical experiments*."

The Institution bought a house on Albermarle Street, where it is still located, and hired an architect to transform it to the uses of the institution. Rumford hired a Thomas Garnett as a lecturer, whose lectures proved to be very popular, but whose independent way of operating annoyed the Count who forced him out after the first year. Rumford hired in his stead a young 23-year-old biochemical technician, Humphry Davy. Although Davy was to mature to be one of the greatest chemists of the 19th century, his only claim to Rumford's attention was a highly questionable[29] paper "On Heat Light and the Combination of Light."[30] Davy was to spend the rest of his life at the Royal Institution. Shortly thereafter Rumford looked around for another lecturer and found Thomas Young who had just announced his theory of light for which he is so justly famous.

In the autumn of 1801 Count Rumford left his labors at the Royal Institution to return to Munich. Elector Carl Theodor had died and the Count wanted to be assured of the continuing patronage (and pension) from the new Elector Maximilian. Finding all was well, he vacationed for two happy months in Paris at a time when Napoleon was trying to be especially nice to the British and he made a great show of friendliness toward the Count. Napoleon put Rumford under the special care of the famous mathematician Laplace, and Lagrange, Delambre, Monge, Berthollet, Fourcroy, and Prony were their constant companions. But mostly Rumford fell very much in love with the widow of the famous chemist Lavoisier. Lavoisier had been guillotined during the Terror of the French Revolution but with the rise of Napoleon, Mme. Lavoisier became the center of scientific, social and political activity as one of the wealthiest women in all of France. Rumford found it difficult to return to London. When he did around Christmas 1801 he found that the Royal Institution was not fulfilling his dreams. His plans for a technical school, his museum of models, his showplace kitchens and his journal publication schemes had all failed and the Institution was facing financial ruin because of the Count's insistence that the poor rather than the rich should be catered to. The Count gave up and sailed from England in May, 1802, never to return. The Royal Institution turned to the fashionable nobility, erasing all signs of Rumford's dreams, and established itself as a continuing and permanent scientific organization.

Count Rumford spent three months with Mme. Lavoisier in Paris and then returned to Munich. There he found everyone glad to see him, but nothing to do. He turned his attention to the phenomenon of radiational cooling and designed an instrument for doing experiments in this area which is shown in Fig. 7. He reported a large number of experiments with this instrument, checked to make sure that the rate of cooling was logarithmic, and discovered that the shinier a surface was, the slower it cooled. Not only was this conclusion of great theoretical value, but the Count was quick to point out its practical significance when applied to clothing, dishes, and pots and pans. These experiments got him involved in a full scale priority battle with John Leslie[31] who had been doing the same experiments in Edinburgh. The controversy hit the public press and accusations flew back and forth for several months. Almost certainly the two men worked completely independently but national honors were involved, the results of which still exist today. Textbooks in English still only credit Leslie with these experiments, and those in German mention only Rumford.

By May 1803, France and England were at war again; Rumford as a British citizen, with a pension from the British military establishment, was effectively barred from France, so Mme. Lavoisier came to him in Munich. They went travelling in Switzerland and on the Mer de Glace above Chamonix they were fascinated by the cylindrical holes in the ice which were to be found on the glacier surface. Rumford hastened to write a paper on this phenomenon,[32] attributing it to the principle he had reported on in 1797[23] that water goes through a point of maximum density at 41°F according to his measurements (it is actually around 39°F). Since the warmer water sank to the surface of the ice, it melted out these holes.

After a happy time together, Rumford and Mme. Lavoisier parted company. The Count still could not go to Paris, but Mme. Lavoisier went back to France to persuade Napoleon to let him in. By November French officialdom was convinced that Rumford was harmless and with assurances from him that he would not indulge in any

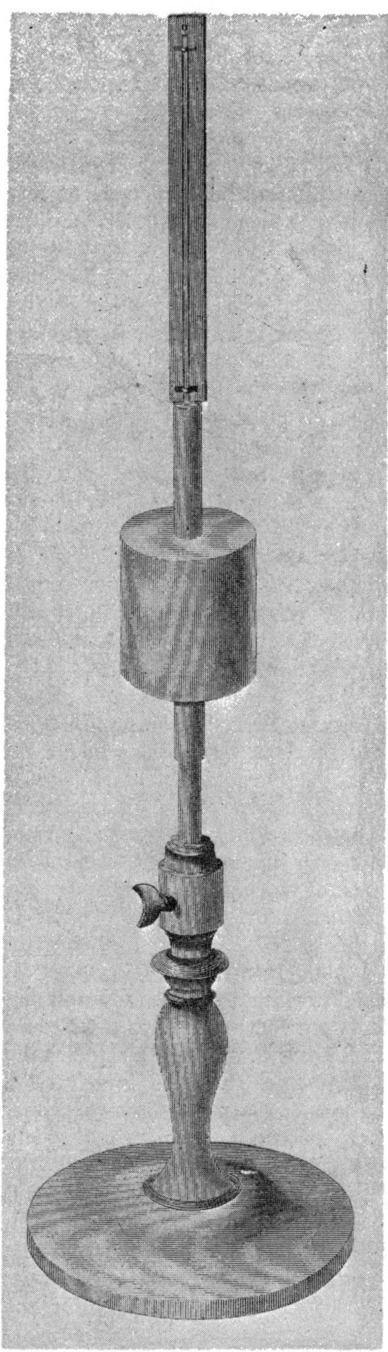

Fig. 7. Apparatus for measuring heat radiation from surfaces.

political activities and apply himself strictly to personal matters and natural philosophy, they allowed him to take up residence in Paris.

When Rumford arrived in Paris he announced that he and Mme. Lavoisier were to be married but legal complications arose almost immediately. The Count discovered that he had to produce his birth certificate, death certificate of his former wife (who had died without any contact from her husband), and of his father. He also needed his mother's permission to marry. With war going on and the common uncertainty of transatlantic travel it was not until October 1805 that all the documents were in order and the ceremony was performed. He wrote to his daughter in Concord, New Hampshire, "I have the best-founded hopes of passing my days in peace and quiet in this paradise of a place, made what it is by me — my money, skill and direction."[33]

It is clear that the Count was busy "Rumfordizing" the house where his new wife had lived for the previous ten years. As a lover and fiancee, Madame could look on with a mixture of amusement and a degree of flattered pride at one of the world's leading authorities improving her heating, her cooking, and her lighting; but after he moved in and proceeded to take over, the sparks began to fly. Mme. Lavoisier-Rumford (she did not give up her former husband's name) was famous for her social events. But these hardly seemed compatible with remodeling kitchens, rebuilding fireplaces, and experimenting with illumination. Also, from all indications, Count Rumford was a bore. He held forth at great length on the minutia of his scientific improvements, and he was inept at the literary repartee which characterized French intellectual social events. His wife was reported to have said,[34] "My Rumford would make me very happy could he but keep quiet."

On the first anniversary of his marriage Rumford characterized his wife as a "female dragon" and it was clear that the arrangement was not working. After one episode when he locked out a group of her callers who had come to attend one of her soirées and she retaliated by pouring boiling water over his flower gardens of which he was very proud, a separation was inevitable and he moved to the suburb of Auteuil to turn his attention entirely to his science and technology.

During the last phase of Rumford's life his writings became voluminous. Let us mention just a few. Having discovered that the taste in coffee was in its volatile oils, he invented what we now call a "dripolator" so that the aromatic oils would be sealed up as the coffee was being brewed.[35] He wrote a long paper "On the Adventures of Employing Wheels with Broad Felloes for Travelling and Pleasure Carriages"[36] for which he devised an ingenious dynamometer connected to the whippletree of his carriage to measure the drawing force exerted by the horses. Rumford was convinced that cold was different from heat, and that "frigorific" rays were best reflected by shiny white cloth; so whenever he went out in winter he was completely dressed from head to foot, including his hat, in white. He measured the heat of combustion of various woods and charcoals[37] and designed a very good combustion calorimeter.[38] And he worked long and hard on improving lamps and illumination, which got him involved in a spirited law suit with some Parisian lamp-makers (he won).

Napoleon's empire was crumbling and Englishmen in France stayed as inconspicuous as possible. In spite of that the last recorded episode of Rumford's life was a visit in November 1813 from his English protégé Humphry Davy who brought along his scientific secretary Michael Faraday to visit the failing Count before he died. Because of the chaos of war we know almost nothing about Rumford's last few months. On March 31, 1814, Czar Alexander's army marched into Paris and ten days later Napoleon abdicated. It was during the social and political uncertainties of the first few months under Louis XVIII that Rumford died of a "nervous Fever" on August 21, 1814. Only a handful of people gathered at the graveside when the Count was buried in the cemetary at Auteuil three days later, where one may still visit his tomb.

Although his scientific contributions are almost unknown today, his name lives on among physicists because his Rumford Medals are still given every two years by the American Academy of Arts and Sciences, and the Royal Society for outstanding contributions to the fields of heat and light; and the fact that he left the residue of his estate to Harvard College assures that there is still at Harvard a Rumford Professor of the Physical and Mathematical Sciences as applied to the Useful Arts.

References

1. A biography written by the same author at the request of the Physical Science Study Committee for their Science Study Series was entitled *Count Rumford, Physicist Extraordinary* (Anchor-Doubleday, Garden City, 1962), p. 178.
2. Owned and operated by the Rumford Historical Association. In celebration of the U.S. bicentennial this association has put on an exhibit at its house in North Woburn a model of Rumford's famous cannon boring experiment.
3. M. A. Pictet, *Science et Arts* XIX Aug. 15, 1801.
4. John Johnston, "Sketch of the Early History of Count Rumford," *Silliman's American Journal of Science* 33, 21 (1838).
5. Thompson's diary is owned by the New Hampshire Historical Society, Concord, NH.
6. Peter Sanborn to G. B. Perry, Reading, MA, Jan. 22, 1841. ALS New Hampshire Historical Society.
7. B. Thompson,"Account of a remarkable Child born at Woburn, N.E. April 1772, with exact drawing," Proc. Amer. Phil. Soc. 1744-1838 (April 2, 1773).
8. Proc. Amer. Phil. Soc. 1744-1839, p. 79.
9. N. Bouton, *The History of Concord* (Concord, N.H. 1856) p. 556.
10. Letters from Thompson to his mother are owned by the Pierce family and were made available through the kindness of William C. Pierce of New York City.
11. Sanborn C. Brown, Elbridge Stein, "Benjamin Thompson and the First Secret-Ink Letter of the American Revolution," Amer. J. of Police Science 40, 627 (1950).
12. B. Thompson, "Miscellanius Observations upon the state of the Rebel Army," Nov. 4, 1775; AM, Clements Library, University of Michigan, Ann Arbor, MI.
13. B. Thompson, Phil. Trans. **71**, 229-328 (1781). This paper is reproduced in the *Collected Works of Count Rumford* (referred to in the following as the *Collected Works*) edited by Sanborn C. Brown (Harvard University, Cambridge, 1970) Vol. 4, pp. 293-394.
14. The paper was not published although it was read to meetings of the Royal Society. It is printed in the *Collected Works* Vol. 3, pp. 433-454.
15. Benjamin F. Thompson, *The History of Long Island*, 1843, Vol. 1, p. 211.
16. B. Thompson, Phil. Trans. **76**, 273 (1786). This paper is reproduced in the *Collected Works* Vol. 1, p. 53.
17. B. Thompson, Vollstandiger *Bericht und Abrechnung uber den Erfolg der neu-eingefuhrten Einrichfungen bey dem churpfalzbaierischen Militar* Munich 1792. A translation of this pamphlet is included in the *Collected Works* Vol. 5, p. 394.
18. B. Thompson, Essay V, London 1796. This paper is reprinted in the *Collected Works* Vol. 5, p. 97.
19. B. Thompson, Phil. Trans. **84**, 67 (1794). This paper is reprinted in the *Collected Works* Vol. 4, p. 97.
20. Rumford, "An Account of some Experiments upon coloured Shadows," Phil. Trans. **84**, 107 (1794). This paper is reprinted in the *Collected Works* Vol. 4, p. 53.
21. Rumford, Essay III, London, 1796. This essay is reprinted in the *Collected Works* Vol. 5, p. 169.
22. Rumford, Essay IV, London 1796. This essay is reprinted in the *Collected Works* Vol. 2, p. 221.
23. Rumford, "On the Propagation of Heat in Fluids," Essay VII, London 1797. This essay is reprinted in the *Collected Works* Vol. 1, p. 119.
24. William Prout, *Bridgewater Treatises* **8**, 65 (1834).
25. Sanborn C. Brown, "The Discovery of Convection Currents by Benjamin Thompson, Count of Rumford," Amer. J. Phys. **15**, 273 (1947).
26. Rumford to Lady Palmerston ALS 20 Sept. 1797. Private collection of S. C. Brown.
27. Sanborn C. Brown, "The Caloric Theory of Heat," Amer. J. Phys. **18**, 367 (1950).
28. Rumford to Lady Palmerston ALS Feb. 2, 1799, archives of the Royal Institution, London.
29. E.N.daC. Andrade, Nature **135**, 359 (1935).
30. H. Davy, *Contributions to Physical and Medical Knowledge principally from the West of England*, edited by T. Beddoes (1799), p. 1.
31. Sanborn C. Brown, Amer. J. Phys. **22**, 13 (1954).
32. Rumford, "an Account of a Curious Phenomenon Observed on the Glaciers of Chamouny," Phil. Trans. **94**, 23 (1804). This paper is reprinted in the *Collected Works* Vol. 2, p. 31.
33. Rumford to Sarah Thompson, Oct. 25, 1805. George E. Ellis, *Memoir of Sir Benjamin Thompson, Count Rumford*, Boston, n.d. (1871) p. 548.
34. Ellis, p. 562.
35. Rumford, "Of the Excellent Qualities of Coffee and the Art of Making it in the Highest Perfection," London 1812. This paper is reprinted in the *Collected Works* Vol. 5, p. 265.
36. Read to the Institute of France, April 15, 1811. This paper is reprinted in the *Collected Works* Vol. 3, p. 461.
37. Rumford, "Account of Some New Experiments on Wood and Charcoal" read before the Institute of France, Dec. 30, 1811. This paper is reprinted in the *Collected Works* Vol. 2, p. 162.
38. Rumford, "Description of a New Calorimeter" read before the Institute of France, Feb. 24, 1812. This paper is reprinted in the *Collected Works* Vol. 2, p. 80.

"[I] am persuaded that a habit of keeping the eyes opened to everything that is going on in the ordinary course of the business of life has oftener led . . . to useful doubts and sensible schemes for investigation and improvement than all the more intensive meditations of philosophers in the hours especially set apart for study."

COUNT RUMFORD

Hans Christian Oersted—Scientist, Humanist and Teacher [1]

J. Rud Nielsen
Department of Physics, University of Oklahoma, Norman, Oklahoma

OERSTED is known to the world as the discoverer of electromagnetism. In Denmark he is remembered with equal gratitude as a great teacher and exponent of physical science, and as the founder of the Royal Technical College and other institutions which have contributed to the enlightenment and welfare of the country. His greatness, not least as an educator, depended on the fact that his interests extended over the entire range of human culture. He was a humanist as well as a scientist. "None of our scientists," he was once told, "regard art, science and men from such a comprehensive point of view as you." The wide scope of Oersted's writings makes them valuable as source material for anyone who wishes to familiarize himself with the status of physical science and philosophy in the early part of the nineteenth century.

Education

Hans Christian Oersted was born in 1777 at Rudkøbing in one of the smaller Danish islands. His early education was rather irregular. The wife of an old German wigmaker taught him to read and to write; the wigmaker taught him German and as much arithmetic as he knew himself, namely to add and to subtract. Multiplication he learned from an older boy and division, from the parson. A former university student taught him Latin and other subjects, and the town judge gave him and a one-year younger brother lessons in French and English. In addition, the brothers "seized with avidity all other means of gaining knowledge. . . ."[2] From his eleventh year Hans Christian helped in his father's pharmacy; and the laboratory work together with the reading of chemical books early aroused his interest in natural science.

In 1794 the two brothers went to Copenhagen to prepare for admission to the University, which they entered the following year. Both of them soon became ardent exponents of Kant's new critical philosophy. Hans Christian heard lectures on mathematics and physical science; his brother took up law and later became a famous jurist. They led a studious life and lived together in *Elers Collegium*, in the very rooms in which the writer of the present article lived more than a century later. The poet Oehlenschläger described them as follows:

> As in a dim monastic cell the Oersteds sat here, grave, silent, at their studies. . . . To all their fellow students they shone resplendent like the Dioscuri, and even ripe scholars soon noticed what was in them.

In 1797 Oersted graduated as a pharmacist with high honors. The previous year he had won the university gold medal for an essay on esthetics. He was awarded the same prize, in 1798, for a medical paper on the origin and function of the amniotic fluid. In addition to the two prize essays, he published in 1798 two "Chemical Letters." In the opening paragraph of the first of these, he wrote:

> I promised . . . to give you an account in letters of the systematic parts of chemistry. . . . I keep my promise with pleasure, both for your sake and for that of science, which you know I find so much pleasure in communicating to others.

This desire to write for the general public stayed with him all his life, and there were few years when he did not write one or more popular papers. In the same year, Oersted wrote the first of many essay book reviews. These were

[1] This article is based largely on Oersted's collected scientific papers, edited by Dr. Kirstine Meyer, née Bjerrum, and published in 1920 by the Royal Danish Academy of Sciences and Letters, under the title *H. C. Ørsted, Naturvidenskabelige Skrifter*, ved Kirstine Meyer f. Bjerrum (Andr. Fred. Høst & Søn, Copenhagen). This work, consisting of three large volumes, contains also two essays by Doctor Meyer. The first of these, entitled "The Scientific Life and Works of H. C. Ørsted," is printed in English; the other, dealing with Oersted's varied activities in the Danish commonwealth, is in Danish. The first two volumes contain 61 scientific papers of which only nine are in Danish. The second volume contains, in addition, 66 papers (in Danish) read before the Royal Danish Academy of Sciences and Letters. In the third volume are reprinted some 50 popular papers on physical science (in Danish). The work does not include Oersted's textbooks, nor his philosophic and poetic works. The writer is indebted to the Danish Academy of Sciences and to Professor Martin Knudsen of the University of Copenhagen for kindly placing this and other material at his disposal.

[2] Since most of the sources quoted are Danish, no detailed references are given. All of the quotations are from the writings of Oersted or his contemporaries. The writer of this article is responsible for the translation of most of them.

rather lengthy and, as he explained, written "in such manner that they have a content of their own, i.e., that they may be read with interest and profit independently of the books which they review and appraise."

In 1798, a new journal was started in Copenhagen for the purpose of promoting the Kantian philosophy, and Oersted became a member of the editorial staff. He contributed a paper on Kant's *Metaphysical Foundations of Natural Science*. This was elaborated into a dissertation, written in Latin, for which, in 1799, he won the degree of Doctor of Philosophy. The thorough study of the critical philosophy gave Oersted a sharp sense for systematic thinking and formed an excellent background for his later scientific work, although it made him an opponent of the atomic theory for the greater part of his life. In these early writings Oersted displayed a great enthusiasm for physical science and a remarkable power to express himself in clear and elegant language.

First Teaching Experience and Physical Research

It was now time to look for a position, and Oersted had his heart set on an academic career. The prospects, however, were far from good. In fact, the state of physics at the University of Copenhagen was a sorry one in those days. Physics had been a subject in its own right since the beginning of the sixteenth century. However, when Pietism flourished in the early part of the eighteenth century, the chair of physics was discontinued to make possible an expansion of the divinity school. Physics was taught by the professors of mathematics or medicine. In 1800, the medical professor who had taught physics and chemistry died, and Oersted, who was then substituting as manager of a well-known pharmacy, applied for that part of the position which concerned physical science. The patron of the University, the Duke of Augustenborg, and several of the older professors were opposed to Oersted because of his advocacy of the newfangled Kantian system; but the pressure of his many friends was so strong that he was finally appointed "adjunct," without salary, with the duty of lecturing two hours a week to the pharmaceutical students. This was hardly an enviable position, but Oersted accepted it with enthusiasm and apparently added a lecture and a laboratory for graduates. Laboratories were then unknown in Denmark, as in most other places, and Oersted could arrange the laboratory only by using the facilities and space of the pharmacy which he managed.

In the year 1800 Volta constructed his galvanic pile, and with its aid Nicholson and Carlisle discovered the electrolytic decomposition of water. This stimulated Oersted's first physical research. He constructed a small battery of novel design and invented a gas voltameter, by means of which, he said, "we shall be able to measure galvanism even more accurately than electricity." He also discovered that "syrup of violets" is stained green by the negative and red by the positive electrode, and found that the colors disappear on shaking. He did not follow up these observations, either for lack of time or because he did not fully recognize their importance.

Oersted as a young man. [From a copper-print made by Crétien, Paris, 1803.]

Travel Abroad

In 1801 Oersted, then nearly 24 years old, received a traveling fellowship and set out on a trip to Germany and France which lasted two and a half years. Although not all the influences to which he was subjected on this journey were of equal value, the trip was of great importance for his later scientific work. He traveled by wagon from town to town, visiting universities, factories, mines and museums. Wherever there was an opportunity he attended lectures, worked in laboratories, or "galvanized" with the small battery that he carried with him. In Berlin he associated mainly with the philosophers, Fichte, A. W. Schlegel and his brother Friedrich, and made a thorough study of Schelling's philosophy of nature. Although he was not blind to the dangers of the romantic movement, which he recognized more clearly later, he was deeply influenced by it. In Oberweimar and, later, in Jena, Oersted visited Ritter who shortly afterwards discovered the electrolytic polarization. Ritter and Oersted became close friends and performed a number of galvanic experiments together. Ritter was a clever experimenter and the creator of several rather fantastic theories that appealed to Oersted at the time. During his first stay with Ritter, Oersted became acquainted with a book obscurely written in Latin by the Hungarian chemist, Winterl. Its leading idea, namely, that all the forces of nature arise from the same fundamental causes, appealed greatly to Ritter and Oersted. They succeeded in showing, to their own satisfaction, the connection between electricity and heat, light, and chemical effects; but with magnetism they had difficulty, although the researches of Coulomb pointed to a fundamental similarity between electricity and magnetism. Oersted decided to adapt Winterl's work for German readers. His book was published in 1803 under the title *Materialien zu einer Chemie des neunzehnten Jahrhunderts*. It was not well received, especially outside Germany where the romantic philosophy of nature had not penetrated; and its publication caused Oersted some embarrassment after his return to Denmark.

In 1802 Oersted went to Paris, where he stayed more than a year. Although he missed the philosophic atmosphere, he gradually learned to appreciate the high development that physical science had reached there. He published two papers on Ritter's galvanic discoveries and did his best to secure for Ritter a large prize offered by Napoleon. He attended lectures and made detailed notes, not only of their contents, but also on their form:

> I learn here daily much about the art of lecturing: from Charles, the way lectures should be delivered; from Vauquelin, how they should not be delivered. . . . I realize how much I have lacked of this art, or rather that I have not known it at all; but when I have completed the training I have now begun I hope to return the wiser. . . .

About Cuvier's lectures he writes:

> These belong to the most interesting of those I have had opportunity to attend. It is the philosophy of natural history that he is here occupied with. . . . It is the spirit of science that he depicts. . . . As to his delivery, it is fluent and beautiful without being embellished with the empty rhetorical phrases of the French.

About Berthollet he writes:

> He speaks with difficulty and hence somewhat slowly; but this very manner fits well the profound ideas that he teaches.

Oersted was enthusiastic about the *École Polytechnique*, especially about the student laboratories:

> The mere dry lectures such as they are given in Berlin without the art of experimentation do not please me; for, after all, all scientific advances must start from experimentation.

Teaching and Textbook Writing

In January, 1804, Oersted returned to Denmark. The enthusiasm for the romantic philosophy of nature revealed in his writings had not pleased the University administration, so he had no hope of obtaining a salaried position. An influential friend advised him to go into applied chemistry, but he replied that while he would be glad to lecture occasionally on technical problems, he was primarily interested in pure chemistry and physics, and rather than seek financial success he would live in accordance with his ideas. After obtaining charge of a collection of physical and chemical instruments belonging to the king, he issued a printed invitation to private lectures on physics and chemistry for which he charged admission. These lectures were a great success, and the attendance increased from year to year. In 1805 he writes in a letter:

> My lectures on chemistry are so strongly attended this year that not all could find room. These lectures are also attended by five or six ladies. You will readily imagine that I make no change in my lectures for their sake.

In another letter, he says:

> These lectures are attended by women as well as by men, although I have only five women. As an introduction to these lectures, I gave in the first three hours a survey of the difference between the older and the newer status of physics. These three lectures were open to the public. My lecture room was far too small to hold all who desired to come, and I won much applause. . . .

A student who attended Oersted's lectures a few years later described his way of lecturing in the following words:

> He usually began very quietly with a few observations and instructive remarks; sometimes also with derivations and definitions of particular terms, whereby he wished to make sure that he would be understood when he proceeded further; once in a while he would also pause to discuss the translation of chemical or physical terms into the Danish language. He then pursued a definite series of phenomena and ideas which were closely related to each other and to the definitions first given. In the beginning his lecture was distinguished almost solely by sharp reasoning, but little by little the separate objects united into larger groups and these in turn joined into a greater unity which he presented vividly to our imagination. Thus his speech became ever mightier, like a stream which grows and is joined by many tributaries, and finally it acted with such a power that at least the younger ones, who were not yet bound by preconceived notions and were susceptible to the new and unusual in his delivery, could with difficulty resist him.

The success of these private lecture courses could not fail to impress the University administration, and in 1806 Oersted was appointed *professor extraordinarius* in physics with the duty to examine candidates in philosophy and to teach physics and chemistry to medical and pharmaceutical students. Although the salary was miserable, Oersted was happy: "I obtain hereby the privilege of being able to found a school of physics in Denmark, for which I hope to find some talented persons among the many young students I shall now have."

With Oersted's appointment, physics was again recognized as a science in its own right rather than merely as a service discipline for medicine and pharmacy. However, only after a long struggle did this recognition become more than a gesture. Oersted worked out detailed proposals for a reform of the study of physics, but they were shelved by the administration. In one of these plans he mentions that on a trip abroad he had made a study of the influence which the experimental sciences had upon a country's welfare. He continues:

> The first question I asked myself was if the chemists and physicists really are right when they claim that their science has such an important influence upon the welfare of the state and if the conviction which I had myself in this regard was based on sufficiently solid reasons. I found that this question must really be answered with a "Yes!" In a country where the scientific knowledge has really penetrated there is soon formed among all educated people a clear idea of what science is able to do and what must be left to practice. . . . But the most important advantage of the diffusion of chemistry and physics among all classes is this, that the practicians acquire theory. . . . If the value of the experimental sciences is considered solely from the point of view of national economics, it may be said that the state needs theorists only to teach the practical people those parts of the theory which are most important to them and to enrich science with new theories which always, sooner or later, will be useful to the practicians.

With his great ability as a writer and his interest in teaching, it was natural that Oersted should write several textbooks. His largest work, *The Science of the General Laws of Nature, Part I,* which appeared in 1809, dealt with the mechanics of solids and fluids and with sound. A second part was to treat "chemical physics"—heat, electricity, magnetism and optics, in addition to what we now call chemistry; but it was never completed, although several of the subjects were treated in separate books. *Part I* appeared in several editions and was used in Denmark for more than fifty years. According to the preface, it was written in such a manner that it could be used both by beginners and by advanced students:

> It is my wish and in part my hope that the students who once, through this work and through my lectures, have acquired a good background in physics should continue this study during the rest of their lives and use this book, with the aid of which they have made their first step in science, as a companion on their further path. . . . I readily admit that I have not

labored for those who merely wish to obtain the knowledge necessary to earn their daily bread in some position.

The introductory chapter, entitled "General Remarks about Science," although now naturally out of date, is vastly superior to the corresponding chapter in most present-day textbooks. In this introduction, as in many other places, Oersted emphasizes the importance of studying the history of science. "By such a study . . . one gains an insight into the development of the whole human mind. . . ." An interesting feature of this book and of several later writings is Oersted's attempt to coin short and natural Danish words to take the place of awkward foreign terms. Many of his new words, such as "ilt" for oxygen (from "ild" = fire) and "brint" for hydrogen (from "brænde" = burn), have become a permanent part of the Danish language. In the preface, Oersted promised to publish an annual supplement in order to bring the book up to date. Time did not permit him to keep this promise. Instead, he delivered, for the rest of his life, a monthly lecture devoted to recent advances in physics and chemistry.

Oersted also wrote papers on the teaching of science with titles such as "On the Manner in Which a Textbook of Physics Should be Written," and "The Briefest Way of Presenting the Theory of Electricity through a Series of Experiments." In this connection it may be mentioned that he, some years later, invented a peculiar grading system (with the scale, 8, 7, 5, 1, −7, −23) which is used in Denmark to this day.

In 1815, the king presented his collection of physical instruments to the University, and increased appropriations were made available for experimental work. Oersted spared no efforts to add to this collection and to find the best available quarters for it. It was of the utmost importance both for his research and for his teaching. Once he wrote in a letter: "I have now beautiful instruments and can fortunately make any experiments whatsoever." In connection with the problem of housing the instrument collection, Oersted succeeded in establishing the first chemical laboratory at the University of Copenhagen. At first, this laboratory consisted merely of a kitchen, but it had the grand name of Royal Chemical Laboratory. A few years later,

Oersted in 1822. On the table stands the compass needle; in his hand is a metal disk on which an acoustic figure is formed; in the background may be seen an early form of his piezometer and also a large galvanic battery. [From a painting by Eckersberg.]

Oersted's pupil, W. C. Zeise, the discoverer of mercaptan and xanthogenic acid, became the first Professor of Chemistry. Oersted's growing reputation as a scientist and his remarkable ability as a lecturer gradually broke down the disfavor with which he was looked upon by the administration and, in 1817, he was made *professor ordinarius* and member of the governing board of the university.

Scientific Work

Oersted's teaching load was heavy and for many years economic difficulties forced him to add to his income by extra work. At times he was discouraged, but somehow he managed to continue his experimental and theoretical researches. There are indications that his teaching gave him stimulation for scientific work. Thus he begins one paper with the words: "My physical researches go in part parallel with my lectures." On the other hand, he began many very promising investigations which his other duties prevented him from carrying to completion.

In 1807 he finished an extensive and important investigation of acoustic figures to which he was led by the hope of finding electrical effects accompanying the oscillations. The introduction to his paper contains this interesting reference to Chladni:

> It follows from the infinity of nature that no observer can discover all that is in an experiment. To understand an experiment quite completely would be the same as finding the key to all of nature. Hence one cannot reproach the ingenious discoverer of the acoustic figures if he has not observed all that really lies in his experiments.

The last part of the paper consists of philosophic speculations on the "profound incomprehensible reason of nature which speaks to us through the flow of music." These ideas were later elaborated in a paper, "On the Cause of the Pleasure Produced by Music."

In 1812 Oersted went abroad again, spending more than a year, mainly in Berlin and Paris, It is interesting to note that he is now rather critical toward the German romantic philosophers about whom he was so enthusiastic in his youth:

> It is also my firm conviction . . . that a great fundamental unity permeates all nature, but just when we have become convinced of this it is doubly necessary that we turn our attention to the world of the manifold where this truth will find its only corroboration. If we do not, unity itself becomes a barren and empty thought leading to no true insight.

On this trip he completed a theoretical work on chemistry which was first published in German and shortly afterwards adapted for French readers under the title, *Recherches sur l'identité des forces électriques et chymiques.* The main conclusion of this book is that all chemical affinities, as well as heat and light, are produced by the positive and negative electricities. The book was well received but, because of its philosophic form and the qualitative nature of its many ingenious observations, it is difficult to ascertain what influence it had. Oersted is still true to his Kantian convictions in opposing the atomistic theory. An interesting feature of the book is the development of an electrical wave theory of light which was elaborated in later papers.

Oersted's researches in the next decade covered a wide range of fields. Volta, and later Simons, had obtained results that seemed to disprove Coulomb's inverse square law for electrostatic forces; so Oersted repeated Coulomb's experiment. He verified the inverse square law for moderate distances but found deviations from this law for very small and very great distances. However, he apparently was not quite convinced about their reality. In the years 1818 and 1819, he investigated the minerals of the island of Bornholm and, in 1820, he discovered the alkaloid piperine.

The Discovery of Electromagnetism

The belief in a connection between electricity and magnetism had taken a firm hold of Oersted's mind during the time of his association with Ritter. It was nourished by his study of the romantic philosophy of nature, although this philosophy otherwise retarded rather than furthered his scientific development. In 1808, Oersted had proposed the problem of the relation between electricity and magnetism for the prize essay of the Danish Academy. In his *Researches on the Identity of Electrical and Chemical Forces*, he had attempted to show that the magnetic effects are produced by electricity; but, as he later put it himself, "he was well aware that nothing in the whole work was less satisfactory than the reasons he alleged for this." In 1817, he constructed, together with Esmarch, a large galvanic battery of small internal resistance with which he made a number of electrolytic experiments. No remarkable results were forthcoming until in April, 1820, when, during an evening lecture, he discovered the effect of an electric current upon a magnetic needle. In the *Edinburgh Encyclopedia*, Volume 18 (1830), Oersted gives the following account of his discovery:

> Electromagnetism itself was discovered in the year 1820 by Professor Hans Christian Oersted of the University of Copenhagen. . . . In the winter of 1819–20, he delivered a course of lectures upon electricity, galvanism and magnetism before an audience that had been previously acquainted with the principles of natural philosophy. In composing the lecture, in which he was to treat of the analogy between magnetism and electricity, he conjectured that if it were possible to produce any magnetical effect by electricity, this could not be in the direction of the current, since this had been so often tried in vain, but that it must be produced by a lateral action. This was strictly connected with his other ideas, for he did not consider the transmission of electricity through a conductor as an uniform stream but as a succession of

interruptions and re-establishments of equilibrium in such a manner that the electrical powers in the current were not in quiet equilibrium but in a state of continual conflict. As the luminous and heating effect of the electrical current goes out in all directions from a conductor which transmits a great quantity of electricity, so he thought it possible that the magnetical effect could likewise eradiate. The observations, above recorded, of magnetical effects produced by lightning in steel-needles not immediately struck confirmed him in his opinion. He was nevertheless far from expecting a great magnetical effect of the galvanical pile; and still he supposed that a power sufficient to make the conducting wire glowing might be required. The plan of the first experiment was to make the current of a little galvanic trough apparatus, commonly used in his lectures, pass through a very thin platina wire which was placed over a compass covered with glass. The preparations for the experiments were made but, some accident having hindered him from trying it before the lecture, he intended to defer it to another opportunity; yet, during the lecture, the probability of its success appeared stronger, so that he made the first experiment in the presence of the audience. The magnetical needle, though included in a box, was disturbed; but as the effect was very feeble and must, before its law was discovered, seem very irregular, the experiment made no strong impression on the audience. It may appear strange that the discoverer made no further experiments upon the subject during three months; he himself finds it difficult enough to conceive it; but the extreme feebleness and seeming confusion of the phenomena in the first experiment, the remembrance of the numerous errors committed upon this subject by earlier philosophers, and particularly by his friend Ritter, [and] the claim [i.e., demand that] such a matter has to be treated with earnest attention may have determined him to delay his researches to a more convenient time. In the month of July, 1820, he again resumed the experiment, making use of a much more considerable galvanical apparatus. The success was now evident, yet the effects were still feeble in the first repetitions of the experiment because he employed only very thin wires, supposing that the magnetical effect would not take place when heat and light were not produced by the galvanical current; but he soon found that conductors of a greater diameter give much more effect, and he then discovered, by continued experiments during a few days, the fundamental law of electromagnetism, *viz.*, *that the magnetical effect of the electrical current has a circular motion round it.*

On July 21, 1820, Oersted announced his discovery in a paper entitled *Experimenta circa effectum conflictus electrici in acum magneticam*, which was sent to learned societies and scholars in the various European countries. An English translation appeared in Thomson's *Annals of Philosophy* (London, 1820). In this paper, which is rather too brief to be perfectly intelligible, Oersted describes some of his experiments and gives a simple rule for finding the direction of the force upon the magnetic pole. In stating his findings he says that the effect passes through all the various mediums which he placed between the conductor and the magnet, and that the force depends upon the nature of this medium, as well as upon the distance from the conductor and the strength of the battery. He mentions that needles of brass and other materials were unaffected by the "electric conflict," as he calls the current. He concludes that:

> It is sufficiently evident from the preceding facts that the electric conflict is not confined to the conductor but [is] dispersed pretty widely in the circumjacent space. From the preceding facts we may likewise collect that this conflict performs circles. . . . This I think will contribute very much to illustrate the phenomena to which the appellation of polarization of light has been given.

Immediately afterwards Oersted published another paper which appeared in the July issue of Schweigger's *Journal für Chemie und Physik* and, in English translation, in Thomson's *Annals of Philosophy*. In this paper he shows that "the magnetic effects do not seem to depend upon the intensity of the electricity but solely on its quantity"; hence, a greater effect is produced by a single large cell than by a large battery of small cells. He also shows that a suspended circuit behaves like a magnet.

On the basis of notes found after Oersted's death, Doctor Kirstine Meyer has succeeded in reconstructing in considerable detail the series of experiments which he carried out in July, 1820. Her findings corroborate Oersted's own account.

The importance of Oersted's discovery was immediately recognized. Ampère wrote: "M. Oersted . . . has forever attached his name to a new epoch." Schweigger expressed the same opinion by beginning a new series of his journal. Faraday wrote of Oersted in 1821,

> . . . his constancy in the pursuit of his subject, both by reasoning and experiment, was well awarded in the winter of 1819 by the discovery of a fact of which not a single person besides himself had the slightest suspicion, but which, when once known, instantly drew the attention of all those who were at all able to appreciate its importance and value.

On a later occasion, he said of the discovery, "It burst open the gate for a domain in science, dark until then, and filled it with a flood of light."

In view of these clear statements of appreciation, it is strange that the stories should have developed that Oersted's discovery was purely accidental and incomplete, that the deflection of the needle was first observed by a janitor, that the discovery was not made by Oersted at all but by Schweigger, and so on. Doctor Meyer has shown that the chief errors, although elaborated by later German writers, originated in an article by the editor of *Annalen der Physik*, Gilbert, who had been unable to understand fully Oersted's Latin paper.

The rapidity with which new discoveries followed Oersted's is well known. Before the end of the year, Ampère had discovered the mutual action of currents, and Arago had constructed the first electromagnet. In 1821 Schweigger invented the "multiplier" or galvanometer and, with its help, Seebeck discovered thermoelectricity. This phenomenon made possible the use of constant electromotive forces and thus led to Ohm's discovery in 1824. An essential part of these advances was the clarification and quantitative definition of such concepts as current, potential difference and electromotive force.

It is symbolic of the great practical importance of Oersted's discovery that the headquarters of the Copenhagen Telephone Company now covers the very spot where he discovered electromagnetism.

Travel and Further Scientific Work

In 1822 Oersted set out on a trip to Germany, France and England, which lasted for nearly a year. Wherever he went, he was honored for his great discovery, for which he had already been awarded the Copley medal by the Royal Society of London and the gold medal of the French Academy. It is again interesting to note the change in his evaluation of the German and the French scientists. He writes:

> In poetry and philosophy I have not noticed that any new shining light has arisen in Germany in recent years. Nor does experimental science fare very well. Berlin has its excellent men in this branch of learning: Seebeck, Erman, Mitscherlich, Heinrich Rose, but from Berlin to Munich, on a journey of about 360 miles during which I have passed through three university towns, I have not found one fairly reliable chemist or experimental physicist. . . . But I found much that was instructive with Fraunhofer at Munich, so that I was able to occupy myself with benefit there for about a fortnight.

From Paris he writes:

> . . . the acquaintances I have made grow every day more cordial and intimate; the benefit I can derive scientifically is thus all the greater. Chevreul, Biot, Fresnel and Pouillet are the men I meet particularly often. . . . Comprehensive science and not only skill in a single branch is now their watchword. . . . If in Germany I am often tempted to protest against the Philosophy of Nature when I see how it is misapplied, in France I feel so much the more called upon to defend it, or rather I feel a fundamental difference in scientific thought which I should not have imagined to be so great if I had not so often felt its vital presence. Still, I am far from falling out with the French on account of this dissimilarity. I now know better than before to appreciate their merit and am therefore on better terms with them.

He had long discussions with Ampère but remained skeptical about the value of the latter's theory. Shortly after his arrival in Paris he gave a report of Seebeck's discovery and with Fourier, as well as alone, made experiments on the new phenomenon for which he proposed the name *thermoelectricity*. With Arago, Oersted discussed the possible connection between light and electromagnetism. In London, Oersted associated especially with Davy, Herschel and Faraday. He also met Wheatstone, who was then a young instrument maker. Oersted performed a number of experiments together with Wheatstone and introduced him to the English scientists.

The most important researches made by Oersted after his discovery of electromagnetism are undoubtedly his preparation of aluminum chloride and metallic aluminum, in 1825, and his extensive series of measurements of the compressibilities of liquids and gases. Since Oersted did not find time to describe his preparation of aluminum in complete form, his name usually is not associated with this achievement. However, a posthumous study of his notes has shown that his claim was well founded. The stimulation to his researches on compressibility came apparently from writing a textbook. These investigations, published in some 20 papers between 1817 and 1845, reveal him as a competent designer of apparatus and a careful and critical experimenter. His piezometer and the methods of measurements which he developed were of basic importance for the work of Despretz, Dulong and Arago. The great difficulty which Oersted and his contemporaries had in determining the correction required because of the compression of the glass vessel containing the liquid being studied is very instructive; it might well give present-day physics teachers food for thought.

Oersted's mind was very fertile in ideas, and his papers are full of statements which show that he was on the track of important discoveries. For example, he early surmised that the tangent of the angle of galvanometer deflection rather than the deflection itself should be taken as a measure of the current; he all but stated Ohm's law before it was announced by Ohm; he developed ideas of electric and magnetic fields, and anticipated in a vague way the electromagnetic theory of light. In all too many cases he did not carry his ideas to fruition. This was due in part to his reluctance or inability to give his ideas mathematical form, but it was also caused by the enormous range of his interests. However, while this great diversity of interests hindered him at times in imposing upon himself that limitation which is required for the complete solution of a scientific problem, it was an essential factor in his greatness as a teacher.

As Oersted became absorbed in the new educational enterprises described in the following pages, he naturally found less and less time for research. Yet he continued to experiment until the end of his life. His last experimental research, completed some two years before his death, dealt with diamagnetism and contained the important result that dia- and paramagnetic rods align themselves differently in a nonhomogeneous magnetic field.

In connection with Oersted's scientific work, it should be mentioned that he served as secretary of the Royal Danish Academy of Sciences and Letters for 36 years. His election to this post was followed by a period of reform of the Academy of Sciences and of great scientific advances in Denmark. The work took a considerable part of his time and energy. Thus, the annual *Proceedings*, containing abstracts of all the papers read at the weekly meetings, were written entirely by Oersted for 27 years. Oersted himself read 66 papers before the Academy, in addition to a large number of brief contributions. As secretary of the Academy of Sciences, he was instrumental in establishing the Meteorological Institute and the Magnetic Observatory at Copenhagen.

The Founding of the Society for the Diffusion of Physical Science

On his return trip from England, in 1823, Oersted conceived a plan for spreading the knowledge of physical science to a larger part of the population, and within a year he had founded the Society for the Diffusion of Physical Science in Denmark, with a membership of 200. In the printed invitation to join this society he emphasizes the importance that a more widespread knowledge of physics would have for the various crafts and industries, and thus for the economic welfare of the state; but he also points out that "the knowledge of the laws of nature form an essential part of man's whole range of knowledge and hence of his culture. As little as we are accustomed to admit this, it is nevertheless true."

The first activity of the Society was the holding of public lectures in Copenhagen by Oersted and two of his colleagues. Each gave two lectures a week, and the arrangement was such that the major parts of physics and chemistry and their applications were treated in a cycle of two winters. The attendance at Oersted's lectures was around 200. At the same time Oersted began to train several young men in the art of giving popular lectures with demonstrations. In order to get a practice school, Oersted offered to furnish a teacher in physics and chemistry to one of the secondary schools in Copenhagen; and the offer was accepted, although without enthusiasm. After half a year the first lecturer, equipped with a collection of instruments, was sent to Aarhus, the largest city in Jutland. He not only gave popular demonstration lectures but also acted as a technical consultant in much the same way as American county agricultural agents. In addition, he sent the Society reports on the condition of the industries in the province. Shortly afterwards three lecturers were sent to other towns. Since artisans worked long hours in those days, Sunday schools were established for them in several towns. In some cases the lecturer managed to get physics introduced in the local high school with himself as unpaid teacher. This method of sending lecturers on physics to the provincial towns was continued for a number of years, although it did not work without difficulties, the chief one being that of securing qualified lecturers.

The Society for the Diffusion of Physical Science did much to encourage the introduction of physics into the schools, especially by lending or donating instruments to schools that were willing to adopt physics as a subject. The outcome was that, in 1845, physics became recognized as a regular part of the curriculum of the secondary schools. After the establishment of the engineering college in 1829, the number of popular lectures in Copenhagen was reduced. However, various new activities were taken up, such as the demonstrations of machines, which began in 1837. During the month of April, 1838, the attendance at these exhibitions was 2185. Printed descriptions and brief oral explanations of the steam engine, "the electromagnetic telegraph," "the electromagnetic motion machine," etc., were given; and the devices were shown in operation. Oersted gave much time and thought to the Society for the Diffusion of Physical Science; and the Society, which is still very active, deserves much credit for the high place that physics and chemistry hold in the interests of the Danish people. Since 1902 it has published the journal, *Fysisk Tidsskrift*, which serves a purpose similar to that of *The American Physics Teacher*. Since 1908, the Society has awarded a gold medal, the Oersted medal, accompanied by a cash prize.

Pioneering in Engineering Education

Although most interested in pure science, Oersted often stressed the benefits which would result from a greater application of physics to practical problems, and he early formed plans for an institution of the type now known as an engineering college. In 1827 one of the officers of the Society for the Diffusion of Physical Science sent plans to the government for the establishment of a trade school similar to a German "Gewerbschule." The proposal was turned over to a committee with Oersted as chairman. It reported that, while such a school might be useful, there was a greater need for an institution on a higher level, for "only a rather high degree of thoroughness can lead to a great and sure application of the natural sciences, and this higher insight is not reached without considerable preparation and prolonged study." Oersted, therefore, proposed that a "polytechnic institute" be formed. He worked out detailed plans whereby it was possible, by a moderate addition to the teaching staff and laboratories already available at the University, to establish such a college without great expense. The entrance requirements, he proposed, should be similar to those of the University. While mathematics, physics and chemistry were to be the chief subjects, he held that:

> . . . the teaching at the Polytechnic Institute must always strive to give the students not only the required knowledge but also such a practice in its application that this knowledge is not a dead treasure when they enter practical life. . . . I believe that no one acquires mathematical ability unless he trains himself in the solution of problems and in applying mathematics everywhere. . . . These lectures [in physics] must be accompanied by experimental exercises conducted in such a way that the students acquire competence in all types of physical experiments. . . . In order to bring about a more perfect cooperation between all the teachers, it is important that each teacher should know what doctrines the others teach; not in order to prevent differences of opinion which stimulate rather than harm when teachers have the requisite wisdom, nor to prevent that repetition of the same truths in different lectures which is inevitable in so related subjects, but in order to enable the teachers to have the proper regard to each other and to correct each other's knowledge in a friendly way. . . . The lectures should be based on printed textbooks. However, these need not be Danish but might be German or French.

Oersted insisted that the teachers should have the same status as University professors.

Oersted's proposal was accepted, and the new institution began its work in the fall of 1829 with Oersted as President and Professor of Physics. In addition, the original faculty consisted of a Professor of Mathematics, two Professors of Chemistry, a Professor of Applied Mechanics, and several teachers of lower rank. At the dedication ceremony Oersted delivered an address "On the Cultural Effects of the Application of the Natural Sciences." The number of engineering students was 22, but several of the lectures were attended also by University students. The course was originally planned to take two years, but soon it was found necessary to increase the time of study.

The building up of the engineering college, then an entirely new type of institution, at a time when the economic conditions in Denmark were very bad was a task that taxed all of Oersted's ability and strength. But he entered upon it with enthusiasm and devotion. He expected a similar devotion from the faculty. In judging the value of a teacher, he would put teaching ability above scientific attainments, a policy for which he was sometimes criticized. He believed in teamwork on the part of the faculty and insisted, for example, that the professors of mathematics have a good knowledge of the subjects to which their students had to apply mathematics. He took a great personal interest in the students and in the apprentices in the shops.

For a number of years, the budget was very inadequate, and the salaries were low. "I have assumed," he said once in connection with the salary problem, "that we all carry this burden because we saw that the Polytechnic Institute would at present be discontinued if we would demand a salary corresponding to our labors." The difficulty of finding laboratory space for the increasing number of students and for research continued for many years. Shortly before his death, his plans for a great expansion of the Polytechnic Institute and for a substantial raise in the salaries were approved by the government; but he did not live to see them carried out.

Another problem which demanded Oersted's attention for many years was how to place the graduates in suitable positions. He had expected that the industries would be anxious to employ such well-trained men but had failed to reckon with the opposition of the guilds to their new competitors. He took up the fight with the guilds with his usual energy, and with considerable success, although the final victory came only in 1849, when a political upheaval gave Denmark a free constitution. Oersted also had to fight for the academic rights of the Polytechnic graduates. After some years they obtained the same status as University graduates, permitting them to hold fellowships and, on certain conditions, to dispute for the doctor's degree.

As the prestige of the Polytechnic Institute grew, the government referred more and more technical problems to it. Oersted himself wrote many important reports which bear witness of the broad understanding and clear vision that characterized his other activities. For example, when a foreign engineer applied for a monopoly on an improved telegraph, Oersted advised "that the construction of electric telegraphs either be undertaken by the government or at least be made a large public enterprise."

The Polytechnic Institute, now called the Royal Technical College, continues to enjoy a high reputation among European engineering schools. Besides the Physical Instrument Collection, it now has an Oersted Museum in which Oersted's instruments, including the famous compass needle, and many of his private possessions may be seen.

Poetic and Philosophic Writings

As previously mentioned, Oersted's first scholarly work was a paper on esthetics, and throughout his life he cultivated poetry and the fine arts. "I philosophize, experiment and write poetry," he once wrote in a letter. In 1836 he published a large poem, *The Airship*, in which he attempted to describe the rich and peaceful life that may be attained through the proper cultivation and application of the natural sciences. He sent a copy to the Swedish chemist, Berzelius, asking him to give it to his wife, and wrote in the accompanying letter: "I know that you do not like that one should thus scatter his strength in several fields; but that much I dare say in my defense that it could not be written except by a physicist." Berzelius replied: "You were, like Davy, born a poet, and your poetic bent has exerted its right in your old age. It is now thirty years since our first acquaintance and I remember, as if it were yesterday, how you then with delightful rapture read to me several of the shorter poetical works of Goethe." In the preface to *The Airship* Oersted wrote: "Amongst other things the sciences would seem able to a great extent to act as a guide to us in the investigation of the nature of the beautiful," an idea which he elaborated in a series of papers on "The Natural Philosophy of the Beautiful."

As Oersted grew old, he turned again to philosophy. However, his long scientific life had given him a sense for reality and a tempered judgment, which he had not always possessed in his youth. Having lived a harmonious and happy life, he never lost his intellectual enthusiasm and optimism or his strong faith in the value of physical science. "It belongs to my plans as a physicist," he wrote in 1836, "to make it evident in every way that the natural sciences should form an essential part of general culture." Some years later he wrote about his work in the decade following the establishment of the Polytechnic Institute:

> My scientific activity in this period is especially marked by an effort to work for my language and my people through popular papers. I have aimed here both at the better educated and the less educated classes. If I am not mistaken, there are certain influences upon human society which ought to issue from the scientists proper, and it is still my opinion that especially the type of thought which is developed by the cultivation of natural science should be extended to a wider circle. Without doubt, the greatest fault of our present culture is an inclination to settle the most important matters according to certain abstract principles without entering into the true relations of these principles to reality. The extremes to which one thereby is misled, be they political, religious, or metaphysical, are all based upon a lack of sense for the truth and reason which lie in the real. That visionary inanity Natural Science is particularly fitted to counteract. I have attempted to apply the scientific spirit in the treatment of religious, esthetic and philosophic topics. That I am far from having produced the effect which I desired is very clear to me; but that I hereby have sown some seeds among the young people who usually flock to my lectures and conversations I dare believe, and I venture to hope for fruits in the future. . . .

His respect for reality is also revealed in the words:

> Should we not feel ashamed in our innermost soul if we caught ourselves in desiring another truth than the real? . . . Let us honor truth! With it the good is inseparably connected. . The full truth carries itself its consolation with it.

Statue of Oersted in the Oersted Park at Copenhagen.

Oersted's conception of the laws of nature is indicated by the following statement:

> The train of ideas through which empirical science comes to realize that the laws of nature are laws of reason is not based on any consideration of the wisdom of these laws . . . but depends upon our seeing that which reason perceives confirmed in nature. It is true that we often come to recognize the agreement of the laws of nature with reason only after finding these laws in experience; but often thought runs ahead of experience, and we find that which is thought verified by nature. Hence, we may say in numerous cases: what reason promises, nature keeps.

The idea of unity is again prominent in Oersted's later philosophic writings:

> The laws of nature in the bodily world are laws of reason, the revelation of one reasonable will; if thus we figure to ourselves the whole bodily world as the continual work of eternal reason, we cannot abide by the consideration of this but are carried on to perceive in our thinking also the same laws of the universe. In other words, spirit and nature are one, viewed under two different aspects. Thus we cease to wonder at their harmony.

About two years before his death Oersted completed a large work, *The Spirit in Nature,* which reached several editions and was translated into a number of languages. In this book he attempts to show that religion, art and philosophy must all be based upon that conception of an order in nature to which we are led by natural science. "The world drawn by the poet, with all its freshness and daring, must after all obey the same laws that our spiritual eye discovers in the real world," he once wrote. Scientific work was to Oersted a sort of religious worship. "God's will can never deviate from the laws of nature for the simple reason that the laws of nature are God's will." He had little use for the dogmas of the Church and on occasions engaged in public controversy with a bishop. He held that Christ has taught us great truths, but has given us no system.

In order to reach a greater part of the population, he often expressed his thoughts in the form of aphorisms or maxims, such as these:

> It is a common spiritual disease to like a new delusion better than an old truth.
>
> When a philosopher scorns Nature because it puts his concepts to shame, he acts like a child who spanks the object that hurt it.
>
> Logic is a dangerous weapon in the hands of passion.
>
> Be stricter to yourself than to others; your egotism is sure to make up for this lack of fairness.

Personality

Oersted possessed a radiant and harmonious personality. He was an optimist by nature and continued to have throughout his life a strong, almost naïve, faith in science and in his fellowmen. He was happy in his work. He was happily married and was a thoughtful husband and father. A friend once came to visit him in his study and found him busy clearing some things from a table. "My wife," Oersted explained, "is going to send my little boy in here, and I put the things away before he comes because I don't like to forbid him to touch them." His home was an intellectual center in Copenhagen and a gathering place for writers, philosophers and scientists. He had many friends whom he was always ready to help by word and deed. "There is nothing in him which one needs conceal or put into a better light before showing it to the world," wrote C. Hauch in 1852, "hence one may well say of him that he was not only a great scientist and a rare thinker but also that he was a great and rare man."

To Hans Christian Andersen, as well as to Zeise and his other students, Oersted was a fatherly and faithful friend. He was the first to appreciate the merits of Andersen's fairy tales and was always ready to console and encourage the poet. Andersen wrote:

> I always returned so clear in thought and rich in mind from the lovable and glorious Oersted; and in the darkest hours of misjudgment and despair it was he who supported me and promised me a better time. . . . One day when I had left him, sick in my soul from the injustice and hardness shown me by others, he could not rest before he, the older man, had looked me up at my house, late at night too, and there once more expressed his sympathy and consolation. It affected me so deeply that I forgot all my sorrow and pain and wept my fill in gratitude and bliss over his infinite goodness; I won again strength and courage to write and work. . . . his wealth of knowledge, experience, and genius, his charming naiveté, something innocent, unconscious as in a child, a rare character with the stamp of divinity were here revealed. . . . He was so mild and good. A child at heart and yet a deep philosopher.

Oersted worked happily to the very end. In March, 1851, he died, after less than a week's illness. He was mourned by his family, by his friends, and by the Danish people.

The Oersted medal of the American Association of Physics Teachers. In 1937 it was awarded to the late Professor Edwin H. Hall and in 1938, to Professor A. Wilmer Duff. A blank of the medal has been sent to Denmark for deposit in a collection commemorating Oersted's work. Professor Frederic Palmer, Jr. suggested the motif appearing on one face of the medal; namely, Oersted, scientist and teacher, discovering electromagnetism in the presence of his assembled pupils. The scene depicted is based on information obtained for the Committee on Awards by Professor J. Rud Nielsen, formerly of Copenhagen, and Professor F. K. Richtmyer, and is believed to be highly authentic. The design of the medal was carried out by the firm of Dieges and Clust.

The search for electromagnetic induction

Samuel Devons

Background:[1]

Hans Christian Oersted's (1777-1851) sensational discovery in 1820 that a magnetic compass experienced a force in the vicinity of an electric current marked the end of a long history of speculation about the *possible* relationship between electricity and magnetism. Oersted's experiments were simple, unambiguous and decisive: not only was the relationship between the electric (Voltaic) current and magnetism established, its extraordinary form was vividly portrayed and reported in a manner which made its speedy verification by others absurdly easy. Within days of the announcement, Oersted's results were reproduced in the major scientific centers of Europe; within a few weeks or months they were rapidly extended. Quickly it was evident that not only could the electric current exert forces on a magnet, but could in fact turn a piece of (nonmagnetized) iron into a magnet as shown by Arago and Davy. Electricity could create magnetism. Ampère, almost from the outset, went much further; magnetism was not only made by electricity – it *was* nothing but electricity, in some special configuration of (molecular) electric currents.

Electricity and magnetism were clearly and firmly related, but the relation was one way – electricity produced magnetism. Yet surely there must be a converse – in some way magnetism must be able to create electricity! It was a belief amply reinforced by other reciprocal relations of electricity – with chemistry and, soon to be discovered (1822), with heat [by Thomas Johann Seebeck (1770-1831)]. It was a belief nourished by the prevalent philosophic faith in the interconnectedness of all of "Nature's innumerable workings." The issue was not so much whether magnetism could produce electricity, but how. There were known parallels that suggested an answer – of sorts.

"Induction" or "influence" was a common enough notion when used to describe familiar electric and magnetic phenomena. A body charged with ordinary (electrostatic) electricity induces electrification in a nearby (conducting) body. Similarly, a magnet induces magnetism in some neighboring magnetic material. The extension to current electricity seemed obvious. The supposition that, somehow, an electric current must induce an electric current (in a nearby conductor) was almost irresistible.

For more than a decade the leading investigators – Ampère and Faraday in particular – repeatedly made deliberate attempts to discover such effects, all with results that were negative, ambiguous, or at best unconvincing. The methods used were by no means unsuited to the purpose, nor were the instruments always lacking in sensitivity. Yet these experiments failed. But at the same time some remarkable new magnetic phenomena, which were discovered "by chance" (Arago) and then thoroughly examined (Christie, Herschel, Babbage, and others), were in fact blatantly displaying the very phenomena being sought:

This work is part of the development of the History-of-Physics Laboratory, and is supported by a grant from the National Science Foundation, Division of Science Education Development and Research: Grant No. HES. 74-17738-A01.

Samuel Devons *received his doctorate from Cambridge University and did research at the Cavendish Laboratory under Rutherford. He came to the U.S. in 1959 as visiting professor of physics at Columbia University and has remained there since then. Physics teachers probably know him best as the director of the Barnard College History of Physics Laboratory. (Department of Physics, Columbia University, New York, NY 10027)*

Fig. 1. Portrait of M. Faraday December 1831

Fig. 2. Portrait of A. M. Ampère c 1820.

striking manifestations of electromagnetic-induction, but, incredible as it must now seem, not recognized as such. In one case the effects were strenuously sought – and were absent or not perceived; in the other they were fully manifest – and not recognized! It would be absurd to put this down simply to ineptness in one case, or blindness in the other. Some of the most perceptive minds and brilliant experimenters of the time were involved. Like so many discoveries, electromagnetic induction looks so simple *afterwards*, if only because experiments can be arranged to make it appear so. It is certainly not difficult – even today when the basic principles are thoroughly established – to display these same phenomena in a setting where their proper interpretation can be quite challenging. When we examine the early experiments in some detail – or even better, rework them – we shall not find all that we look for neatly separated from "irrelevant" distraction; and we may then better appreciate how much experimental observation is determined by expectation, both collective and individual. And how different is the interpretation of the *same* experiment before and after discovering the principle that guides the eyes as well as the mind.

Attempts by Ampère and others:

The first to propose – and to claim to have observed – the reciprocal phenomena was Augustin Fresnel (1788-1827) in one of his rare excursions from the field of optics. In November 1820, only three months after Oersted's announcement, he reported to the Academie des Sciences in Paris that he had succeeded in decomposing water – an incontrovertible test for galvanic electricity – by means of the current generated by a helical wire with a magnet inside. This experiment was suggested by the already familiar magnetization of iron by a current helix.

The experimental results were spurious, however, and the claim soon retracted. Meanwhile, Fresnel's colleague, André Marie Ampère (1775-1836), already deeply involved in what were to be his marathon labors in electrodynamics, was encouraged by Fresnel's claim to add his own – that he too had observed signs, albeit unclear, of the same phenomenon. Ampère's conviction seemed to ebb with Fresnel's but his interest was alerted. A few months later, early in 1821, after some more careful and deliberate investigations – this time based on the analogy with induction – Ampère returned a negative verdict from experiment:[2]

> The proximity of an electric current does not induce another current in a metallic conductor made of copper, even under the most favorable conditions for its influence to be made effective.

This conclusion marks the end of the first phase in Ampère's cycle of belief and disbelief in induced electric currents, and the beginning of a long period of uncertainty and vacillation, and of interest and disinterest. For whether true or not, these induced currents were not, for Ampère, a central issue. His real concern was his doctrine of the essentially electrical (Voltaic) nature of magnetism, his exploration of the structure of the dynamical interaction of electric currents, and his attempt to build on this basis a comprehensive mathematical theory from which all observed phenomena could be "deduced." This theory did not require (or predict) induced currents, although no doubt these could be encompassed within it. On the whole it might be simpler if they did not exist: there was certainly no shortage of *un*ambiguous phenomena for the theory to explain.

But the experiments did continue. The arrangement Ampère used in 1821 was well conceived to detect any induced current, if such existed. A closed circular loop of copper was suspended by a torsionless silk thread so as to lie in the plane of, and close to, a slightly larger fixed circle

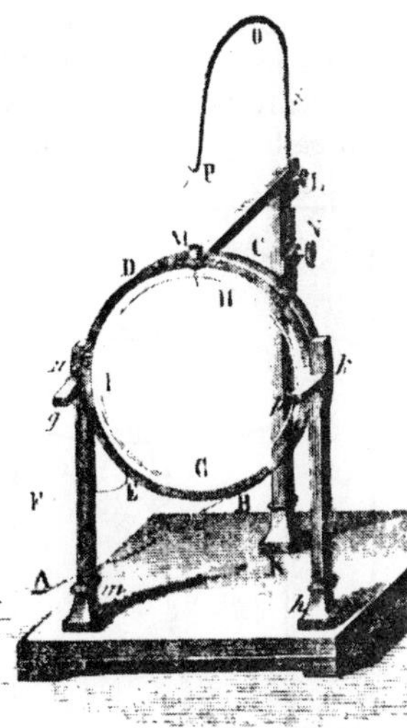

Fig. 3. Apparatus used in original Ampère-de la Rive experiment 1831. (Letter to Van Beck. Reproduced in Journal de Phys. XCIII p. 447 Oct. 1821.)

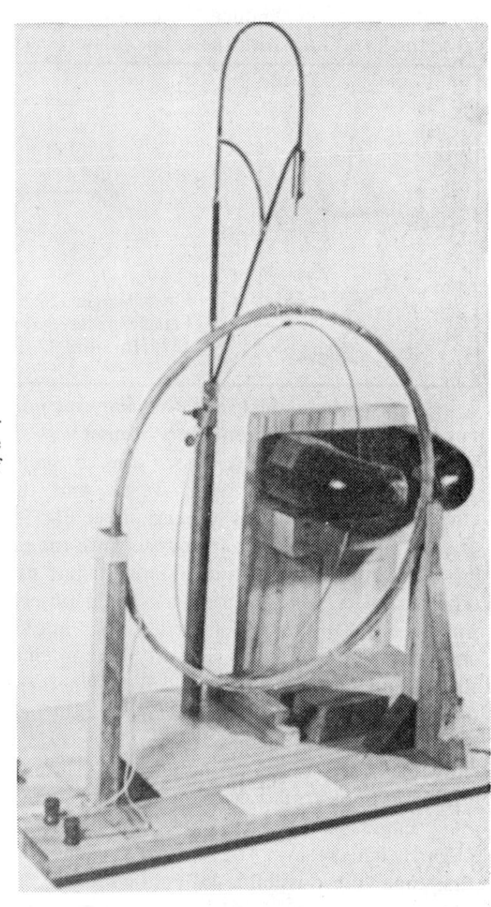

Fig. 4. Reproduction of Ampère-de la Rive experiment in the Barnard-Columbia History of Physics Laboratory.

comprising many turns of insulated copper wire, through which a large current could be passed (Figs. 3 and 4). If current-induction occurred, then the current in the fixed coil would cause some – presumably small – current to flow in the suspended loop, and this latter could be detected by bringing up a strong magnet: the suspended loop should be moved from its position of rest. Ampère's original failure in 1821 to detect such movement could then be attributed to lack of sensitivity in the "instrument" (as well as lack of any real induction!). Needless to say these experiments were conducted in the spirit of a true analogy to familiar electric and magnetic induction: i.e., a steady current in one conductor should induce a steady current in another placed nearby. There was no suggestion of any transient effects, and no hint that such were looked for or observed.

Early in the following year, 1822, in the course of a visit to Geneva, and taking advantage of the availability there of a more powerful magnet, Ampère repeated this experiment with the aid of a young Swiss collaborator, Auguste de la Rive (1801-1873). This time a positive result was obtained. It seems that the suspended ring was observed to move when the current was set up, and then returned to its original position when the current was disconnected. This was interpreted – according to an explication which Ampère gave *very much later* – as showing steady induced current persistently whilst the main current passed, bringing the suspended coil to a new position of equilibrium in which the force of the magnet on the induced current was balanced by the *torsion* of the suspension. When the main current was turned off, this torsion restored the loop to the original equilibrium. The results of this investigation were deemed interesting enough to be reported by Ampère to the Academie later in the year,[3] but not apparently of sufficient importance – or sufficiently certain – to warrant publication, at least by Ampère. There were, however, the other accounts – a brief one by the young de la Rive,[4] which refers to

> the effect which at first M. Ampère believed to be nonexistent [but which now] has been verified by him very definitely while in Geneva.

The "effect" is characterized as the ability of conductors "not otherwise able to acquire permanent magnetism" being able to "at least acquire a sort of temporary magnetism whilst they are under the influence of the current."

The experiments begun in Paris and continued in Geneva did not end with the 1822 version. In 1825, another young member of the Geneva circle and an associate of de la Rive, Jean Daniel Colladon (1802-1893), repeated the attempt made earlier by Fresnel to "induce" current in a helix by a magnet, but now using a galvanometer as a far more sensitive test of induced current than decomposition of water. To obviate any *direct* influence which moving the magnet might have on the galvanometer, this was placed some distance away in an adjacent room and connected to the helix by long wires. Unfortunately, lacking an assistant, Colladon had to move from one room to another to examine the response of the galvanometer to changes in the position of the magnet. By the time he arrived, all transient effects had, needless to say, disappeared: and not suspecting these, he naturally regarded the experiment as having failed to provide evidence for electromagnetic induction.

Attempts by Faraday

About this same time Faraday at the Royal Institution in London was also making sporadic attempts along similar lines to detect electromagnetic induction. Currents induced in wires laid close alongside separate current-carrying wires, and inside helixes, were also sought in vain.[5] Stimulated by the observations of Arago and others with *moving* conductors (see below), Faraday also tried an arrangement in which the charged Leyden-jar was suspended so that its two terminals (of opposite polarity) lay close above a rotating copper plate. There was no sign of any interaction (or "induced" electrification) which would cause the jar to turn with the copper plate (Fig. 5):

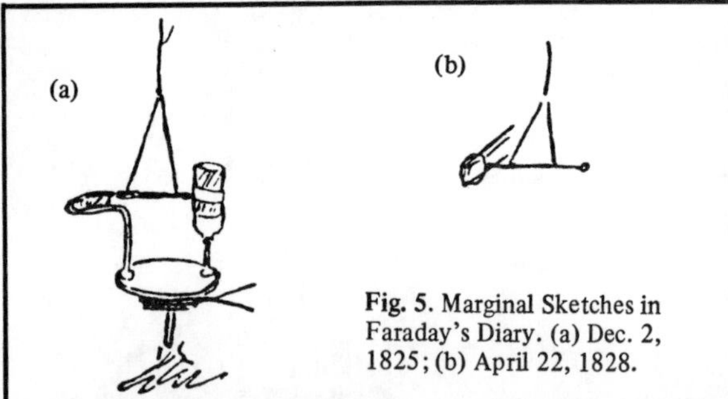

Fig. 5. Marginal Sketches in Faraday's Diary. (a) Dec. 2, 1825; (b) April 22, 1828.

A few years later (1828) his diary records an attempt to observe induced currents in a manner very similar to the Geneva experiments.[6] A closed ring of copper wire is counterbalanced by a small weight and the whole is mounted on a torsionless suspension (see Fig. 5b). The pole of a strong bar magnet is introduced into the ring "Supposing it might exert an influence on it; but upon bringing other magnets near to the wire, could observe no effect whatever . . . ." Faraday also tried rings not soldered (i.e., open circuited), and also rings of platinum and silver – in all cases no signs of induced currents. Whether he regarded these attempts as inconclusive, or of insufficient interest because of their negative outcome, Faraday did not apparently feel they warranted publication. During this period (1824-1830) electricity was not his major preoccupation.

Faraday's dramatic entry into the area of electrical research had been in 1821, when he demonstrated for the first time the possibility of continuous rotation based on Oersted's discovery. Faraday, like others, and especially his senior colleague William Wollaston (1776-1828), was struck by the strange, unprecedented *circularity* of Oersted's magnetic influence. Attention was quickly riveted on this feature. Wollaston seemed convinced that it should be possible for the electric current to make a (cylindrical) magnet rotate about its own axis; but eventually he abandoned the attempt to demonstrate this. Faraday tenaciously persisted and was eventually rewarded by success – a simple, direct, practical discovery (or invention) demonstrating continuous rotation, at first of the current-carrying wire about a "magnetic pole," and later the converse. This circularity of magnetism is a theme which pervades Faraday's later electromagnetism. This achievement led to a correspondence with Ampère in which, whilst ostensibly exchanging views, each presented and maintained his own. In style and background and outlook the divergence between the two could hardly be greater; but both became deeply immersed in the new puzzles (for Faraday) and problems (for Ampère) of electromagnetism. Yet unlike Ampère, for whom electromagnetism became the overwhelming preoccupation, Faraday's interest and energies were, after a year or two, turned in other directions. In 1822/23 there were many entries in his diary showing concern with different forms of electromagnetic rotation – a phenomenon which seemed hard to reconcile with Ampère's theory of action *along* the line joining currents elements. But after that – apart from the occasional entry in 1825, and again in 1828 (as mentioned above) – there was no concern with electromagnetism. For most of a decade chemistry, optics and acoustics were the domains of Faraday's scientific activities. When he did return to the subject in 1831, it was with incredible force and resolution, as if his ideas and intentions which had germinated but remained pent-up in his mind suddenly burst forth. The contrast with his own (and others) earlier furtive attempts could hardly have been more dramatic.

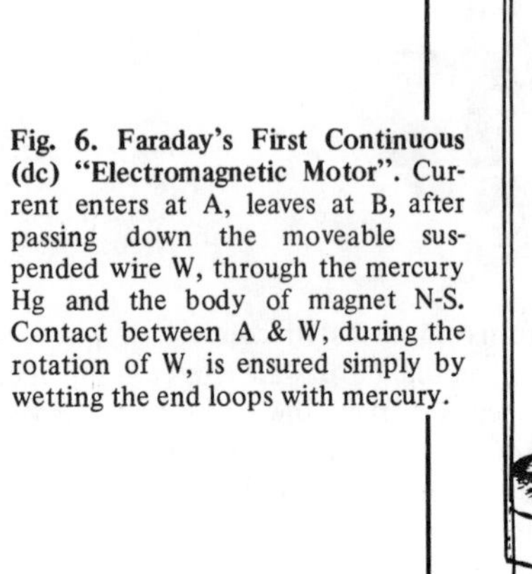

Fig. 6. Faraday's First Continuous (dc) "Electromagnetic Motor". Current enters at A, leaves at B, after passing down the moveable suspended wire W, through the mercury Hg and the body of magnet N-S. Contact between A & W, during the rotation of W, is ensured simply by wetting the end loops with mercury.

"Magnetism of Rotation": Dominique Francoise Jean Arago (1786-1853) and others

Whilst Ampère, Faraday and others were pursuing the indecisive search for "induced" electricity, a phenomenon – the "Arago Effect" – was "accidentally" discovered involving just such currents, but despite its thorough examination was not recognized as such for seven or eight years until Faraday in 1831/32 recognized it as an example

Fig. 7. Apparatus for Demonstrating "Arago Effect." Barnard-Columbia History-of-Physics Laboratory.

of his comprehensive principle of electromagnetic induction.

The first announcement of the phenomenon – an increased damping without change of frequency of a magnet oscillating above, and close to the surface of a metal, or even a nonmetal – was made by Arago in a very brief report to the French Academie in 1824.[7] The following year Arago again reported briefly to the Academie. By 1826, when Arago gave a fuller account of his original investigation and its extension, the subject had stimulated investigations in England, Germany, Switzerland, and Italy – as well as in France – and had become a lively topic of controversy and of the inevitable rival claims for priority.

One of the most extensive inquiries into the nature of the new phenomena was made by Charles Babbage (1792-1871, of calculating-machine fame) and J.F.W. Herschel (1792-1871, the famous astronomer, then Secretary of the Royal Society), and reported in 1825 as "an imperfect and hasty note justified by the great interest [regarding] the curious experiments of M. Arago described by M. Gay-Lussac during his visit to London in the spring of the present year."[8]

By this time, presumably, Arago had observed not only the damping of the oscillations by a stationary metal, but also the deflection, and even the continuous rotation of the compass needle, by a rotating copper disc placed beneath it. These effects with copper and some other metals were confirmed by Babbage and Herschel – experimenting at Babbage's home in London. Then the experiment was "reversed": a powerful (20 lb) magnet was rotated under a copper disc (6 in. diam, 0.05 in. thick) freely suspended by a silk thread, and the disc was observed to deflect (or rotate) in the same sense as the magnet. Sheets of different materials (paper, glass, wood, copper, lead, and tinned iron) were interposed between the suspended disc and the rotating magnet. All had no influence, *except* the iron! Exactly as one would expect if one were observing some form of magnetism induced by the magnet in the metal! But not the ordinary induced magnetism – the new dynamical effect was only observed when there was a relative motion (rotation) of the magnet and the metal. This is the view Babbage, Herschel, and others adopted almost from the outset as a guide to further experiments, for their interpretation, and as a basis for a general "theory" of the phenomena.

They continued by measuring the deflection by the rotating magnet of suspended discs of various metals; the angle was greatest for copper, and successively less for zinc, tin, lead, antimony, mercury, and bismuth. It was zero for nonmetals – except perhaps for a small effect in carbon (from coal gas retorts). This correlated, more or less, with the effectiveness of the materials in damping the oscillations of a compass needle. The best conductors showed the largest "induced magnetism"! Then – as if they were picturing some currents flowing in the disc (but they were not)– they examined the effect of cutting successively more slots in the suspended (or rotating) metal discs:

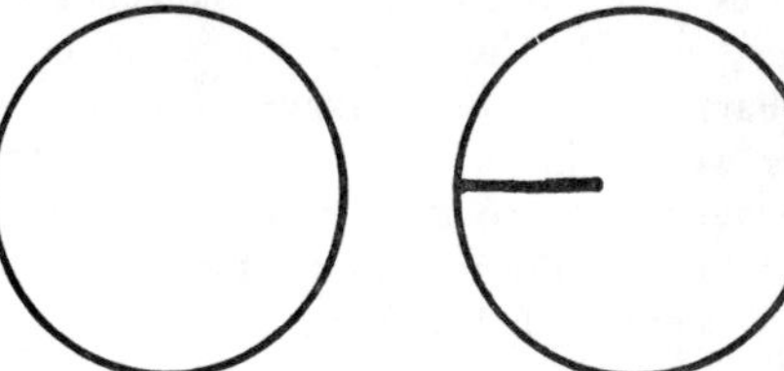
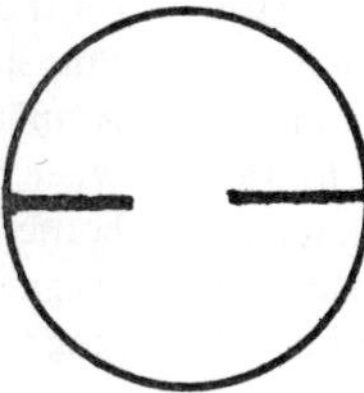

The "induced magnetism" was progressively suppressed. But by filling the slots with tin (or even bismuth – which displays little induced magnetism itself) the induced magnetism was wholly restored. Powdered metals also showed greatly reduced effects. All this was interpreted as an inhibition by the slots of the propagation of *magnetism* from one point to another in the disc: a familiar enough effect in ordinary induced magnetism. For the new dynamical induction, metallic conduction seemed equivalent to magnetic contact.*

In an attempt at a basic theory to embrace all the phenomena they proposed a principle; a

> *postulatum*, viz, that in the (dynamical) induction of magnetism, time enters as an essential element, . . . that no finite degree of magnetic polarity can be communicated to, or taken from any body whatever susceptible of magnetism in an *instant*.

Time is required to lose, as well as to gain, magnetism. It seems as if almost against their will they have hit upon the key factor in electromagnetic induction. But, alas, it was still magnetism only that they were thinking of. Their picture was of a retentive or coercive power of induced magnetism, and of a process in which magnetization consists of separation of two magnetic fluids by infinitesimal distances. Motion, plus the delay in magnetic fluid separation or recombination, enables these fluids to be separated by *finite* distances, and so to exhibit magnetic attractions and repulsions. In Paris, Simeon Denis Poisson (1781-1840) was elaborating a formal theory – a generalization of his earlier (1824) mathematical theory of magnetism – which also introduced the principle of a time-dependence.

Whilst it is no surprise to find the experts in magnetism striving to extend the orthodox theory of magnetic-fluids to embrace the strange new phenomena, and inventing their magnetism of rotation, surely Ampère himself was under no such spell. Had he not already witnessed induced currents – rather than induced magnetism – in his own experiments? Certainly he was not unaware of the work of his colleague (and editor of Annales) Arago. Indeed in 1826

> Ampère was approached by Arago himself who wished to make use of the Voltaic pile and other equipment belonging to the College de France for a continuation of his researching. Arago had in mind substituting a solenoidal electromagnet for the magnetic needle in his original experiment.[9]

With Ampere's approval the experiment (similar to one by Georg Frederick Pohl,[10]) was mounted, and after

* Radial slots are *not* the most effective way of inhibiting the induced currents, which are, where most intense, themselves flowing radially. As Faraday showed in his first paper on electromagnetic induction (Nov. 1831) concentric slots are more effective. A tangential spread of induced magnetism may have seemed more natural: and in any case a disc with radial slots is easier to make than one with concentric slots.

some mishaps was successfully completed by Colladon (who was then in Paris) under Ampère's supervision. Ampère apparently was satisfied that it was just one more piece of evidence in support of his theory: the identity of magnetism with electricity, in this case the equivalence of a bar-magnet with a solenoid. How did he picture the induced magnetism (or whatever else) in the rotating metal, – if he contemplated this question at all? It seems that at this time his faith in his own demonstration of the reality of currents induced by currents had weakened. He seemed content to accept Colladon's (apparent) failure to observe induced currents; at least he offered no encouragement to continue this search. Moreover, Antoine César Becquerel[11] referring to the Ampère-de la Rive experiment in connection with his own investigations of weak magnetism, remarks that although this experiment seemed to demonstrate current-current induction "Monsieur Ampère has subsequently become convinced that this is not so!"

Was Ampère familiar with the experiments with the slotted discs – so strikingly suggestive of currents? Or of the correlation of the Arago effect in different materials with electrical conductivity (which was at least qualitatively recognized)? It is certainly ironic that faced with the most spectacular – albeit complex – demonstration of electromagnetic induction, Ampère failed to recognize or had lost interest in the phenomenon he had long sought. Ampère was not alone.

After a couple of years the great surge of interest in the Arago phenomena waned. The puzzle remained unsolved. All who theorized about induced magnetism of rotation could only succeed by closing their eyes to the facts – especially the existence of repulsive as well as attractive force–or by taking refuge in unfulfilled and unrealizable dreams and speculations. Undoubtedly the phenomena were too complex to provide a striking and convincing demonstration of the simple but revolutionary new principle that was needed. In any event it was not until 1832, after Faraday had taken up the whole issue of electromagnetic induction afresh, that the meaning of the Arago effect was revealed.

Retrospect

In 1822 Faraday had entered in his day book, amongst other "notes, hints, suggestions and objects of pursuit," the prophetic exhortation to: "Convert magnetism into Electricity."

Within ten years the prophecy was fulfilled and the title headings of the beginning of his very first, definitive paper (Nov. 1831)[12] show just how, in doing so, he had found the key to all the mysteries:

> 1. On the Induction of Electric Currents. 2. On the Evolution of Electricity from Magnetism. 3. On a New Electrical Condition of Matter. 4. On Arago's Magnetic Phenomena.

After observing some first feeble indications of momentarily induced currents when currents in neighboring wires were started or stopped (a possibility he may have ruminated on for years) Faraday persisted with extraordinary energy and speed born of confident intuition. In a matter of days, he demonstrated by dozens of experiments and tests both the essential principle of the new, transient, induced currents, and the great variety of circumstances in which it can be expressed: changing currents, changing magnetism, movement of inducing currents or magnets, or of conductors in which currents are induced. Without pause he went on to the first climax of this triumph of experimentation: the demonstration of the *continuous* generation of electricity from magnetism. What a remarkable echo of his earlier triumph – the generation of continuous rotation by electromagnetism, and, incidentally, the secret of the mysterious Arago effect!

This was just one by-product of his new discoveries, which were more significant for the future development of electromagnetism which they opened up than in clearing away some long-standing riddles. Nevertheless, some backward glances and heart-searching questions were inevitable. Why, now that the true answers lay revealed, had they eluded all the earlier probings of Ampère, Arago, and others, and even of Faraday himself? Or had they? The perennial issue of priorities – not untinged with national pride – inevitably arose. Had electromagnetic induction really been discovered earlier, even if only dimly perceived and unconvincingly reported?

For Arago himself, the matter was relatively simple. His own claims had never been more than to have discovered the new phenomena and to insist on their correct description. Insofar as he ventured to explain them, it was only the conjectures of others that he reported (in fact he attributed the first suggestion of induced magnetism to his young colleague Duhamel), and not necessarily with conviction. Reviewing these events in retrospect, Faraday praised Arago's honesty and open-mindedness when confronted with the new phenomena:

> What an education Arago's mind must have received in relation to philosophic reservation . . . what a fine example he has left us of that condition of judgment to which we should strive to attain.[13]

"Philosophic reservation" and freedom from preconceived ideas can hardly explain Ampère's indifferent success, and in his case the phenomena he examined were hardly of the same circumstantial complexity. Ampère had, like Faraday, seemingly posed the right questions; were his experiments capable, sensitive, and direct enough to give the replies? After the event there seemed little doubt; so little that Ampère was rashly tempted to claim that not only had his experiments with de la Rive demonstrated electromagnetic induction, but that this was in essence predicted by his own "theory that traced all magnetic phenomena to the production of electricity in motion." Late in 1831, on learning of Faraday's success, but without full knowledge of all its particulars, Ampère hastily and imprudently published a revised version of his, by then, almost forgotten experiments with de la Rive.[14] If before there had been some doubt about what was observed (which was not lessened by Ampère's own vacillations as to its implication), now Ampère made matters more explicit, but unfortunately not more correct. Absent from the new account is any hint that the induced current is a transient effect, or any clear statement about its *direction* (with respect to the primary current), and still less of the extreme importance of these features. Recriminations and misunderstandings, not surprisingly, ensued. The genial intercourse between Ampère and Faraday took on a sharper tone. But later, when passions were calmed, Ampère wrote to Faraday confessing his failure to discover the true nature of electromagnetic induction and the reasons for his failure. Referring to the experiments with de la Rive,

> At that time (he wrote) I had but one aim in making these experiments . . . to resolve this question . . . [of] the pre-existence of molecular currents in metals able to be magnetized.

Ampère's experiments were, typically, directed at answering "Yes" or "No" to a preconceived question. Faraday's, by contrast, were exploratory: a succession of questions and answers in which experiments (or "Nature's") answer to one question immediately prompted others. For Ampère, experiment was a judge: for Faraday, an indispensable and trusted guide.

Unknown to the scientific world of Ampère and Faraday, in remote Albany, New York, another investigator – Joseph Henry (1797-1878) – was at this time observing or recording puzzling electrical phenomena of a quite different sort. They were also unrecognized manifestations of the principle of electromagnetic-induction soon to be discovered and formulated by Faraday. Henry recognized this after even a brief announcement of Faraday's discovery. Had Henry's work been known in Europe, surely it would have earned the same recognition and commendation as Arago's.

It is much easier to recognize the idiosyncracies of a particular individual than the influence of the character of a period in the development of science. Ampère was not alone in failing to recognize the essential features of time dependence in electromagnetic processes. For him, as for his contemporaries, it was, consciously or not, as much a matter of avoiding these as exploring them. And when such considerations could – apparently without danger, since no Voltaic discharges were involved – be introduced, as in the Arago experiments, the time dependence introduced was, alas, an irrelevant one. Today we have a whole domain of "quasi-stationary" electrical current phenomena with proper criteria for establishing an appropriate time-scale. But such prescience was not vouchsafed for those who struggled with the puzzling phenomena of the 1820s.

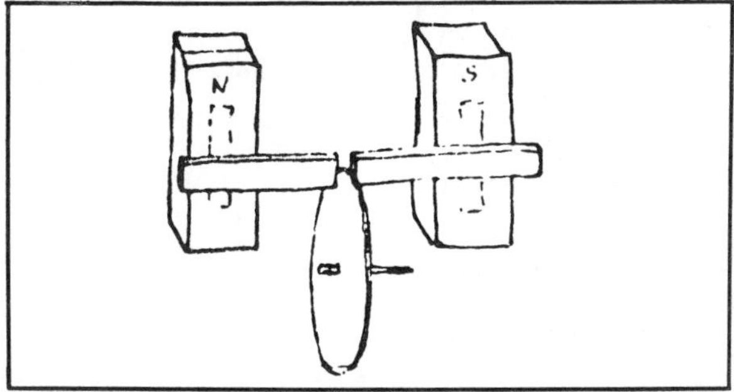

Fig. 8. Diagram from Faraday's diary for Aug. 29, 1831: Recording the first evidence for electromagnetic-induction.

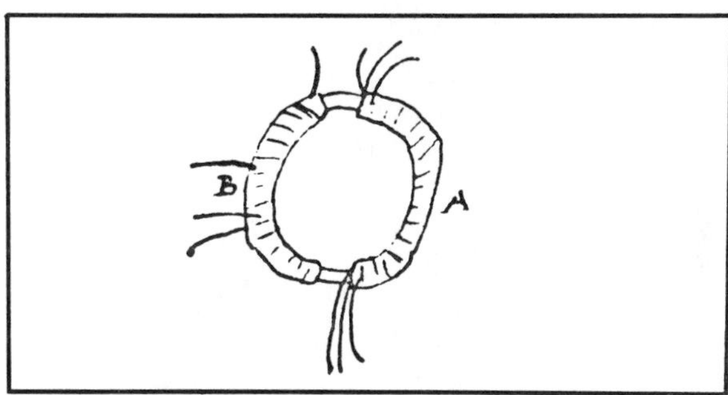

Fig. 9. Diagram from Faraday's diary for Oct. 28, 1831: First production of steady currents by electromagnetic induction.

References

1. Sydney Ross, *The Search for Electromagnetic Induction*, 1820-1831. Notes and Records of the Royal Society **20**, p. 184 (1965).
2. M. A. Ampère, Journal de Physique **93**, p. 447 (1821). Collection de mémoires relative á la physique, publics par la Societe francaies de Physique. Paris 1885-1887, Vol. 2, p. 212.
3. M. A. Ampère. Letter sent to the Académie Royal das Sciences 16 September 1822, Collection de mémoires relative á la physique, publics par la Societé francaies de Physique. Paris 1885-1887. Vol. 2, pp. 332-334.
4. de la Rive, Annales de Chimie et de Physique (2) **21**, p. 24, 1822. See reference 1, p. 187.
5. M. Faraday, Diary Vol. I, London, 1932, pp. 279-280. Entries for Nov/Dec 1825.
6. M. Faraday, Diary, Vol. I, London, 1932, p. 310. Entry for April 22 (1828).
7. Arago, Annales de Chimie et de Physique (2), **27**, p. 363 (1824).
8. Babbage and Herschel, Philosophical Transactions of the Royal Society No. 115, pp. 467-497, June (1825).
9. P. Barlow, Philosophical Transactions of the Royal Society, No. 115, pp. 317-328, May (1825).
10. G. F. Pohl, Pogg. Annalen. 84 (8), pp. 369-396 (1826).
11. A. C. Becquerel, Annales de Chimie et de Physique (2) **25**, p. 269 (1824).
12. M. Faraday, Philosophical Transactions of the Royal Society, Nov. 1831. Also in reference 5.
13. See reference 1, p. 194.
14. M. A. Ampère. Annales de Chimie et de Physique (2) 48, pp. 405-412 (1831).

AMERICAN JOURNAL of PHYSICS

A Journal Devoted to the Instructional and Cultural Aspects of Physical Science

VOLUME 25, NUMBER 6 SEPTEMBER, 1957

On the Relations between Light and Electricity

E. C. WATSON
California Institute of Technology, Pasadena, California
(Received February 5, 1957)

The purpose of this article is to celebrate the centenary of the birth of Heinrich Rudolf Hertz by reprinting his famous lecture on the relations between light and electricity delivered before the German Association for the Advancement of Natural Science and Medicine at Heidelberg on September 20, 1889.

IN the course of only a few years immediately following 1887 a young German physics professor at the Technical High School in Karlsruhe, while still in his early thirties, put the electromagnetic theory of Maxwell upon a firm experimental basis and began a new era of electric wave physics and astounding applications which is continuing at the present time.

Heinrich Rudolf Hertz was born on the 22nd of February, 1857, and died on New Year's Day, 1894. His life work, accomplished in a life span of less than thirty-seven years, has been collected in three separate volumes, each of which fortunately has been translated into English by D. E. Jones. These are *Electric Waves* (Macmillan and Company, Ltd., London, 1894 and 1900), *Miscellaneous Papers* (Macmillan and Company, Ltd., London, 1896), and *The Principles of Mechanics* (Macmillan and Company, Ltd., London, 1899).

As a result, Hertz's important work should be readily available to modern readers. Unfortunately, this is not the case, however, because these volumes have already become rare and only the *Mechanics* is available in a modern reprint (Dover Publications, New York, 1956). It seems desirable, therefore, to celebrate the centenary of Hertz's birth by reprinting the admirable account of the most significant of his discoveries which he gave in a public lecture

HEINRICH RUDOLF HERTZ
(From the frontispiece to *Miscellaneous Papers* engraved from a photograph by R. Krewaldt of Bonn.)

before the German Association for the Advancement of Natural Science and Medicine in Heidelberg on September 20, 1889. This lecture was published by Emil Strauss in Bonn, but was also included in the *Miscellaneous Papers*. The reprint which follows is the English translation by D. E. Jones. It is a classic of clear, simple exposition.

An intimate account of Hertz's life has been given by his daughter, Johanna Hertz, in her *Heinrich Hertz, Erinnerungen: Briefe: Tagebücher* (Leipzig, 1927). Many biographical details as well as discussions of the significance of his work will also be found in the prefaces (by Lord Kelvin, Philipp Lenard, and Helmholtz) to the three volumes mentioned above, as well as in *A History of the Theories of Aether and Electricity*, by Edmund Whittaker, pp. 319–330, in *Great Men of Science*, by Philipp Lenard (translated by H. Stafford Hatfield, Macmillan and Company, Ltd., London, 1933), pp. 358–371, in *Portraits of Famous Physicists with Biographical Accounts*, by Henry Crew (Scripta Mathematica, New York, 1942), in the Introductory Essay by R. S. Cohen in the Dover reprint of *The Principles of Mechanics*, in *Heinrich Rudolf Hertz: Rede zu seinem Gedächtnis*, by Max Planck (Leipzig, 1894), and in J. Zenneck's paper, "Heinrich Hertz," in the *Deutsches Museum Abhandlungen und Berichte* I, Heft 2, pp. 1–36 (1929). Hertz himself in the introduction to his *Electric Waves* has described the manner in which his investigations came to be undertaken and has discussed their bearing upon electrical theory. And finally a fascinating eye-witness account of Helmholtz's announcement of Hertz's discoveries to the Physical Society in Berlin in 1887 will be found in *From Immigrant to Inventor* by Michael Pupin (Charles Scribner and Sons, New York, 1924), pp. 263–274.

On the Relations between Light and Electricity

A lecture delivered at the Sixty-Second Meeting of the German Association for the Advancement of Natural Science and Medicine in Heidelberg on September 20, 1889. [Reprinted from *Miscellaneous Papers* by H. R. Hertz, translated from the German by D. E. Jones (Macmillan and Company, Ltd., London, 1896), by permission of the publishers. The German original was published in Bonn by Emil Strauss.]

WHEN one speaks of the relations between light and electricity, the lay mind at once thinks of the electric light. With this the present lecture is not concerned. To the mind of the physicist there occur a series of delicate mutual reactions between the two agents, such as the rotation of the plane of polarization by the current or the alteration of the resistance of a conductor by the action of light. In these, however, light and electricity do not directly meet; between the two there comes an intermediate agent—ponderable matter. With this group of phenomena again we shall not concern ourselves. Between the two agents there are yet other relations—relations in a closer and stricter sense than those already mentioned. I am here to support the assertion that light of every kind is itself an electrical phenomenon—the light of the sun, the light of a candle, the light of a glow-worm. Take away from the world electricity, and light disappears; remove from the world the luminiferous ether, and electric and magnetic actions can no longer traverse space. This is our assertion. It does not date from today or yesterday; already it has behind it a long history. In this history its foundations lie. Such researches as I have made upon this subject form but a link in a long chain. And it is of the chain, and not only of the single link, that I would speak to you. I must confess that it is not easy to speak of these matters in a way at once intelligible and accurate. It is in empty space, in the free ether, that the processes which we have to describe take place. They cannot be felt with the hand, heard by the ear, or seen by the eye. They appeal to our intuition and conception, scarcely to our senses. Hence we shall try to make use, as far as possible, of the intuitions and conceptions which we already possess. Let us, therefore, stop to inquire what we do with certainty know about light and electricity before we proceed to connect the one with the other.

What, then, is light? Since the time of Young and Fresnel we know that it is a wave motion. We know the velocity of the waves, we know their length, we know that they are transversal waves; in short, we know completely the geometrical relations of the motion. To the physicist it is inconceivable that this view should be refuted; we can no longer entertain any doubt about the matter. It is morally certain that the wave theory of light is true, and the conclusions that necessarily follow from it are equally certain. It is therefore certain that all space known to us is not empty, but is filled with a substance, the ether, which can be thrown into vibration. But whereas our knowledge of the geometrical relations of the processes in this substance is clear and definite, our conceptions of the physical nature of these processes is vague, and the assumptions made as to the properties of the substance itself are not altogether consistent. At first, following the analogy of sound, waves of light were freely regarded as elastic waves, and treated as such. But elastic waves in fluids are only known in the form of longitudinal waves. Transversal elastic waves in fluids are unknown. They are not even possible; they contradict the nature of the fluid state. Hence men were forced to assert that the ether which fills space behaves like a solid body. But when they considered and tried to explain the unhindered course of the stars in the heavens, they found themselves forced to admit that the ether behaves like a perfect fluid. These two statements together land us in a painful and unintelligible contradiction, which disfigures the otherwise beautiful development of optics. Instead of trying to conceal this defect let us turn to electricity; in investigating it we may perhaps make some progress towards removing the difficulty.

What, then, is electricity? This is at once an important and a difficult question. It interests the lay as well as the scientific mind. Most people who ask it never doubt about the existence of electricity. They expect a description of it—an enumeration of the peculiarities and powers of this wonderful thing. To the scientific mind the question rather presents itself in the form—Is there such a thing as electricity? Cannot electrical phenomena be traced back, like all others, to the properties of the ether and of ponderable matter? We are far from being able to answer this question definitely in the affirmative. In our conceptions the thing conceived of as electricity plays a large part. The traditional conceptions of electricities which attract and repel each other, and which are endowed with actions-at-a-distance as with spiritual properties—we are all familiar with these, and in a way fond of them; they hold undisputed sway as common modes of expression at the present time. The period at which these conceptions were formed was the period in which Newton's law of gravitation won its most glorious successes, and in which the idea of direct action-at-a-distance was familiar. Electric and magnetic attractions followed the same law as gravitational attraction; no wonder men thought the simple assumption of action-at-a-distance sufficient to explain these phenomena, and to trace them back to their ultimate intelligible cause. The aspect of matters changed in the present century, when the reactions between electric currents and magnets became known; for these have an infinite manifoldness, and in them motion and time play an important part. It became necessary to increase the number of actions-at-a-distance, and to improve their form. Thus the conception gradually lost its simplicity and physical probability. Men tried to regain this by seeking for more comprehensive and simple laws—so-called elementary laws. Of these the celebrated Weber's law is the most important example. Whatever we may think of its correctness, it is an attempt which altogether formed a comprehensive system full of scientific charm; those who were once attracted into its magic circle remained prisoners there. And if the path indicated was a false one, warning could only come from an intellect of great freshness—from a man who looked at phenomena with an open mind and without preconceived opinions, who started from what he saw, not from what he had heard, learned, or read. Such a man was Faraday. Faraday, doubtless, heard it said that when a body was electrified something was introduced into it; but he saw that the changes which took place only made themselves felt outside and not inside. Faraday was taught that forces simply acted across space; but he saw that an important part was played by the particular kind of matter filling the space across which the forces were supposed to act. Faraday read that electricities certainly existed, whereas there was much contention as to the forces exercised by them; but he saw that the effects of these forces were clearly displayed, whereas he could perceive nothing of the electricities themselves. And so he formed a quite different, an opposite conception of the matter. To him the electric and magnetic forces became the actually present, tangible realities; to him electricity and magnetism were the things whose existence might be disputable. The lines of force, as he called the forces independently considered, stood before his intellectual eye in space as conditions of space, as tensions, whirls, currents, whatever they might be—that he was himself unable to state—but there they were, acting upon each other, pushing and pulling bodies about, spreading themselves about and carrying the action from point to point. To the objection that complete rest is the only condition possible in empty space he could answer—Is space really empty? Do not the phenomena of light compel us to regard it as being filled with something? Might not the ether which transmits the waves of light also be capable of transmitting the changes which we call electric and magnetic force? Might not there conceivably be some connection between these changes and the light waves. Might not the latter be due to something like a quivering of the lines of force?

Faraday had advanced as far as this in his ideas and conjectures. He could not prove them, although he eagerly sought for proof. He delighted in investigating the connection between light, electricity, and magnetism. The beautiful connection which he did discover was not the one which he sought. So he tried again and again, and his search only ended with his life. Among the questions which he raised there was one which continually presented itself to him—Do electric and magnetic forces require time for their propagation? When we suddenly excite an electromagnet by a current, is the effect perceived simultaneously at all distances? Or does it first affect magnets close at hand, then more distant ones, and lastly, those which are quite far away? When we electrify and discharge a body in rapid succession, does the force vary at all distances simultaneously? or do the oscillations arrive later, the further we go from the body? In the latter case the oscillation would propagate itself as a wave through space. Are there such waves? To these questions Faraday could get no answer. And yet the answer is most closely connected with his own fundamental conceptions. If such waves of electric force exist, traveling freely from their origin through space, they exhibit plainly to us the independent existence of the forces which produce them. There can be no better way of proving that these forces do not act across space, but are propagated from point to point, than by actually following their progress from instant to instant. The questions asked are not unanswerable; indeed they can be attacked by very simple methods. If Faraday had had the good fortune to hit upon these methods, his views would forthwith have secured recognition. The connection between light and electricity would at once have become so clear that it could not have escaped notice even by eyes less sharp-sighted than his own.

But a path so short and straight as this was not vouchsafed to science. For a while experiments did not point to any solution, nor did the current theory tend in the direction of Faraday's conceptions. The assertion that electric forces could exist independently of their electricities was in direct opposition to the accepted electrical theories. Similarly the prevailing theory of optics refused to accept the idea that waves of light could be other than elastic waves. Any attempt at a thorough discussion of the one or the other of these assertions seemed almost to be idle speculation. All the more must we admire the happy genius of the man who could connect together these apparently remote conjectures in such a way that they mutually supported each other, and formed a theory of which every one was at once bound to admit that it was at least plausible. This was an Englishman—Maxwell. You know the paper which he published in 1865 upon the electromagnetic theory of light. It is impossible to study this wonderful theory without feeling as if the mathematical equations had an independent life and an intelligence of their own, as if they were wiser than ourselves, indeed wiser than their discoverer, as if they gave forth more than he had put into them. And this is not altogether impossible: it may happen when the equations prove to be more correct than their discoverer could with certainty have known. It is true that such comprehensive and accurate equations only reveal themselves to those who with keen insight pick out every indication of the truth which is faintly visible in nature. The clue which Maxwell followed is well known to the initiated. It had attracted the attention of other investigators: it had suggested to Riemann and Lorenz speculations of a similar nature, although not so fruitful in results. Electricity in motion produces magnetic force, and magnetism in motion produces electric force; but both of these effects are only perceptible at high velocities. Thus velocities appear in the mutual relations between electricity and magnetism, and the constant which governs these relations and continually recurs in them is itself a velocity of exceeding magnitude. This constant was determined in various ways, first by Kohlrausch and Weber, by purely electrical experiments, and proved to be identical, allowing for the experimental errors incident to such a difficult measurement, with another important velocity—the velocity of light. This might be an accident, but a pupil of Faraday's could scarcely regard it as such. To him it appeared as an indication that the same ether must be the medium for the transmission of both electric force and light. The two velocities which were found to be nearly equal must really be identical. But in that case the most important optical constants must occur in the electrical equations. This was the bond which Maxwell set himself to strengthen. He developed the electrical equations to such an extent that they embraced all the known phenomena, and in addition to these a class of phenomena hitherto unknown—electric waves. These waves would be transversal waves, which might have any wavelength, but would always be propagated in the ether with the same velocity—that of light. And how Maxwell was able to point out that waves having just these geometrical properties do actually occur in nature, although we are accustomed to denote them, not as electrical phenomena, but by the special name of light. If Maxwell's electrical theory was regarded as false, there was no reason for accepting his views as to the nature of light. And if light waves were held to be purely elastic waves, his electrical theory lost its whole significance. But if one approached the structure without any prejudices arising from the views commonly held, one saw that its parts supported each other like the stones of an arch stretching across an abyss of the unknown, and connecting two tracts of the known. On account of the difficulty of the theory the number of its disciples at first was necessarily small. But every one who studied it thoroughly became an adherent, and forthwith sought diligently to test its original assumptions and its ultimate conclusions. Naturally the test of experiment could for a long time be applied only to separate statements, to the outworks of the theory. I have just compared Maxwell's theory to an arch stretching across an abyss of unknown things. If I may carry on the analogy further, I would say that for a long time the only additional support that was given to this arch was by way of strengthening its two abutments. The arch was thus enabled to carry its own weight safely; but still its span was so great that we could not venture to build up further upon it as upon a secure foundation. For this purpose it was necessary to have special pillars built up from the solid ground, and serving to support the center of the arch. One such pillar would consist in proving that electrical or magnetic effects can be directly produced by light. This pillar would support the optical side of the structure directly and the electrical side indirectly. Another pillar would consist in proving the existence of waves of electric or magnetic force capable of being propagated after the manner of light waves. This pillar again would directly support the electrical side, and indirectly the optical side. In order to complete the structure symmetrically, both pillars would have to be built; but it would suffice to begin with one of them. With the former we have not as yet been able to make a start; but fortunately, after a protracted search, a safe point of support for the latter has been found. A sufficiently extensive foundation has been laid down: a part of the pillar has already been built up; with the help of many willing hands it will soon reach the height of the arch, and so enable this to bear the weight of the further structure which is to be erected upon it.

At this stage I was so fortunate as to be able to take part in the work. To this I owe the honor of speaking to you today; and you will therefore pardon me if I now try to direct your attention solely to this part of the structure. Lack of time compels me, against my will, to pass by the researches made by many other investigators; so that I am not able to show you in how many ways the path was prepared for my experiments, and how near several investigators came to performing these experiments themselves.

Was it then so difficult to prove that electric and magnetic forces need time for their propagation? Would it not have been easy to charge a Leyden jar and to observe directly whether the corresponding disturbance in a distant electroscope took place somewhat later? Would it not have sufficed to watch the behavior of a magnetic needle while some one at a distance suddenly excited an electromagnet? As a matter of fact these and similar experiments had already been performed without indicating that any interval of time elapsed between the cause and the effect. To an adherent of Maxwell's theory this is simply a necessary result of the enormous velocity of propagation. We can only perceive the effect of charging a Leyden jar or exciting a magnet at moderate distances, say up to ten meters. To traverse such a distance, light, and therefore according to the theory electric force likewise, takes only the thirty-millionth part of a second. Such a small fraction of time we cannot directly measure or even perceive. It is still more unfortunate that there are no adequate means at our disposal for indicating with sufficient sharpness the beginning and end of such a short interval. If we wish to measure a length correctly to the tenth part of a millimeter it would be absurd to indicate the beginning of it with a broad chalk line. If we wish to measure a time correctly to the thousandth part of a second it would be absurd to denote its beginning by the stroke of a big clock. Now the time of discharge of a Leyden jar is, according to our ordinary ideas, inconceivably short. It would certainly be that if it took about the thirty-thousandth part of a second. And yet for our present purpose even that would be a thousand times too long. Fortunately nature here provides us with a more delicate method. It has long been known that the discharge of a Leyden jar is not a continuous process, but that, like the striking of a clock, it consists of a large number of oscillations, of discharges in opposite senses which follow each other at exactly equal intervals. Electricity is able to simulate the phenomena of elasticity. The period of a single oscillation is much shorter than the total duration of the discharge, and this suggests that we might use a single oscillation as an indicator. But, unfortunately, the shortest oscillation yet observed takes fully a millionth of a second. While such an oscillation is actually in progress its effects spread out over a distance of three hundred meters; within the modest dimensions of a room they would be perceived almost at the instant the oscillation commenced. Thus no progress could be made with the known methods; some fresh knowledge was required. This came in the form of the discovery that not only the discharge of Leyden jars, but, under suitable conditions, the discharge of every kind of conductor, gives rise to oscillations. These oscillations may be much shorter than those of the jars. When you discharge the conductor of an electrical machine you excite oscillations whose period lies between a hundred-millionth and a thousand-millionth of a second. It is true that these oscillations do not follow each other in a long continuous series; they are few in number and rapidly die out. It would suit our experiments much better if this were not the case. But there is still the possibility of success if we can get only two or three such sharply-defined indications. So in the realm of acoustics, if we were denied the continuous tones of pipes and strings, we could get a poor kind of music by striking strips of wood.

We now have indicators for which the thirty-thousandth part of a second is not too short. But these would be of little use to us if we were not in a position to actually perceive their action up to the distance under consideration, *viz.*, about ten meters. This can be done by very simple means. Just at the spot where we wish to detect the force we place a conductor, say a straight wire, which is interrupted in the middle by a small spark-gap. The rapidly alternating force sets the electricity of the conductor in motion, and gives rise to a spark at the gap. The method had to be found by experience, for no amount of thought could well have enabled one to predict that it would work satisfactorily. For the sparks are microscopically short, scarcely a hundredth of a millimeter long; they only last about a millionth of a second. It almost seems absurd and impossible that they should be visible; but in a perfectly dark room they *are* visible to an eye which has been well rested in the dark. Upon this thin thread hangs the success of our undertaking. In beginning it we are met by a number of questions. Under what conditions can we get the most powerful oscillations? These conditions we must carefully investigate and make the best use of. What is the best form we can give to the receiver? We may choose straight wires or circular wires, or conductors of other forms; in each case the choice will have some effect upon the phenomena. When we have settled the form, what size shall we select? We soon find that this is a matter of some importance, that a given conductor is not suitable for the investigation of all kinds of oscillations, that there are relations between the two which remind us of the phenomena of resonance in acoustics. And lastly, are there not an endless number of positions in which we can expose a given conductor to the oscillations? In some of these the sparks are strong, in others weaker, and in others they entirely disappear. I might perhaps interest you in the peculiar phenomena which here arise, but I dare not take up your time with these, for they are details—details when we are surveying the general results of an investigation, but by no means unimportant details to the investigator when he is engaged upon work of this kind. They are the peculiarities of the instruments with which he has to work; and the success of a workman depends upon whether he properly understands his tools. The thorough study of the implements, of the questions above referred to, formed a very important part of the task to be accomplished. After this was done, the method of attacking the main problem became obvious. If you give a physicist a number of tuning-forks and resonators and ask him to demonstrate to you the propagation in time of sound waves, he will find no difficulty in doing so even within the narrow limits of a room. He places a tuning fork anywhere in the room, listens with the resonator at various points around, and observes the intensity of the sound. He shows how at certain points this is very small, and how this arises from the fact that at these points every oscillation is annulled by another one which started subsequently but traveled to the point along a shorter path. When a shorter path requires less time than a longer one, the propagation is a propagation in time. Thus the problem is solved. But the physicist now further shows us that the positions of silence follow each other at regular and equal distances: from this he determines the wavelength, and, if he knows the time of vibration of the fork, he can deduce the velocity of the wave. In exactly the same way we can proceed with our electric waves. In place of the tuning fork we use an oscillating conductor. In place of the resonator we use our interrupted wire, which may also be called an electric resonator. We observe that in certain places there are sparks at the gap, in others none; we see that the dead points follow each other periodically in ordered succession. Thus the propagation in time is proved and the wavelength can be measured. Next comes the question whether the waves thus demonstrated are longitudinal or transverse. At a given place we hold our wire in two different positions with reference to the wave: in one position it answers, in the other not. This is enough—the question is settled: our waves are transversal. Their velocity has now to be found. We multiply the measured wavelength by the calculated period of oscillation and find a velocity which is about that of light. If doubts are raised as to whether the calculation is trustworthy, there is still another method open to us. In wires, as well as in air, the velocity of electric waves is enormously great, so that we can make a direct comparison between the two. Now the velocity of electric waves in wires has long since been directly measured. This was an easier problem to solve, because such waves can be followed for several kilometers. Thus we obtain another measurement, purely experimental, of our velocity, and if the result is only an approximate one, it at any rate does not contradict the first.

All these experiments in themselves are very simple, but they lead to conclusions of the highest importance. They are fatal to any and every theory which assumes that electric force acts across space independently of time. They mark a brilliant victory for Maxwell's theory. No longer does this connect together natural phenomena far removed from each other. Even those who used to feel that this conception as to the nature of light had but a faint air of probability now find a difficulty in resisting it. In this sense we have reached our goal. But at this point we may perhaps be able to do without the theory altogether. The scene of our experiments was laid at the summit of the pass which, according to the theory, connects the domain of optics with that of electricity. It was natural to go a few steps further, and to attempt the descent into the known region of optics. There may be some advantage in putting theory aside. There are many lovers of science who are curious as to the nature of light and are interested in simple experiments, but to whom Maxwell's theory is nevertheless a seven-sealed book. The economy of science, too, requires of us that we should avoid roundabout ways when a straight path is possible. If with the aid of our electric waves we can directly exhibit the phenomena of light, we shall need no theory as interpreter; the experiments themselves will clearly demonstrate the relationship between the two things. As a matter of fact such experiments can be performed. We set up the conductor in which the oscillations are excited in the focal line of a very large concave mirror. The waves are thus kept together and proceed from the mirror as a powerful parallel beam. We cannot indeed see this beam directly, or feel it; its effects are manifested in exciting sparks in the conductors upon which it impinges. It only becomes visible to our eyes when they are armed with our resonators. But in other respects it is really a beam of light. By rotating the mirror we can send it in various directions, and by examining the path which it follows we can prove that it travels in a straight line. If we place a conducting body in its path, we find that the beam does not pass through—it throws shadows. In doing this we do not extinguish the beam but only throw it back; we can follow the reflected beam and convince ourselves that the laws of its reflection are the same as those of the reflection of light. We can also refract the beam in the same way as light. In order to refract a beam of light we send it through a prism, and it then suffers a deviation from its straight path. In the present case we proceed in the same way and obtain the same result; excepting that the dimensions of the waves and of the beam make it necessary for us to use a very large prism. For this reason we make our prism of a cheap material, such as pitch or asphalt. Lastly, we can with our beam observe those phenomena which hitherto have never been observed excepting with beams of light—the phenomena of polarization. By interposing a suitable wire grating in the path of the beam we can extinguish or excite the sparks in our resonator in accordance with just the same laws as those which govern the brightening or darkening of the field of view in a polarizing apparatus when we interpose a crystalline plate.

Thus far the experiments. In carrying them out we are decidedly working in the region of optics. In planning the experiments, in describing them, we no longer think electrically, but optically. We no longer see currents flowing in the conductors and electricities accumulating upon them: we only see the waves in the air, see how they intersect and die out and unite together, how they strengthen and weaken each other. Starting with purely electrical phenomena we have gone on step by step until we find ourselves in the region of purely optical phenomena. We have crossed the summit of the pass: our path is downwards and soon begins to get level again. The connection between light and electricity, of which there were hints and suspicions and even predictions in the theory, is now established: it is accessible to the senses and intelligible to the understanding. From the highest point to which we have climbed, from the very summit of the pass, we can better survey both regions. They are more extensive than we have ever before thought. Optics is no longer restricted to minute ether waves a small fraction of a millimeter in length; its dominion is extended to waves which are measured in decimeters, meters, and kilometers. And in spite of this extension it merely appears, when examined from this point of view, as a small appendage to the great domain of electricity. We see that this latter has become a mighty kingdom. We perceive electricity in a thousand places where we had no proof of its existence before. In every flame, in every luminous particle we see an electrical process. Even if a body is not luminous, provided it radiates heat, it is a center of electric disturbances. Thus the domain of electricity extends over the whole of nature. It even affects ourselves closely: we perceive that we actually possess an electrical organ—the eye. These are the things that we see when we look downwards from our high standpoint. Not less attractive is the view when we look upwards towards the lofty peaks, the highest pinnacles of science. We are at once confronted with the question of direct actions-at-a-distance. Are there such? Of the many in which we once believed there now remains but one—gravitation. Is this too a deception? The law according to which it acts makes us suspicious. In another direction looms the question of the nature of electricity. Viewed from this standpoint it is somewhat concealed behind the more definite question of the nature of electric and magnetic forces in space. Directly connected with these is the great problem of the nature and properties of the ether which fills space, of its structure, of its rest or motion, of its finite or infinite extent. More and more we feel that this is the all-important problem, and that the solution of it will not only reveal to us the nature of what used to be called imponderables, but also the nature of matter itself and of its most essential properties—weight and inertia. The quintessence of ancient systems of physical science is preserved for us in the assertion that all things have been fashioned out of fire and water. Just at present physics is more inclined to ask whether all things have not been fashioned out of the ether? These are the ultimate problems of physical science, the icy summits of its loftiest range. Shall we ever be permitted to set foot upon one of these summits? Will it be soon? Or have we long to wait? We know not: but we have found a starting point for further attempts which is a stage higher than any used before. Here the path does not end abruptly in a rocky wall; the first steps that we can see form a gentle ascent, and amongst the rocks there are tracks leading upwards. There is no lack of eager and practised explorers: how can we feel otherwise than hopeful of the success of future attempts?

Germanium Available Through Apparatus Committee

In June, 1952, the Bell Telephone Laboratory sponsored a summer school for teachers of transistor physics. The experiments used during these sessions were published in the *American Journal of Physics* [W. J. Leivo, Am. J. Phys. 22, 622 (1954)]. Since that time, many people have tried to introduce these experiments into their undergraduate laboratory courses, but have found difficulty in obtaining the proper germanium samples. Because of this difficulty, the Bell Telephone Laboratory has made available through the AAPT Committee on Apparatus for Educational Institutions samples of germanium rods and germanium wafers suitable for these experiments. Members of the AAPT who wish to obtain N-type and P-type germanium rods and N-type germanium wafers for carrying out the first three experiments described in the above paper, as well as writeup of these experiments and details on finishing the samples, are invited to write for them to Dr. W. C. Kelly, American Institute of Physics, 335 East 45th Street, New York 17, New York.

Michelson–Morley Experiment*

R. S. SHANKLAND
Case Institute of Technology, Cleveland, Ohio
(Received 16 July 1963)

The Michelson–Morley experiment, performed in Cleveland in 1887, proved to be the definitive test for discarding the Fresnel aether hypothesis which had dominated physics throughout the 19th century. The experiment had been suggested to Michelson by his study of a letter of James Clerk Maxwell, and a preliminary but inconclusive trial had been made at Potsdam in 1881. It seems certain that the experiment would never have been repeated except for the urging of Kelvin and Rayleigh at the time of Kelvin's Baltimore Lectures in 1884, which Michelson and Morley attended. The conclusive null result of the Cleveland experiment was decisive in its influence on Lorentz, FitzGerald, Larmor, Poincaré, and Einstein in developing their theories of the electrodynamics of moving bodies, which culminated in the special theory of relativity. The present account contains material from extensive notes and correspondence related to the work of Michelson and Morley which the writer has assembled during the past years.

I. INTRODUCTION

THE revival and development of the wave theory of light at the beginning of the nineteenth century, principally through the contributions of Young and Fresnel, raised a problem which proved to be of major interest for physics throughout the entire century. The question was on the nature of the medium in which light is propagated. This medium was called the aether and an enormous amount of experimental and theoretical work was expended in efforts to determine its properties. On the experimental side, a long series of electrical and optical investigations was carried out attempting to measure the motion of the earth through the aether medium. For many years, the experimental precision permitted measurements only to the first power of the ratio of the speed of the earth in its orbit to the speed of light ($v/c \simeq 10^{-4}$), and these "first-order experiments" uniformly gave null results. It became the accepted view that the earth's motion through the aether could not be detected by laboratory experiments of this sensitivity. With the development of Maxwell's electromagnetic theory of light, and especially with its extensions by Lorentz in his electron theory, theoretical explanations for the null results obtained in the first-order aether-drift experiments were provided. This situation was in harmony with the Galilean–Newtonian principle of relativity in mechanics, which asserts that the essential features of all uniform motions are independent of the frame of reference in which they are described. In Maxwell's electromagnetic theory, however, the situation was different when quantities of the second order in (v/c) were considered. According to the Maxwell theory, effects depending on $(v/c)^2$ should have been detectable in optical and electrical experiments. The presence of these effects would reveal a preferred reference frame for the phenomena in which the aether would presumably be at rest. At first, this feature of Maxwell's theory, which implied that aether-drift effects to the second order in (v/c) might be observed, raised a purely hypothetical question because the accuracy needed for such experiments was about one part in a hundred million, and no experimental techniques then known could attain this sensitivity.[1]

* This paper is the result of three talks given by the the writer. The first was on 19 December 1952 when the Cleveland Physics Society celebrated the centenary of the birth of Michelson. The second was at the New York meeting of the American Association of Physics Teachers on 30 January 1959, and the third was at a symposium on 24 November 1962 organized by The American Physical Society for its Cleveland meeting to commemorate the 75th Anniversary of the Michelson–Morley experiment.

[1] James Clerk Maxwell, article on aether in Encyclopaedia Britannica 9th ed., Vol. 8; also, in *Scientific Papers* (Dover Publications, Inc., New York, 1952), Vol. 2, pp. 763–775. "If it were possible to determine the velocity of light by observing the time it takes to travel between one station and another on the earth's surface, we might, by comparing the observed velocities in opposite directions, determine the velocity of the aether with respect to these terrestrial stations. All methods, however, by which it is practicable to determine the velocity of light from terrestrial experiments depend on the measurement of the time required for the double journey from one station to the other and back again, and the increase of this time on account of the relative velocity of the aether equal to that

Michelson pondered this problem and it led him to invent the Michelson interferometer and to plan the aether-drift experiment, which he carried to completion in collaboration with Edward W. Morley at Cleveland in 1887. This famous optical-interference experiment was devised to measure the motion of the earth through the aether of space by means of an extremely sensitive comparison of the velocity of light in two mutually perpendicular directions. The experiment as carried out in 1887 gave a most convincing null result and proved to be the culmination of the long nineteenth-century search for an aether. At that time, the definitive null result of the Michelson–Morley experiment was a most disconcerting finding for theoretical physics, and indeed for many years repetitions of this experiment and related ones were performed with the hope of finding positive experimental evidence for the earth's motion through the aether. These later experiments, however, have all been shown to be consistent with the original null result obtained by Michelson and Morley.[2] In the years following 1887, their experiment led to extensive and revolutionary developments in theoretical physics. It proved to be the major incentive for the work of FitzGerald, Lorentz, Larmor, Poincaré, and others, leading finally in 1905 to the special theory of relativity of Albert Einstein.

II. ANNAPOLIS AND WASHINGTON

In the years immediately following Michelson's graduation from the U. S. Naval Academy in 1873, his researches in optics were exclusively concerned with measurements of the speed of light. While serving during 1875–1879 as instructor in physics at Annapolis, he made his first determination of this quantity with a demonstration for the students in November 1877 in which he repeated, with essential improvements, the rotating-mirror experiment of Foucault. These simple trials gave such good results that he decided to repeat and extend them with improved apparatus. This led to his transfer in 1879 to the Nautical Almanac Office in Washington, D. C., where Professor Simon Newcomb was director. Newcomb was the leading scientist in Washington, and he obtained ample support for their measurements of the speed of light made between stations at Fort Myer, Virginia, and the Old Naval Observatory and the Washington Monument.

Perhaps the most important event which occurred for Michelson while he was at the Nautical Almanac Office was his opportunity to study a letter dated 19 March 1879 from James Clerk Maxwell to David Peck Todd,[3] then also associated with the Nautical Almanac Office. In this, Maxwell inquired whether the existing observations of the eclipses of Jupiter's satellites made at several epochs of the earth's orbital motion were of sufficient accuracy to permit a determination of the absolute motion of the earth through space by an extension of Roemer's method, which Maxwell had proposed. The essential contents of this letter, for Michelson, were in the final paragraph containing the statement that all terrestrial methods for measuring the velocity of light could not detect the earth's motion through space, since "in the terrestrial methods of determining the velocity of light, the light comes back along the same path again, so that the velocity of the earth with respect to the ether would alter the time of the double passage by a quantity depending on the square of the ratio of the earth's velocity to that of light, and this is quite too small to be observed."

Michelson's interest was keenly aroused by the discussions of this problem with Todd and Newcomb, and especially by Maxwell's belief that no experiment to measure the speed of light could be devised with sufficient sensitivity to make possible a terrestrial measurement of the earth's motion through the aether. This was the challenge which led Michelson to his studies of optical interference methods and to his determination to pursue this problem as the principal objective of his study and research in Europe while on a leave of absence from regular Navy duty, which had been arranged for him by Simon Newcomb.

of the earth in its orbit would be only about one hundred millionth part of the whole time of transmission, and would therefore be quite insensible."

[2] R. S. Shankland, S. W. McCuskey, F. C. Leone, and G. Kuerti, Rev. Mod. Phys. **27**, 167 (1955).

[3] J. C. Maxwell letter to D. P. Todd, reprinted in Nature **21**, 314–317 (1880); Proc. Roy. Soc. (London) **A30**, 109–110 (1880); reply of Todd to Maxwell (19 May 1879) furnished to the writer by his daughter, Mrs. Millicent Todd Bingham, of Washington, D. C.

III. STUDIES IN EUROPE AND THE MICHELSON INTERFEROMETER

When Michelson, Master, U. S. Navy, sailed with his family for Europe in September 1880 for two years of study and research, he already was well-known for his precise measurements of the speed of light. Following brief stays in London and Paris, where he had letters of introduction to leading physicists from Simon Newcomb, he went on to Berlin. His plans for the optical-interference experiment, which he had started in a preliminary way at the Nautical Almanac Office, were continued at von Helmholtz's laboratory in the Physikalisches Institut at the University of Berlin, where he began his studies in the winter semester. This laboratory of von Helmholtz had a distinguished reputation and had attracted many students from abroad. "From America in 1876 had come Henry A. Rowland with all the plans and specifications already on paper, for his now famous 'Berlin Experiment' on the magnetic effect of electric convection. Helmholtz had furnished Rowland a research room, materials with which to construct his apparatus, and then had 'let him alone' to carry out his famous experiment. Later Henry Crew, James S. Ames, Arthur Gordon Webster, Michael Pupin, and D. B. Brace were to study and carry on research in Helmholtz's laboratory. In spite of his brilliance and high position and the awe in which he was held by the whole scientific world of Germany, Helmholtz was in fact a kindly, quiet and benevolent man who showed a deep interest in any proposed plan of study or research of a student. After his 10 a.m. lecture on Experimental Aspects of Physics, Helmholtz would walk around the laboratory talking in a friendly way on the progress of the experiments of his younger colleagues and research students, who at the time of Michelson's visit, included Otto Lummer, Ernst Hagen, and Heinrich Hertz."[4]

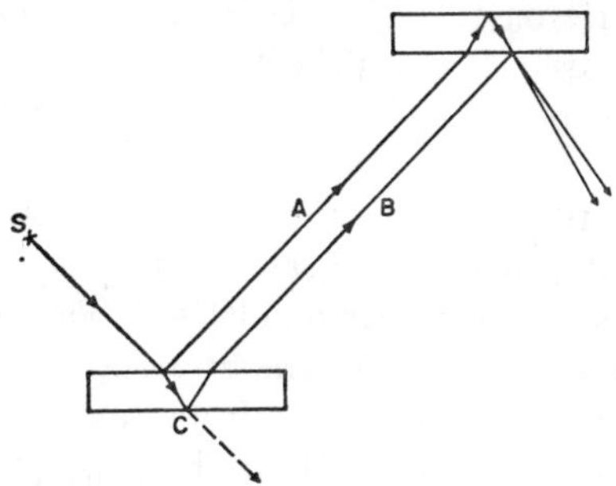

FIG. 1. Jamin interferometer.

It was in this congenial atmosphere that Michelson's studies led him to the invention of his interferometer. It is, of course, impossible to trace all the threads that lead to a great invention, but it is probable that Michelson was influenced by careful consideration of the interference devices developed by Jamin[5] in which the entire wavefront is *divided in amplitude* at a plane parallel plate set at an angle to the incident beam, and then recombined at a second plane parallel plate of exactly equal thickness and set approximately parallel to the first. Figure 1 shows the Jamin form of interferometer.

This device although having the advantage that it produces two coherent beams by *division of wave amplitude*, with correspondingly high intensity, nevertheless suffers from the limitation that the separation of the two beams is small, being limited by the thickness of the glass plates employed. Michelson realized, however, that, by using the coherent light beams (B) and (C) separated at the second surface of the first Jamin plate, he could then reflect these from widely spaced mirrors and then reunite them to produce interference fringes. The great advantage of Michelson's method over the Jamin interferometer is that beams (B) and (C) can be separated to accomodate apparatus of many forms, whereas using the beams (A) and (B) in the Jamin instrument limits the possible experiments to those requiring only a relatively small separation of the two coherent beams. Although in principle the two beams in Michelson's method may travel at various angles,[6] in the usual form of his interferometer the beams are oriented at 90°, as shown in Fig. 2.[7] Michelson's first trials of his interferometer were made by fastening the optical parts to a pier by pieces of wax. It took several hours of continuous

[4] Letters from Professor Henry Crew to the writer (26 Nov. 1950; 7 Aug. 1952).

[5] J. Jamin, Ann. Chim. Phys. **52**, 163, 171 (1858).

[6] A. A. Michelson, (a) *Studies in Optics* (University of Chicago Press, Chicago, 1927), pp. 21–26; (b) *Light Waves and Their Uses* (University of Chicago Press, Chicago, 1903), pp. 35–43.

[7] A. A. Michelson, Am. J. Sci. **22**, 120 (1881).

searching to find the white-light fringes in this way.[8]

It may perhaps seem only a simple step from Jamin's apparatus to the Michelson interferometer and indeed many years later Mascart referred to it as one in which the optical parts, "sont disposées comme dans l'appareil interferentiel de M. Jamin."[9] But, however great may be Michelson's debt to his predecessors for their development of interference methods, it must be remembered that his form of interferometer was devised for the express purpose of measuring an effect of the earth's motion on the speed of light[10] and that his interferometer and the famous experiment for which it was the elegant and simple tool are alike the product of the genius revealed in all his optical researches.

As had been his custom in America, Michelson reported frequently to Simon Newcomb on the progress of his work, and a most interesting letter of this period is the following[11]:

Berlin, November 22, 1880

Dear Sir,

Your very welcome letter has just been received. It will give me much pleasure to let you know how I am progressing.

At present the work in the laboratory is quite elementary, and I am trying to get over that part somewhat hurriedly.

Besides this work I attend the lectures on Theoretical Physics by Dr. Helmholtz, and am studying mathematics and mechanics at home.

I had quite a long conversation with Dr. Helmholtz concerning my proposed method for finding the motion of the earth relative to the ether, and he said he could see no objection to it, except the difficulty of keeping a constant temperature. He said, however, that I had better wait till my return to the U. S. before attempting them, as he doubted if they had the facilities for carrying out such experiments on account of the necessity of keeping a room at a constant temperature.

With all due respect, however, I think differently, for if the apparatus is surrounded with melting ice, the temperature will be so nearly constant as possible.

There is another and unexpected difficulty, which I fear will necessitate the postponement of the experiments indefinitely—namely—that the necessary funds do not seem to be forthcoming.

Dr. Helmholtz was however quite willing to have me make experiments upon light passing through a narrow aperture—but did not give much encouragement. In his opinion the polarization arises purely from reflection from the sides of the slit.

The change in color, he ignores entirely.

With many thanks for your kind interest in my affairs I remain,

Very truly yours,
Albert A. Michelson

Prof. Simon Newcomb, U.S.N.
Supt. Naut. Almanac

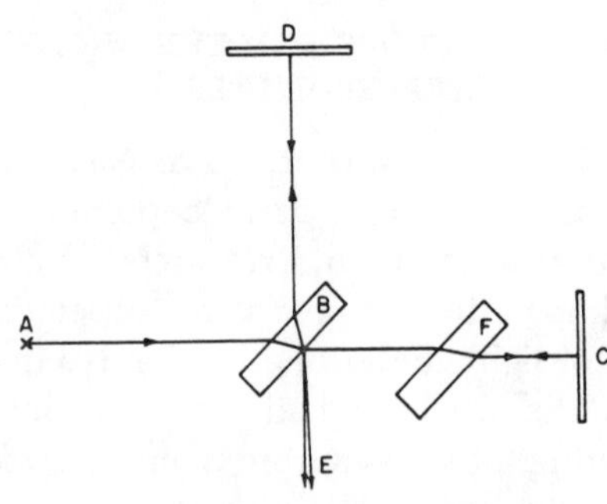

FIG. 2. Michelson interferometer.

The necessary funds (£100) referred to in this letter were furnished by Alexander Graham Bell at the suggestion of Simon Newcomb, and Michelson then selected the firm of Schmidt and Haensch in Berlin to build an instrument. This firm had specialized in the construction of optical-polarimeter equipment, but they could not supply the precise optical flats needed for the interferometer. These Michelson obtained from "Maison Breguet" in Paris, who were well-known suppliers of optical plates for Jamin interferometers.

When the Schmidt and Haensch instrument was ready, Michelson set up the new interferometer in von Helmholtz's laboratory. A perspective drawing of this apparatus is shown in Fig. 3. The light beam from the source (a) was divided in amplitude at the glass plate (b) set at 45° on the axis of the instrument. The two

[8] Dayton C. Miller to writer (10 Apr. 1933).

[9] M. E. Mascart, *Traité d'optique* (Gauthier-Villars et Fils, Paris, 1893), Vol. 3, p. 111.

[10] Reference 6(b); pg. 159. "The experiment is to me historically interesting, because it was for the solution of this problem that the interferometer was devised. I think it will be admitted that the problem, by leading to the invention of the interferometer, more than compensated for the fact that this particular experiment gave a negative result."

[11] The existence of this letter was reported to the writer by Dr. Loyd S. Swenson, Jr., of the University of California, Riverside. The fourth paragraph clearly assumes that Newcomb was acquainted with the general plan of Michelson's experiment, and thus indicates that it had been discussed with Newcomb while Michelson was still in Washington. This paragraph also emphasizes the importance of good temperature control for the experiment—a matter of concern throughout the long history of the repetitions and refinements of the experiment.

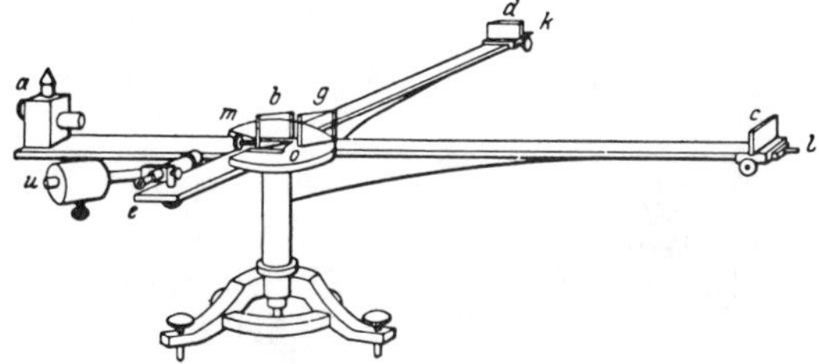

FIG. 3. Michelson interferometer used at Berlin and Potsdam.

coherent beams traveled at right angles to each other along the two arms of the interferometer and were then reflected back by the mirrors (c) and (d). With these set at the extreme ends of the arms, the two optical paths were each 120 cm. The interference fringes were found by first using a sodium light source and after adjustment for maximum visibility, the source was changed to white light and the colored fringes then located. White-light fringes were employed to facilitate observation of shifts in position of the interference pattern. These were viewed and measured on a scale ruled on glass in the small telescope (e), which was focussed on the surface of the mirror (d), where the interference fringes were most distinct. In his first description of this device, Michelson referred to it as an "interferential refractor."

The experiment to observe "the relative motion of the Earth and the luminiferous ether," for which this instrument was devised was planned by Michelson as follows. When the interferometer is oriented, as in Fig. 4, with the arm L_1 parallel to the direction of the earth's velocity v in space, the time required for light to travel from M to M_1 and return to M in its new position is

$$t_{\parallel}^{(1)}=\frac{L_1}{c-v}+\frac{L_1}{c+v}=\frac{2L_1}{c}\frac{1}{1-\beta^2}\quad\left(\beta=\frac{c}{v}\right).$$

The time for light to make the to and fro journey to the mirror M_2 in the other interferometer arm L_2 is

$$t_{\perp}^{(1)}=[2L_2(1+\tan^2\alpha)^{\frac{1}{2}}/c],$$

and since $\tan^2\alpha=v^2/(c^2-v^2)$

$$t_{\perp}^{(1)}=\frac{2L_2}{c}\frac{1}{(1-\beta^2)^{\frac{1}{2}}}.$$

In his first analysis, Michelson incorrectly assumed that the time required for light to travel along the arm at right angles to the earth's motion to mirror M_2 would be unaffected by this motion, thus assuming incorrectly that

$$t_{\perp}^{(1)}=2L_2/c.$$

When the interferometer is rotated through 90° in the horizontal plane so that the arm L_2 is parallel to v, the corresponding times are

$$t_{\parallel}^{(2)}=\frac{2L_2}{c}\frac{1}{1-\beta^2};$$

$$t_{\perp}^{(2)}=\frac{2L_1}{c}\frac{1}{(1-\beta^2)^{\frac{1}{2}}}.$$

Thus, the total phase shift (in time) between the two light beams expected on the aether theory for a rotation of the interferometer through 90° is

$$\Delta t=\frac{2L_1}{c}\left[\frac{1}{1-\beta^2}-\frac{1}{(1-\beta^2)^{\frac{1}{2}}}\right]+\frac{2L_2}{c}\left[\frac{1}{1-\beta^2}-\frac{1}{(1-\beta^2)^{\frac{1}{2}}}\right]=\frac{2(L_1+L_2)}{c}\left[\frac{1}{1-\beta^2}-\frac{1}{(1-\beta^2)^{\frac{1}{2}}}\right].$$

For equal interferometer arms, as used in this experiment,

$$L_1=L_2=L, \text{ and, since } \beta\ll 1,$$

$$\Delta t\simeq\frac{2L}{c}\beta^2.$$

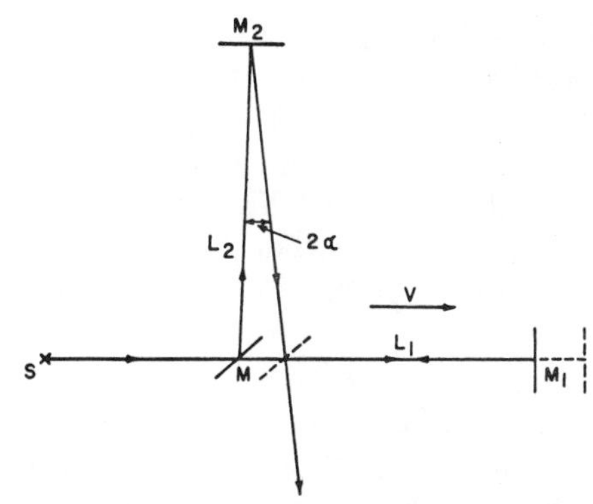

FIG. 4. Michelson experiment.

However, the observations give the positions of the fringes, rather than times, so the quantity of importance for the experiment is the change in optical path in the two arms of the interferometer.

$$A = c\Delta t = 2L(v/c)^2.$$

This is the quantity of the second order in (v/c) referred to by Maxwell,[1] which started Michelson thinking about this problem.

In Michelson's original apparatus, $L = 120$ cm, and in terms of waves of white light ($\lambda \sim 5700$ Å), this distance equals 2×10^6 wavelengths. The orbital speed of the earth around the sun is $v \simeq 30$ km/sec so that $(v/c)^2 = 10^{-8}$. Hence, $A \simeq 4\times10^6 \times10^{-8} \simeq 0.04$ fringe. In neglecting the effect of the earth's motion on the light beam traveling along the interferometer arm L_2, Michelson had anticipated a fringe shift of twice this amount when he rotated the interferometer through 90°.

Michelson made his observations by recording the position of the central black fringe on a graduated scale in the telescope eyepiece when the orientation of the instrument about its vertical axis was set successively at each of the eight points of the compass. When the apparatus was set up on a stone pier in the Physikalisches Institut of the University of Berlin, vibrations due to street traffic made observation of the interference fringes wholly impossible, except during brief intervals after midnight. So von Helmholtz made arrangements with Professor Vogel, Director of the Astrophysicalisches Observatorium at Potsdam, for the experiment to be performed there. It was conducted in the cellar, whose circular walls formed the foundation for the pier of the equatorial telescope. Here, observations were possible, although it was first necessary to return the interferometer to the maker with instructions to make it revolve more easily without a bending of the arms. Michelson finally completed his measurements early in April of 1881. Although he observed shifts in the position of the interference fringes when the apparatus was turned in azimuth, they were smaller than anticipated and, moreover, did not show the proper phase relationship with respect to the earth's motion. Michelson concluded that, "The interpretation of these results is that there is no displacement of the interference bands. The result of the hypothesis of a stationary ether is thus shown to be incorrect."[12]

After completing the Potsdam experiment, Michelson remained for more than a year in Europe. Having written up the Potsdam experiment for publication, he took his family to Heidelberg, where he spent the summer semester attending the lectures of Quincke and Bunsen. Quincke had for years been a leader in optics, especially in researches involving interference phenomena in white light. He was an expert in the use of the Jamin-type interferometer, so it was natural that Michelson should go to Heidelberg to discuss the characteristics of this instrument, as well as his own form of interferometer. Quincke had introduced the practice of silvering the back surfaces of the Jamin plates, thus making the interference fringes much clearer.[13]

It was during this period in Heidelberg that Michelson was appointed to the faculty of the newly organized Case School of Applied Science in Cleveland. Professor George F. Barker, of the University of Pennsylvania, knew Michelson and had strongly urged his appointment in a letter to John N. Stockwell, the first Professor of Astronomy at Case.

March 22, 1881

My dear Dr. Stockwell:

I have received your letters of the 10th and 21st insts. and should have acknowledged their receipt before had I not been very busy. I mailed at once the enclosure to Mr. Michelson and also cabled him to the effect that the intentions of the Cleveland people were good. I have also mailed him your letter to me received today.

I can appreciate his position. He is now drawing his support from the Navy. He cannot afford to resign from that position until he has a positive certainty to fall back on. While it may be very true that the Trustees of the Case School are favorably disposed towards him, that favorable disposition would not, in my opinion, warrant his resigning from the Navy. If he does not do this, however, he will be sent to sea.

The chance to secure him for the Case School I regard as one not to be trifled with. He does not ask any salary and will expect to wait until such time as may be necessary before beginning his duties. But he asks (and I think rightly) that he be elected something, even Instructor in Physics as you and Dr. Taylor are, if not a full Professor

[12] A. A. Michelson, Am. J. Sci. **22**, 120 (1881); Phil. Mag. **13**, 236 (1882); Am. J. Sci. **23**, 395 (1882); J. Phys. (Paris) **1**, 183 (1882).

[13] G. Quincke, Ann. Physik [2] **132**, 321 (1867).

right away, with the understanding that he is not to go on duty until wanted. I can see no reason why some such guarantee cannot be given him, one on which he can have the courage to resign from the Navy and spend another year or more in Europe in Special Study. Then he can select and bring home his apparatus and be ready to go at once to work. I have not heard a word from him since I wrote to you, so I say all this on my own responsibility.

With best wishes,

Cordially yours,
George F. Barker
3909 L
Philadelphia

On 28 March 1881, the Trustees appointed Michelson to the Case faculty: "Resolved, that Albert A. Michelson be and is hereby appointed Instructor in Physics in the Case School of Applied Science, at a salary of $2000 per annum, this appointment, if accepted, to take effect September 1, 1882."[14] Final arrangements were made with the help of Professor Barker who could write to Professor Stockwell:

May 5, 1881

My dear Professor Stockwell:

I am in receipt of a letter from Professor Michelson dated at Heidelberg April 19th announcing the arrival of my telegram and also of a letter from the Secretary of your board of Trustees informing him of his election. He says he has sent his acceptance of the position. So this matter has now been successfully arranged and I believe to the satisfaction of all concerned. I am sure the Case School will never regret this step and I hope they will provide for him the amplest apparatus for experiment and instruction. Physical apparatus is more costly at the outset than chemical, but then the wear and tear is less and the renewals are less frequent. I should not think to judge from my own experience that it would be worthwhile to start with less than $10,000 worth, with the intention of doubling it after 4 or 5 years.

Cordially yours,
George F. Barker

So, on 30 September 1881, Michelson resigned his commission in the Navy. At that time, there was only a freshman class at Case, so Michelson was granted leave of absence to continue his studies and researches in Europe for another year, and later was authorized to purchase the apparatus that he would need in Cleveland.[15]

After the summer semester at Heidelberg was completed, Michelson took his family for a holiday at Schluchsee in the Black Forest. While there in August, he sent a letter to *Nature*[16] criticizing a recent report by Young and Forbes[17] of their measurements of the velocity of light made across the Firth of Clyde in Scotland. These investigators had claimed that the speed of blue light in air exceeded that of red light by 1.8%. Michelson pointed out that, if this were in fact the case, then, in his own measurements made at Annapolis[18] and Washington, the white-light image of the slit as deflected by the revolving mirror would be spread out into a spectrum 2.4 mm in length. Actually, no dispersion was observed although it could easily have been detected in Michelson's apparatus if the velocities of red and blue light in air differed by as much as 0.1%. Michelson's letter to *Nature* attracted the attention of Lord Rayleigh who had already concluded that Young and Forbes must be in error.[19] Thus began the acquaintance and friendship of Rayleigh and Michelson, which continued until Rayleigh's death.

This discussion with Rayleigh about the result of Young and Forbes is especially interesting, for it is part of a recurrent pattern throughout Michelson's career. Each time that he completed an aether-drift experiment (in 1881, 1887, 1897, 1913, 1925, and 1929), he immediately returned to his absorbing passion for optical experiments that would give "numbers"—usually the speed of light. His steady correspondence with Simon Newcomb—from Heidelberg, Schluchsee, and Paris—is almost exclusively concerned with speed-of-light measurements, especially those he was planning for Cleveland. There is not a single mention of the Potsdam experiment in this entire period!

In the autumn of 1881, Michelson moved to Paris to continue his studies and research at Le Collège de France and at l'École Polytechnique. He remained in Paris until his return to America in June 1882. Thus began the very pleasant relationships between Michelson and Mascart, Cornu and Lippmann, who with their great predecessors had made France the leader in optical research for almost a century.

[14] Minutes, Board of Trustees, Case Inst. Technol. (28 Mar. 1881).

[15] Minutes, Board of Trustees, Case Inst. Technol. ($7500 appropriated) (3 Nov. 1881).

[16] A. A. Michelson, Nature **24**, 460 (1881).

[17] J. Young and G. Forbes, Nature **24**, 303 (1881).

[18] A. A. Michelson, Nature **18**, 195 (1878); **21**, 94, 120, 226 (1879).

[19] Lord Rayleigh, Nature **24**, 382 (1881).

Michelson took his interferometer to Paris and demonstrated it to the physicists there. Cornu was not convinced that the fringes were produced as Michelson claimed, contending that they were actually Lloyd fringes produced in the first plate. However, when Michelson showed Cornu that the fringes disappeared when a piece of glass was placed in one of the arms, the latter was at once satisfied.[20]

The friendships Michelson made in Paris continued throughout his life, and led to his return a decade later to determine the length of the standard meter in terms of light waves. While in Paris, he published[21] the theory of his new interferometer ("refractometer") and also devised a very sensitive thermometer.[22]

In addition to the original report on the Potsdam experiment, a revised and shortened account was presented at the 20 February 1882 meeting of the Paris Académie des Sciences.[23] In this paper, which was sponsored by Cornu at a meeting with Jamin in the Chair, Michelson acknowledged his error in neglecting the effect of the earth's motion on light traveling in the interferometer arm set at right angles to the motion. When this effect is included, the expected fringe displacement is reduced by half: in the Potsdam interferometer from 0.08 to 0.04 fringe. Michelson credits A. Potier with calling his attention to this matter, although Potier had in fact concluded that the effect would reduce the expected fringe shift to zero.

Neither Michelson himself nor the scientific world generally ever considered the Potsdam trial conclusive, although Lord Rayleigh and Lord Kelvin (then Sir William Thomson) in England and H. A. Lorentz and others on the continent of Europe gave careful and respectful attention to Michelson's first published result on the aether-drift experiment. However, this interest led to no serious revision of the theories then current and even Lorentz's electron theory, which he was continually developing to adapt it to the growth of experimental fact, was not altered because of the Potsdam result. The situation was emphasized by Robert A. Millikan in a letter to the writer[24] in which he states, "This experiment (Potsdam) must have been a very crude one, and it was only after he got to Case that he set up in connection with Morley the outfit which has since gone under the name of the Michelson–Morley Experiment."

IV. CASE PROFESSORSHIP

Michelson came to Cleveland about 1 July 1882, but several years were to pass before he would repeat his interferometer experiment. Time was required to organize the physics laboratory and courses at Case. His early Case students remembered that "Michelson gave the most elegant lectures they ever heard—absolutely clear, everything finished."[25]

Michelson's lectures on light are of special interest. He apparently never mentioned the Potsdam experiment to his Case students. The result of this experiment was still a subject of controversy, and he was reluctant to go into the arguments in the classroom.

However, his Case students were told that "the luminiferous Ether is to some extent a hypothetical substance and if it consists of matter at all must be very rare and very elastic. *It entirely escapes all our senses of perception.*"

It is also a curious fact that Michelson did not describe his own form of interferometer in his lectures at Case, although his treatment of optical interference was detailed and included a complete description of the Jamin type of "refractometer" and he discussed its use in measuring the refractive indices of gases.

On the subject of the velocity of light, however, Michelson gave his students a complete story from the early determinations of Roemer and Bradley, through the work of the great Frenchmen, Fizeau, Foucault, and Cornu, to his own measurements and those of Newcomb.

It is clear from his correspondence at this time with Willard Gibbs and Lord Rayleigh that he was far from satisfied with the Potsdam

[20] D. C. Miller to writer (10 Apr. 1933); R. A. Millikan, Proc. Natl. Acad. Sci. U. S. **19**, (1938), and letters to writer (31 July 1950; 21 Jan. 1952; 11 Aug. 1952; 18 Mar. 1953).

[21] A. A. Michelson, Am. J. Sci. **23**, 395 (1882); Phil. Mag. **13**, 236 (1882).

[22] A. A. Michelson, J. Phys. (Paris) **1**, 183 (1882); Am. J. Sci. **24**, 92 (1882).

[23] A. A. Michelson, Compt. Rend. **94**, 520 (1882).

[24] Letter from R. A. Millikan to writer (11 Aug. 1952); see, also, Rev. Mod. Phys. **21**, 343 (1949).

[25] Comfort A. Adams (Case Inst. Technol., Class of 1890) to writer (19 May 1950); also, Wm. Koehler (Case Inst. Technol., Class of 1889) (10 Dec. 1950).

result. However, when he again found time for research, he did not take up the aether-drift experiment but returned to his earlier work on precision measurements of the speed of light. An important factor in this decision was the urging and support of Professor Simon Newcomb, who obtained a grant of $1200 from the Bache fund of the National Academy of Sciences which he turned over entirely to Michelson to support his velocity of light experiments in Cleveland. During the two years, 1882–1884, Michelson worked constantly to improve his methods for measuring the speed of light, and here he made three notable contributions.[26] An accurately measured base line along the railroad tracks (N. Y., C. and St. L.) at the rear of the Case campus was prepared by Professor John Eisenmann, and, with an improved optical system and better timing methods than hitherto employed, Michelson obtained 299 850 km/sec for the speed of light reduced to vacuum, a value much superior to his earlier determinations made at Annapolis and Washington. In fact, this measurement of the speed of light made at Cleveland in 1882–1883 was the accepted standard from that time until his own measurements between Mt. Wilson and San Antonio Peak in California were completed in 1927.

The other two important measurements made in Cleveland in 1882–1884 were determinations of the group velocity of light in distilled water and in carbon disulphide. The first of these gave a precision check with the value predicted from the refractive indices of water. Both Foucault and Fizeau had shown that the speed of light in water is less than in air, but Michelson's measurement gave the first accurate value. His measurement of the speed of light in carbon disulphide, however, appeared to be in disagreement with theory. Nevertheless, Michelson reported his result, and it drew the immediate attention of Lord Rayleigh for it provided the first reliable experimental verification of his theory for the difference between wave and group velocities in a dispersive medium.[27] Michelson communicated his results to Lord Rayleigh and, in the latter's presidential address[28] at the Montreal meeting of the British Association in the summer of 1884, he paid special attention to Michelson's measurements. Kelvin also was at the Montreal meeting, as were many of the leading physicists of that time, and thus a much wider acquaintance with Michelson's researches was established.

In October of 1884, Kelvin came to Baltimore to give a series of 20 lectures at The Johns Hopkins University. Professors Rowland and Sylvester had urged this project, and for over two years President Gilman had been in correspondence with Kelvin, inviting him to lecture on any subjects of his choice, emphasizing the strong impulse which such a series of lectures would give to the study of physics in America.

Kelvin lectured on "Molecular Dynamics and the Wave Theory of Light,"[29] dwelling principally on the failures of the wave theory, especially those related to the "Luminiferous Ether." The lectures were attended by a group of twenty-one "coefficients,"[30] including Michelson and Edward W. Morley from Cleveland, Henry A. Rowland, Henry Crew,[31] A. L. Kimball, T. C. Mendenhall, and George Forbes, a visitor from England.

The evening before the first lecture, a grand reception was held in Hopkins Hall; many visitors were present and it was a notable affair. The lectures were given in a small lecture room at Johns Hopkins, seating perhaps thirty people. Lord Rayleigh was there for some of Kelvin's lectures, which bore unmistakable signs of having been prepared in haste. In the midst of the

[26] R. S. Shankland, Am. J. Phys. **17**, 488–489 (1949).

[27] Lord Rayleigh, Proc. London Math. Soc. **9**, 21 (1877); *Theory of Sound*, (The MacMillan Company, London, 1877), Vol. 1, p. 191 and Appendix; see, also, Willard Gibbs, *Collected Works*, (Yale University Press, New Haven, Connecticut, 1948); Vol. 2, pp. 247–254.

[28] Lord Rayleigh, Brit. Assn. Rept. 654; *Scientific Papers* (Cambridge University Press, Cambridge, England, 1884), Vol. 2, p. 348.

[29] Sir Wm. Thomson, "Molecular Dynamics and the Wave Theory of Light," stenographic report by A. S. Hathaway; G. Forbes, Nature **31**, 461, 508, 601 (1884); Lord Kelvin, *The Baltimore Lectures* (Cambridge University Press, Cambridge, England, 1904) (republished in revised form).

[30] So called in suggestion of the twenty-one coefficients by which the most general state of an elastic body is specified.

[31] The writer's picture of the "Baltimore Lectures" has been formed very largely from the privilege of conversation (26 Nov. 1950) and correspondence (29 Nov., 7 Dec-1949; 25 Oct. 1950; 27 Aug. 1952; 17 Sept. 1952) with Professor Henry Crew, who attended Thomson's lectures as a graduate student and Fellow in Physics at Johns Hopkins.

first lecture, Kelvin once appealed to Rayleigh to verify some statement and inform him on the fact of the case; but Rayleigh merely shook his head, as if to say, "These are your lectures, not mine."

Rayleigh and Rowland sat in easy chairs at the end of the lecture table on which were placed Kelvin's various models. Michelson sat in the row next to the front, a keen-eyed, handsome, young chap with jet black hair, quiet in manner and dress. Already at that early date and among a group of students and well-known physicists from all over the country and abroad, Michelson was an outstanding figure; everyone held him in the highest esteem. The basis of Michelson's distinguished reputation in 1884 was without question his determination of the speed of light in a manner vastly superior to that of Fizeau or of Foucault. In his fifth lecture, Thomson alluded to Michelson's work on the group velocity problem, which had greatly interested Rayleigh, and in the eighth lecture referred again to his improved methods for measuring the speed of light.

Rowland was not able to attend the lectures regularly, as in the autumn of 1884 he was a very busy man working on his second ruling engine, finishing up the determination of the *Ohm* for the U. S. Government and designing his new laboratory. As a result, but with no intended disrespect for Kelvin, he often closed his eyes and dozed beside Rayleigh. Professor Crew also felt that Rowland was so certain of the correctness and beauty of Maxwell's then new electromagnetic theory of light (for which his own experiments on the magnetic effect of electric convection[32] provided an important basis) that he considered Kelvin's elaborate attempt to provide a mechanical basis for optical phenomena as a step backwards.

Michelson, however, was intensely interested and, although he asked no questions during the lectures, it is certain that during his stay in Baltimore he discussed the results of his Potsdam experiment with Kelvin and Rayleigh during the ten-minute intervals between the two parts of each day's lecture when Kelvin chatted with his students; and discussions started at the lecture often were continued at the supper table. Kelvin was always enthusiastic about a "sweet" experiment and no doubt urged Michelson to give his interferometer work another trial, especially since both he and Lord Rayleigh were not convinced that the apparatus used at Potsdam had sufficient sensitivity to give a decisive test; and the weight of scientific opinion was tending more and more to the view that Fresnel's hypothesis of a stationary aether was probably correct.

It also seems likely that during these informal discussions Professor Morley was drawn into the problem. Morley was considerably older than Michelson and was well-recognized not only in chemistry, but in physics and mathematics. He was an acknowledged leader in experimental work, and his theoretical insight was an invaluable addition to the great experimental skills of Michelson.

According to Professor Crew, Morley was the "shark" of the Baltimore Lectures, helping Kelvin over many rough spots and working out the hardest "homework" problems proposed to the "coefficients" by Kelvin. One of these was a complete solution by Morley [see Baltimore Lectures by Kelvin, p. 408 (1904)] of the dynamical model of a molecule consisting of seven mutually interacting particles, giving the fundamental periods of vibration, relative displacements, and the energy ratios between the normal modes. Michelson was already well-acquainted with Morley as they had both come to Cleveland in 1882, but the informal and inspiring atmosphere of the Baltimore Lectures without question contributed greatly to their subsequent collaboration in research.

V. REPETITION OF THE FIZEAU EXPERIMENT BY MICHELSON AND MORLEY

The first joint research of Michelson and Morley was not to repeat the Potsdam experiment, as might have been expected, but to carry out in greatly improved fashion Fizeau's 1851 experiment[33] on the speed of light in moving water, as had been urged at Baltimore by

[32] H. A. Rowland, Ann. Physik **158**, 487 (1876); Am. J. Sci. **15**, 30 (1878).

[33] H. L. Fizeau, Compt. Rend. **33**, 349 (1851); Ann. Chim. Phys. **57**, 385 (1859).

both Kelvin and Rayleigh.[34] Many theoretical discussions on the aether problem involved Fizeau's measurement, and it was felt that a new experiment should be performed to subject this question to a decisive test.

During the thirty years between 1851 and Michelson's Potsdam experiment of 1881, Fizeau's measurement had been considered as one of the decisive experimental bases of the validity of Fresnel's hypothesis of a stationary aether, on which he had developed his theory of the influence of the motion of a medium on the propagation of light.

The Fresnel theory predicted that the observed speed of light in a moving transparent medium should be

$$u=(c/n)+v[1-(1/n^2)],$$

where c is the speed of light in vacuum, n is the index of refraction of the transparent medium which is moving at speed v, relative to the observer, while u is the observed speed of light in the moving medium. The speed of light is not altered by the full speed v of the moving medium, but by only a fraction $[1-(1/n^2)]$ called the Fresnel drag coefficient.

In obtaining this coefficient, Fresnel had made rather artificial assumptions about the behavior of aether in a moving transparent medium—namely, that only that part of the aether in the moving body constituting the excess over the aether normally present *in vacuo* partakes of the motion, while the remaining aether remains at rest.

Fresnel's[35] theory had originally been developed to explain the experiment of Arago which demonstrated with reasonable precision that starlight is refracted in a prism by the same amount as is light from a terrestrial source, regardless of the orientation of the prism to the direction of motion of the earth in its orbit. An achromatic prism is employed for the experiment so that the deflection of the light is insensitive to Doppler shifts, caused by the earth's motion and the radial velocity of the stars.

Fresnel had also correctly predicted the result of the experiment of Airy,[36] who had demonstrated at the Greenwich Observatory that the aberration angle of the star γ Draconis determined with a water-filled zenith telescope (35.3-in. tube with special lens) is the same as that originally found by Bradley[37] with an ordinary telescope.

The Fresnel theory had had other notable successes in explaining the null results obtained in experiments by Hoek,[38] Mascart and Jamin,[39] Maxwell,[40] and others, which had failed to detect an influence of the Earth's motion on the propagation of light in transparent media.

It is true that all experiments before Michelson's were capable only of detecting effects to the first order of (v/c), while his method permitted observation of effects to the second order of this ratio; but the preponderance of the evidence supporting Fresnel's hypothesis of a stationary aether in space was so great that little serious attention was given to the Potsdam result. This was especially true after it was known that Michelson had originally overestimated his expected fringe shifts by a factor of two.

However, the Fizeau experiment was not as conclusive as could be desired, for his observed drag coefficient of 0.5 ± 0.1 differed considerably from Fresnel's theoretical value of 0.438 for water. Furthermore, J. J. Thomson[41] had recently obtained a theoretical value of exactly $\frac{1}{2}$ for the drag coefficient by an argument based on Maxwell's electromagnetic theory of light.

[34] Letter of A. A. Michelson to Simon Newcomb: "I have been asked by Sir Wm. Thomson and Lord Rayleigh to repeat Fizeau's experiment for testing the question of the effect of motion of medium on the velocity of light; could I use the remainder of the [Bache] money yet in my possession for that purpose?" (30 Jan. 1885); see also a letter from Michelson to Willard Gibbs dated Cleveland, Ohio, 15 Dec. 1884, in Lynde P. Wheeler, *Josiah Willard Gibbs* (Yale University Press, New Haven, Connecticut, 1951).

[35] A. J. Fresnel, Ann. Chim. Phys. **9**, 57 (1818), including a reprint of Fresnel's famous "Letter to Arago." Arago's result was published much later—D. F. Arago, Compt. Rend. **8**, 326 (1839); **36**, 38 (1853).

[36] G. B. Airy, Proc. Roy. Soc. (London) **20**, 35 (1871); **21**, 121 (1873); *Autobiography* (Cambridge University Press, Cambridge, England, 1896), pp. 240, 286, 291, 294.

[37] J. Bradley, Phil. Trans. Roy. Soc. **35**, 637 (1728).

[38] M. Hoek, Arch. Néerl. Sci. **3**, 180; **4**, 443 (1868); Astron. Nach. **73**, 193 (1869); Compt. Rend. Acad. Sci. Amsterdam **2**, 189 (1868).

[39] M. E. Mascart and J. Jamin, Ann. École Norm. **3**, 336 (1874); *Traité d'Optique* (Gauthier-Villars et Fils, Paris, 1889); Vol. 1, p. 462, Vol. 3, pp. 109–111 (1893).

[40] J. C. Maxwell, Phil. Trans. Roy. Soc. **158**, 532 (1868); *Scientific Papers* (Dover Publication, Inc., New York, 1952), Vol. 2, p. 769.

[41] J. J. Thomson, Phil. Mag. **9**, 284 (1880); repeated with no references to the work of Michelson, Morley, or Lorentz in *Recent Researches in Electricity and Magnetism* (Clarendon Press, Oxford, England, 1893), pp. 543–546.

Lorentz[42] later showed that Maxwell's theory in its original form applied only to vacuum, and that its correct application to the propagation of light in material media required the extensions that he developed in his electron theory. Among the many results obtained by Lorentz in this theory is the same value of the drag coefficient as that given originally by Fresnel.

Thus, the situation was confused, and it was generally agreed that a new trial of the Fizeau experiment was desirable. In the words of Michelson and Morley, it was felt that, "Notwithstanding the ingenuity displayed in this remarkable contrivance (Fizeau's apparatus) which is apparently so admirably adapted for elimating accidental displacement of the fringes by extraneous causes, *there seems to be a general doubt* concerning the results obtained, or at any rate the interpretation of these results given by Fizeau. This together with the fundamental importance of the work must be our excuse for its repetition."[43]

The interference technique used by Fizeau for his experiment had suffered from the basic defect that it employed coherent light beams obtained by *division of wavefront*. The interference fringes obtained by this method are either very faint or are extremely narrow, and any attempt to separate the two beams to accommodate the disposition of the apparatus results in a further narrowing of the interference pattern. The closely spaced fringes can be magnified, but only with a corresponding loss of intensity and distinctness. Fizeau had made a slight improvement in his apparatus by use of a biplate, permitting some increased separation of the two beams without a corresponding increase in the angle between them at interference. However, only minor improvements were possible in this way.

The improved method employed by Michelson and Morley in their repetition of Fizeau's experiment is shown in Fig. 5. They used a form of the

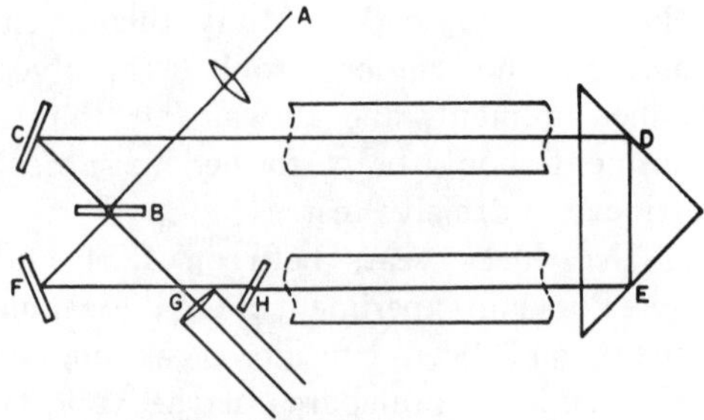

FIG. 5. Michelson–Morley method for the Fizeau experiment.

Michelson interferometer with light source at A, half-reflecting plate at B, and compensating plate at H. The brass tubes (shown only in part in Fig. 5) were 2.8 cm i. d. and in these the distilled water flowed in opposite directions. The two coherent light beams traversed the apparatus in opposite directions; one beam along BCDEHFB to the telescope at G; and the other by BFHEDCB to G. The position of the central fringe in white light was observed with the water flowing in one direction; then its shift was measured when the flow of water was reversed. Fizeau had used a transient flow of water, and, to permit a more accurate observation of the position of the interference fringes, Michelson and Morley employed a steady flow of water from a tank located in the attic of the Case Main Building, 70 ft above their basement laboratory. They also used an ingenious method to determine the velocity distribution of the flow across the diameter of the tube, thus getting an accurate value for the water speed along the optical path.

By employing the modified form of the Michelson interferometer, it was possible to obtain a wide spacing between the optical paths in the tubes carrying the flowing water, while at the same time the distinct and widely spaced interference fringes permitted an accuracy in the optical measurements impossible with Fizeau's apparatus.

A situation which gave Morley the complete responsibility for their experiment during the fall of 1885 was a serious illness of Michelson's. He left Cleveland on 19 September 1885 and did not expect to return. He thought (because of an erroneous diagnosis) that he would never work again, and so asked Morley to complete the experiment, turning over to him his equip-

[42] H. A. Lorentz, *Versuch einer Theorie der electrischen und optischen Erscheinungen in bewegten Körpern* (E. J. Brill, Leiden, 1895). In this work, Lorentz also showed that for dispersive media an additional term $(\nu/n)(dn/d\nu)$ must be added to the Fresnel drag coefficient, a refinement later beautifully confirmed by P. Zeeman, Proc. Amsterdam Acad. **17**, 445 (1914); **18**, 398 (1915).

[43] A. A. Michelson and E. W. Morley, Am. J. Sci. **31**, 377 (1886).

ment and the unexpended funds of the Bache grant.[44]

Fortunately, Michelson's illness proved much less serious then had been feared,[45] and in mid-December of 1885 he returned to Cleveland to finish the experiment with Morley.

Michelson and Morley made, in all, 65 trials of their experiment, varying the length of the tubes carrying the water and the speed of the liquid. They found that the change in the observed velocity of light was accurately proportional to water speed and was altered by almost the exact amount predicted by the Fresnel formula.

Upon completing their measurements, Michelson and Morley reported the result to Kelvin.

Cleveland, Ohio
March 27, 1886

Dear Sir William:

You will no doubt, be interested to know that our work on the effect of the medium on the velocity of light has been brought to a successful termination. The result fully confirms the work of Fizeau. The factor by which the velocity of the medium must be multiplied to give the acceleration of the light was found to be 0.434 in the case of water, with a possible error of 0.02 or 0.03. This agrees almost exactly with Fresnel's formula $1-1/n^2$. The experiment was also tried with air with a negative result. The precautions taken appear to leave little room for any serious error, for the result was the same for different lengths of tube, different velocities of liquid, and different methods of observation. We hope to publish the details within a few weeks. Very respectfully, your obedient servants,

Albert A. Michelson
Edward W. Morley

Kelvin's prompt and enthusiastic reply[46] stated that he would incorporate their results as an appendix in the final publication of his Baltimore Lectures, as he had urged them to undertake this work during their stay at Johns Hopkins.

Michelson also sent a preliminary report to Willard Gibbs.

Cleveland, March 1886

My dear Prof. Gibbs,

Your welcome letter was duly received, and I have delayed answering till my experiments were completed. My result fully confirms the work of Fizeau and the result found for $1-1/n^2$ was 0.434 which is almost exactly the number for this expression when for n we put the index of refraction of water. I had heard that the relationship between maximum and mean velocity of liquids in tubes had been worked out–but have not been able to find it–so I made an experimental determination and found the ratio to be 1.165.

I think my result shows that your estimate of Thomson's [Ref. 47] work is correct. The number 0.434 is correct within 2 or 3% and I can say with a good deal of confidence that it is not one half. I also repeated the experiment with air with a negative result.

I expect to publish details in a few weeks.

Very sincerely yours,
A. A. Michelson

VI. MICHELSON–MORLEY EXPERIMENT

After the publication of the results on the moving-water experiment, there appeared a long article by Lorentz[48] in which he attempted to reconcile Michelson's Potsdam result with a combination of the aether theories of Fresnel and Stokes, and Michelson and Morley's new determination of the Fresnel drag coefficient. Stokes had shown that the aberration of light as observed in either an ordinary telescope[37] or a water-filled telescope[36] can be explained on the wave theory by "supposing that the aether close to the earth's surface is at rest relatively to that surface, while its velocity alters as we recede from the surface, till at no great distance, it is at rest in space."[49] Stokes' theory carried a condition that the motion of the aether must be irrotational, a condition later shown by Lorentz to be inconsistent with other characteristics of motion through the aether. However, in 1886 Lorentz accepted the Stokes theory, but modified it by assuming that the earth

[44] Letter from A. A. Michelson to Simon Newcomb from New York (28 Sept. 1885); letter from E. W. Morley to his father (27 Sept. 1885).

[45] Letters from A. A. Michelson to E. W. Morley (19, 23 Oct. 1885); letter from A. A. M. to Willard Gibbs (13 Dec. 1885).

[46] S. P. Thomson, in *Life of Lord Kelvin* (The MacMillan Company, London, 1910), Vol. 2, p. 857.

[47] This refers to J. J. Thomson (Ref. 41) in which he had obtained a theoretical value of ½ for the Fresnel drag coefficient.

[48] H. A. Lorentz, Arch. Néerl. **21**, 103–176 (1886). It is of interest to note that in this article Lorentz states that "M. Michelson a réalisé l'interférence au moyen d'un appareil qui' présente quelque analogie avec le réfracteur interférential de Jamin."

[49] G. G. Stokes, Phil. Mag. **27**, 9 (1845); **28**, 76 (1846); **29**, 6 (1846); *Mathematical and Physical Papers* (Cambridge University Press, Cambridge, England, 1880), Vol. 1, pp. 134, 141, 153.

imparts only that fraction of its own motion to the aether within it given by the Fresnel drag coefficient for transparent media, as measured by Michelson and Morley, instead of the full velocity as originally assumed by Stokes. Since the velocity of the aether near the earth's surface would, on this view, be less then half the earth's velocity, the expected displacement of the interference fringes in Michelson's Potsdam apparatus would be correspondingly reduced, and would be less than the experimental accuracy. In this same paper, Lorentz also recomputed the expected fringe shift for the Potsdam experiment, showing it to be only half that originally calculated by Michelson.

Lord Rayleigh wrote to Michelson calling his attention to Lorentz's paper and urged him to repeat his experiment on the relative motion of the earth and the aether. Michelson replied as follows:

Cleveland, March 6, 1887

My dear Lord Rayleigh,

I have never been fully satisfied with the results of my Potsdam experiment, even taking into account the correction which Lorentz points out.

All that may be properly concluded from it is that (supposing the ether were really stationary) the motion of the earth thro' space cannot be very much greater than its velocity in its orbit.

Lorentz' correction is undoubtedly true. I had an indistinct recollection of mentioning it either to yourself or to Sir. W. Thomson when you were in Baltimore.

It was first pointed out in a general way by M. A. Potier of Paris, who however was of the opinion that the correction would entirely annul any difference in the two paths; but I afterwards showed that the effect would be to make it one half the value I assigned, and this he accepted as correct. I have not yet seen Lorentz' paper and fear I could hardly make it out when it does appear.

I have repeatedly tried to interest my scientific friends in this experiment without avail, and the reason for my never publishing the correction was (I am ashamed to confess it) that I was discouraged at the slight attention the work received, and did not think it worth while.

Your letter has however once more fired my enthusiasm and it has decided me to begin the work at once.

If it should give a definite negative result then I think your very valuable suggestion concerning a possible influence of the vicinity of a rapidly moving body should be put to the test of experiment; but I too think the result here would be negative. [This experiment, yielding a negative result, was later performed by Sir Oliver Lodge.[50]]

[50] Sir Oliver Lodge, Nature **46**, 501 (1892); Phil. Trans. Roy. Soc. (London) **184**, 727 (1893); *Past Years* (Charles Scribner's Sons, New York, 1932), Chap. 15.

But is there not another alternative?

Suppose for example that the irregularity of the earth's surface be crudely represented by a figure like this:

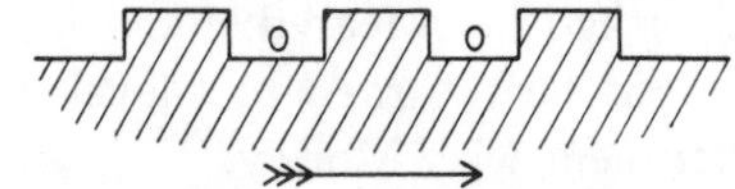

If the earth's surface were in motion in the direction of the arrow, would not the ether in 00 be carried with it?

This supposes of course, contrary to Fresnel's hypothesis, that the ether does not penetrate the opaque portions, or if it does so penetrate, then it is held prisoner. Fizeau's experiment holds good for transparent bodies only, and I hardly think we have a right to extend the conclusions to opaque bodies.

If this be so and the ether for such slow motions be regarded as a frictionless fluid—it must be carried with the earth in the depression.

Would this not be partly true, say in a room of this shape?

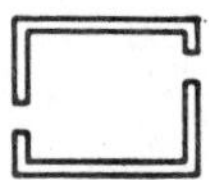

If this is all correct then it seems to me the only alternative would be to make the experiment at the summit of some considerable height, where the view is unobstructed at least in the direction of the earth's motion.

The Potsdam experiment was tried in a cellar, so that if there is any foundation for the above reasoning, there could be no possibility of obtaining a positive result.

I should be very glad to have your view on this point.

I shall adopt your suggestion concerning the use of tubes for the arms, and for further improvements shall float the whole arrangement in mercury; and will increase the theoretical displacement by making the arms longer, and doubling or tripling the number of reflections so that the displacement would be at least half a fringe.

I shall look forward with great pleasure to your article on "Wave Theory" (hoping however, that you will not make it too difficult for me to follow).

I can hardly say yet whether I shall cross the pond next summer. There is a possibility of it, and should it come to pass I shall certainly do myself the honour of paying you a visit.

Present my kind regards to Lady Rayleigh and tell her how highly complimented I felt that she should remember me.

Hoping soon to be able to renew our pleasant association, and thanking you for your kind and encouraging letter,

I am,
Faithfully yours,
Albert A. Michelson

A definitive test of the Potsdam experiment had been the ultimate objective of Michelson and Morley for which their repetition of Fizeau's moving-water experiment had been only an

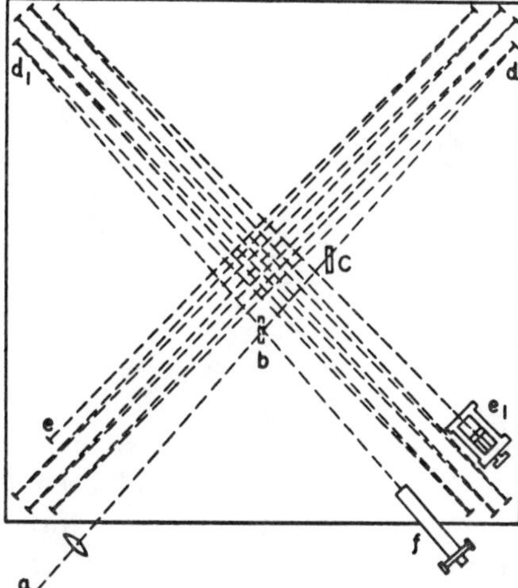

FIG. 6. Optical paths in the Michelson–Morley interferometer.

important preliminary. Their accurate result for the moving-water experiment was based on observed shifts in the interference pattern of about half the distance between fringes, and they were greatly encouraged by this positive result to devise an aether-drift apparatus of comparable sensitivity. Professor Morley's keen interest in the new experiment is evident in the following letter to his father:

Cleveland, April 17, 1887

"Michelson and I have begun a new experiment. It is to see if light travels with the same velocity in all directions. We have not got the apparatus done yet and shall not be likely to get done for a month or two. Then we shall have to make observations for a few minutes every month for a year. We have a stone on which the optical parts of the apparatus are to be fixed. The stone is five feet square, and about fourteen inches thick. This we shall have to support so that it can be turned around and used in different positions. Now since a strain of half a pound would make our observations useless, we have to support it so that its axis of rotation is rigorously vertical. My way to secure this was to float the stone on mercury. This we accomplished by having an annular trough full of mercury, with an annular float in it on which the stone is placed. A pivot in the center makes the float keep concentric with the trough. In this way, I have no doubt, we shall get decisive results."[51]

Michelson's Potsdam experiment had been greatly hampered by strains produced in the apparatus as it was turned in azimuth, and by vibrations transmitted to the optical parts which made the interference fringes unsteady, and often disappear entirely. The new Cleveland interferometer devised by Michelson and Morley overcame these difficulties as mentioned in Morley's letter by mounting the optical parts on a heavy sandstone slab 5 feet square and a foot thick and placing this on an annular wooden float supported by mercury contained in an annular cast-iron trough. An essential feature of the float design was that it permitted a comparatively small amount of mercury to support the heavy stone. This arrangement permitted the interferometer to be continuously rotated in the horizontal plane so that observations of the interference fringes could be made at all azimuths with respect to the earth's orbital velocity. When set in motion, it would rotate slowly (about once in 6 min) for hours at a time. No starting and stopping was necessary, and the motion was so slow that accurate readings of fringe positions could be made while the apparatus rotated.

It was most natural for Professor Morley to suggest the use of mercury to support the apparatus, as he had already used large quantities of mercury in his own work and in 1884 had obtained new apparatus of the type invented by Professor Wright of Yale for distilling mercury in vacuum, and which Morley himself had considerably improved. The sandstone slab used by Michelson and Morley was the one used later by Morley for the pier on which he supported his barometer and reading microscopes for his chemical researches.[52]

The optical paths in the Michelson–Morley interferometer are shown in plan in Fig. 6. Light from (a) is divided into two coherent beams at the half-reflecting, half-transmitting rear surface of the optical flat (b). These two beams travel at 90^0 to each other and are multiply reflected by two systems of mirrors $d-e$ and d_1-e_1. On returning to (b) part of the light from $e-d$ is reflected into the telescope at (f), and light from e_1-d_1 is also transmitted to (f). These two coherent beams produce interference

[51] Letter in Biography of E. W. Morley by H. R. Williams, Ph.D. thesis, Western Reserve University (1942).

[52] E. W. Morley, "On the Densities of Oxygen and Hydrogen and on the Ratio of their Atomic Weights," Smithsonian Inst. Publ. (Contributions to Knowledge) No. 980 (1895), pp. 22–23 and Fig. 8.

fringes. These are formed in white light only when the optical paths in the two arms are exactly equal, a condition produced by moving the mirror at e_1 by a micrometer. (C) is the usual compensating plate. The effective optical length of each arm of the apparatus was thus increased to 1100 cm by the repeated reflections, as compared to the 120 cm optical paths of the Potsdam interferometer.

Figure 7 is a perspective drawing of the Michelson–Morley interferometer showing the optical system mounted on the sandstone slab. The slab is supported on the annular wooden float, which in turn fitted into the annular cast-iron trough containing the mercury. On the outside of this tank can be seen some of the numbers 1 to 16 used to locate the position of the stone in azimuth. The trough was mounted on a brick pier, which in turn was supported by a special concrete base. The height of the apparatus was such that the telescope was at eye level to permit convenient observation of the fringes when the instrument was rotating. While observations were being made, the optical parts were covered with a wooden box to reduce air currents and temperature fluctuations.

Figure 8 is a cross section through the sandstone slab and its supports. The wooden float, cast-iron trough, and brick pier are shown, and also the centering pin which prevented the float from bumping into the sides of the cast-iron trough. The pin was engaged only while the interferometer was being set into rotation, and, once started, the apparatus would continue to turn freely for hours at a time.

With this new interferometer, the magnitude of the expected shift of the white-light interference pattern was 0.4 of a fringe as the instrument was rotated through an angle of 90° in the horizontal plane. (The corresponding shift in the Potsdam interferometer had been 0.04 fringe.) From their recent experience with the Fizeau moving-water experiment, Michelson and Morley felt completely confident that fringe shifts of this order of magnitude could be determined with high precision.

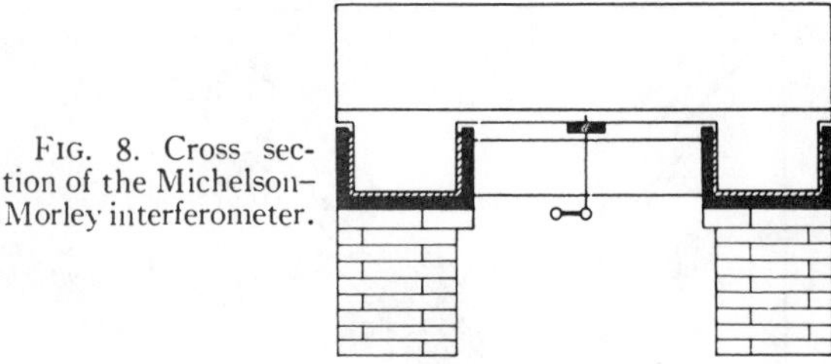

Fig. 8. Cross section of the Michelson–Morley interferometer.

The interferometer was set up in the southeast corner basement laboratory of the Case Main Building, a room having heavy stone walls and rather constant temperature conditions. Here, Michelson and Morley carried on the preliminary work of their experiment, but were prevented from making final observations for, on 27 October 1886, the Case Main Building suffered a disastrous fire.[53] A large part of all Michelson's physics equipment which he had purchased in Europe in 1881–1882, was destroyed in this fire, but the apparatus for the Michelson–Morley experiment was rescued by students living in the nearby Western Reserve University dormitory known as Adelbert Hall, a building later called Pierce Hall, which was razed in 1961. The Michelson–Morley equipment was moved to the southeast corner of the basement of Adelbert Hall.[54]

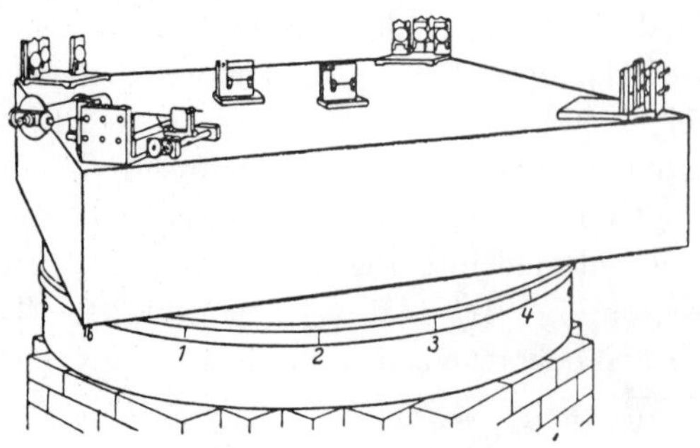

Fig. 7. Michelson–Morley interferometer used at Cleveland in 1887.

[53] *The Cleveland Leader and Herald* (28 Oct. 1886).

[54] Letter from Sidney S. Wilson (Western Reserve Univ. Class of 1888) to Frank N. Shankland (Apr. 1949):

Twenty-Nine Palms, California
April, 1949

Dear Frank,

The night Case burned there were about ten of us Adelbert fellows living in the Old Dorm. We were the first to arrive . . . 15 minutes at least before the first fire engine (horse-drawn) . . . the building was all ablaze, the roof had fallen . . . the Morley-Michelson salvage from their apparatus was moved from the ruins to the Dorm basement and not into the Adelbert College basement as stated in your letter . . . their Lab. was then set up in the Southeast quarter of the basement directly under my room . . . Morley and Michelson were there much of the time . . . We were forbidden and threatened not to molest or even to inspect the Morley-Michelson apparatus. We didn't! . . . however, as words of caution turned to orders from Prexy Haydn for

In addition to the great efforts needed to reestablish, in Adelbert Hall, the Case physics laboratories and lectures for teaching, many weeks were required to set up the Michelson–Morley equipment for trial observations on the interferometer, improvements in the optics of the experiment, and mechanical refinements which would permit the stone to be rotated freely without introducing strains or vibrations. Finally, in July of 1887, Michelson and Morley were able to make their definitive observations. The experiments which gave their published data were conducted at noon and during the evening of the days of 8, 9, 11, and 12 July 1887. Instead of the expected shift of 0.4 of a fringe, they found "that if there is any displacement due to the relative motion of the earth and the luminiferous ether, this cannot be much greater than 0.01 of the distance between the fringes."[55]

Upon completion of the July 1887 experiments, Michelson sent a preliminary report of the work to Lord Rayleigh as follows:

New York, August 17, 1887

My dear Lord Rayleigh,

The Experiments on the relative motion of the earth and ether have been completed and the result decidedly negative. The expected deviation of the interference fringes from the zero should have been 0.40 of a fringe—the maximum displacement was 0.02 and the average much less than 0.01—and then not in the right place.

As displacement is proportional to squares of the relative velocities it follows that if the ether does slip past the relative velocity is less than one sixth of the earth's velocity.

I enclose a poor photograph of the apparatus—which consists of a stone five feet square and one foot thick which floats on mercury and which holds the optical parts. Light from an argand lamp falls on *a*, part going to bcbcbcbaf and part to dedededaf.

I hope to be able to send you a copy of the paper within a month.

With kind regards to Lady Rayleigh,

Very sincerely yours,
Albert A. Michelson

No longer was it possible to believe that a positive result might be hidden in the errors of observation, and the doubts which had hung over Michelson's 1881 Potsdam experiment were now entirely removed by the Cleveland experiment. All explanations which had attempted to reconcile the Potsdam result with existing theories must now be abandoned, and new explanations for the behavior of moving optical and electrical systems had to be found.

At first, however, the full significance of this situation was not appreciated, and as late as 18 August 1892 Lorentz wrote to Lord Rayleigh:

I have read your note with much interest and I gather from it that we agree completely as to the position of the case. Fresnel's hypothesis, taken conjointly with his coefficient $1-1/n^2$, would serve admirably to account for all the observed phenomena were it not for the interferential experiment of Mr. Michelson, which has, as you know, been repeated after I published my remarks on its original form, and which seems decidedly to contradict Fresnel's views. I am totally at a loss to clear away this contradiction, and yet I believe that if we were to abandon Fresnel's theory, we should have no adequate theory at all, the conditions which Mr. Stokes has imposed on the movement of the aether being irreconcilable to each other.

Can there be some point in the theory of Mr. Michelson's experiment which has as yet been overlooked?

And Lord Kelvin, as late as 1900, in a lecture at the Royal Institution on 27 April of that year still referred to the Michelson–Morley experiment as one of the two "19th century clouds" (the other being the Maxwell–Boltzmann doctrine regarding the partition of energy) which dimmed the otherwise brilliant sky of the new scientific century. Kelvin had never fully accepted either the Maxwell electromagnetic theory or its extensions by Lorentz, and, as time passed, the older aether school which he represented fell farther and farther behind the new advances. Kelvin hoped to the end that the aether concepts could be saved, and in the preface to his Baltimore Lectures when reprinted in revised form in 1904 he would say ". . . that two of ourselves, Michelson and Morley, have by their great experimental work on the motion

---

less noise and cause for jarring the basement stairs, our prejudice turned into a feeling of partnership and sympathy, later of pride. . . .

Yours,
Sid.

Also, letters from W. W. Coblentz (Case, Inst. Technol., Class of 1900) to R. S. Shankland (31 Oct. 1952) (quotes D. C. Miller); Comfort A. Adams (Case Inst. Technol., Class of 1890) to R. S. Shankland (3 Nov. 1952); and conversations with William Koehler (Case Inst. Technol., Class of 1889) (10 Dec. 1950; 11 Nov. 1952); and discussions with Prof. Hippolyte Gruener (17 Feb. 1950; 21 Dec. 1950; 28 Nov. 1952).

[55] A. A. Michelson and E. W. Morley, Am. J. Si. **34**,c 333 (1887); Phil. Mag. **24**, 449 (1887); J. Phys. (Paris) **1**, 444 (1888); Sidereal Messenger **6**, 306 (1887).

of ether relatively to the earth, raised the one and only serious objection against our dynamical explanations;. . . ." And at the 1900 International Congress of Physics in Paris, Kelvin had urged Morley and D. C. Miller (Michelson had left Case in 1889) to make another trial of the experiment, which they did with an even more conclusive null result than that obtained by Michelson and Morley.[56]

It is of interest to note that, although Michelson and Morley had planned to repeat their observations at regular intervals throughout the calendar year so that all epochs related to the earth's motion through space would be encompassed, they in fact made no further trials of this experiment after July 1887. The reason for this is probably the following.

The laboratory quarters in the basement of Adelbert Hall had been only temporary, and, at the beginning of the autumn term in September 1887, Michelson and Morley transfered their research activities to a basement laboratory in the Main Building of Adelbert College. Had they again set up the Michelson–Morley aether-drift interferometer there, it would have been a simple matter to repeat their observations at regular intervals during the year. Instead, they soon became involved in new lines of research that absorbed all their interests and efforts for the remaining two years that Michelson was a member of the Case faculty. In fact, Michelson, in his address on optical research as vice-president of section B of the A.A.A.S. in Cleveland in August 1888, failed to mention the Michelson–Morley experiment![57]

It was at this time that they developed the method which proved the feasibility of using light waves as the standard of length.[58] The files of the Warner and Swasey Company of Cleveland (for whom Professor Morley acted as consultant) still have a number of working shop drawings of the "Metre Subdividing Machine" which was built for Morley and Michelson in 1888. This machine was a highly engineered double-screw interferometer and has all of the essential features of the apparatus later used by Michelson in Paris to determine the length of the standard meter in light waves.

One of the most notable features of this work was their discovery of fine structure in the spectrum of hydrogen and hyperfine structure in the spectra of mercury and thallium. These optical multiplets were established with the Michelson interferometer by using the new Warner and Swasey instrument with optical arms of variable length and observing the changes in the visibility of the interference fringes as a function of optical-path difference between the two arms. This method was the natural outgrowth of their technique of moving the mirror e_1 (see Fig. 6) by a micrometer screw in order to improve the visibility of sodium-light fringes, as an aid in finding the white-light fringes in their aether-drift interferometer. The fine structure of the red hydrogen line was measured accurately in this way and shown to have a "doublet" structure.[58] Their early measurements of this "doublet" separation was the standard determination for many years and is in close agreement with the presently accepted value. The correct theoretical explanation of fine structure and hyperfine structure in spectrum lines, of course, had to wait many years until the development of quantum mechanics showed their relationship to electron spin and nuclear spin.

VII. CONCLUSION

It has already been emphasized that the Michelson–Morley experiment of 1887 could never be lightly considered, as had been the Potsdam result, as being due to some experimental inadequacy. Their result has always been accepted as definitive and formed an essential base for the long train of theoretical developments that finally culminated in the special theory of relativity.

The first important suggestion advanced to explain the null result of Michelson and Morley was G. F. FitzGerald's hypothesis, made to Sir Oliver Lodge[59] in his study at Liverpool, that the length of the interferometer is contracted in the direction of its motion through the aether by the exact amount necessary to compensate

[56] E. W. Morley and D. C. Miller, Phil. Mag. **8**, 753 (1904); **9**, 680 (1905).

[57] A. A. Michelson, Proc. AAAS **37**, 3 (1888).

[58] A. A. Michelson and E. W. Morley, J. Assn. Engrg. Soc. (May 1888); Am. J. Sci. **38**, 181 (1889).

[59] Sir Oliver Lodge, Nature **46**, 165 (1892); G. F. FitzGerald, *Scientific Writings* (Dublin University Press, Dublin, 1902), Vol. 34.

for the increased time needed by the light signal in its to and fro path. This contraction hypothesis was made quantitative by H. A. Lorentz in further development of his electron theory. In the original form of this theory used to deduce the Fresnel dragging coefficient, and other effects, Lorentz[60] had obtained equations for a moving electrical system by applying a transformation in which terms of higher order than (v/c) were neglected, and which showed that to this approximation the relative velocity of medium and observer had no influence on the phenomena. After Larmor[61] had shown how these transformations could be extended to include quantities of order $(v/c)^2$, Lorentz[62] gave the general solution, exact to all orders of (v/c), by introducing what has since been known as the "Lorentz transformation," under which Maxwell's equations for the electromagnetic field in empty space are covariant. This treatment provided a more general explanation for the null result of the Michelson–Morley experiment than that previously given by FitzGerald and Lorentz. In his development, Lorentz not only obtained new measures of length and mass, but also employed a new method of specifying time which he called "local time," needed for the description of the properties of moving systems.

During the same period, Poincaré[63] had contributed both to the philosophical and mathematical development of the subject. As early as 1899, he had asserted that the result of Michelson and Morley should be generalized to a doctrine that absolute motion is in principle not detectable by laboratory experiments of any kind, and, in lectures at international congresses held at Paris in 1900 and at St. Louis in 1904,[64] he gave to this generalization the name "The Principle of Relativity," whereby "the laws of physical phenomena must be the same for a 'fixed' observer as for an observer who has a uniform motion of translation relative to him: so that we have not, and cannot possibly have any means of discerning whether we are, or are not, carried along with such a motion." From the experimental evidence, Poincaré also concluded that "there must arise an entirely new kind of dynamics, which will be characterized above all by the rule, that no velocity can exceed the velocity of light."

In June 1905, Poincaré[65] again cited the result of the Michelson–Morley experiment and asserted that "it seems that this failure to demonstrate absolute motion must be a general law of nature." He then proceeded to complete the theory of Lorentz by obtaining the equations of transformation of the electric-charge density and the electric current when the time and space coordinates are changed by the Lorentz transformation (so named here by Poincaré). Poincaré put the Lorentz transformation into the symmetrical form now universally used in which they form a mathematical group, and also gave the relativistic formulation for the addition of velocities.

In September 1905, Einstein[66] published his famous paper on the "Electrodynamics of Moving Bodies," which developed the special theory of relativity from two postulates: (1) the principle of relativity was accepted as a postulate asserting the impossibility of detecting uniform motion as defined for an inertial system in Newtonian mechanics, and (2) the constancy of the speed of light as contained in Maxwell's equations was generalized to a postulate stating that light is always propagated in empty space with a velocity c which is independent of the state of motion of the emitting body. Both postulates could, of course, be considered as having a close relationship to the Michelson–Morley experiment, but actually Einstein arrived at his theory by a less direct route, becoming aware of the observational material principally through the writings of Lorentz which he began to study as a student in 1895. He was also keenly aware of the phenomena of stellar aberration and the experiment of Fizeau on the

[60] H. A. Lorentz, Arch. Néerl. Sci. **25**, 363 (1892); *Versuch einer Theorie . . .* (E. J. Brill, Leiden, 1895); Proc. Amsterdam Acad. **1**, 427 (1899).

[61] J. Larmor, *Aether and Matter* (Cambridge University Press, Cambridge, England, 1900), Chap. 11.

[62] H. A. Lorentz, Proc. Amsterdam Acad. **6**, 809 (1904); Encykl. Math. Wiss. Leipzig (1904).

[63] H. Poincaré, *Électricité et Optique* (Carré et Naud, Paris, 1901).

[64] H. Poincaré, Bull. Sci. Math. **28**, 302 (1904).

[65] H. Poincaré, Compt. Rend. **140**, 1504 (1905); Circ. Mat. Palermo Rend. **21**, 129 (1906), presented 23 July 1905.

[66] A. Einstein, Ann. Physik **17**, 891 (1905) [also in English transl. (Dover Publications, Inc., New York)]; Poincaré and Einstein seem to have arrived at their results independently.

speed of light in moving water.[67] In his great paper, Einstein made the two postulates compatible by means of the Lorentz transformations for the conversion of coordinates and times of events between two inertial systems. He provided direct and convincing explanations for the classic experimental facts, including new treatments of aberration and the transverse Doppler effect. Einstein's approach led him to the relativity of the simultaneity of events, and so made the aether concept superfluous while also demonstrating that the "local time" of Lorentz is in fact the only meaningful time for the description of physical phenomena. His paper is generally considered as the definitive exposition of the special relativity principle and the climax of the century-long developments which had begun with Young and Fresnel to understand electrical and optical phenomena in moving media.

In the years following the acceptance of the theory of relativity, the Michelson–Morley experiment was subject to continual scrutiny, leading to a much deeper understanding of its significance. During the same period, the relativity theory was gradually presented on a more sophisticated basis, and, in more abstract form, less dependent on the special relationship with the optics and electromagnetism from which it had its origin.[68] The special theory in its turn led the way to the general theory of relativity and so to Einstein's theories of gravitation and cosmology.

[67] R. S. Shankland, Am. J. Phys. **31**, 47 (1963).

[68] Especially through the work of H. Minkowski, who introduced the four-dimensional formulation of the theory in terms of the geometry of space–time, and also the use of tensor calculus; Gött. Nach. 53 (1908); Math. Ann. **68**, 526 (1910); H. A. Lorentz, A. Einstein, and Minkowski (*Das Relativätsprinzip* Leipzig, 1910) [also in English transl. (Dover Publications, Inc., New York)].

In closing this account of the Michelson–Morley experiment, it may be appropriate to do so with the statement which Professor Einstein sent for a special meeting of the Cleveland Physics Society held on 19 December 1952 honoring the centenary of Michelson's birth.[69]

"I always think of Michelson as the artist in Science. His greatest joy seemed to come from the beauty of the experiment itself, and the elegance of the method employed. But he has also shown an extraordinary understanding for the baffling fundamental questions of physics. This is evident from the keen interest he has shown from the beginning for the problem of the dependence of light on motion.

The influence of the crucial Michelson–Morley experiment upon my own efforts has been rather indirect. I learned of it through H. A. Lorentz's decisive investigation of the electrodynamics of moving bodies (1895) with which I was acquainted before developing the special theory of relativity. Lorentz's basic assumption of an ether at rest seemed to me not convincing in itself and also for the reason that it was leading to an interpretation of the result of the Michelson–Morley experiment which seemed to me artificial. What led me more or less directly to the special theory of relativity was the conviction that the electromotive force acting on a body in motion in a magnetic field was nothing else but an electric field. But I was also guided by the result of the Fizeau-experiment and the phenomenon of aberration.

There is, of course, no logical way leading to the establishment of a theory but only groping constructive attempts controlled by careful consideration of factual knowledge."[70]

ACKNOWLEDGMENTS

It is a pleasure to express my sincere thanks to Professor Sidney W. McCuskey, Professor Martin J. Klein, and Professor Leslie L. Foldy for constructive criticisms in the preparation of this paper.

[69] R. S. Shankland, Nature **171**, 101 (1953).

[70] Professor Einstein also sent the German original, but as the English translation is his, it is given here.

Radioactivity before the Curies*

Lawrence Badash
Department of History of Science and Medicine, Yale University, New Haven, Connecticut
(Received 15 September 1964)

A scientific discovery of great significance need not lead inexorably to widespread and more penetrating investigation. Immediately after Becquerel's discovery of radioactivity and his study of some physical properties connected to the phenomenon there was very little interest evident among his colleagues. In fact, radioactivity lay essentially moribund until the Curies began their investigations using a new and fertile approach. The story of the two years between Becquerel's discovery and Marie Curie's first research is illustrative of the manner in which a major physical phenomenon may remain neglected, even after it was first observed, because of the confusion of related studies and a failure to recognize its relevant properties.

FROM a vantage point of almost seventy years we may look back upon the discoveries of x rays and radioactivity and assert that the latter field has been more fruitful and of greater consequence. In the last years of the 19th century, however, few scientists thought radioactivity had much of a future. Even Becquerel, who discovered the phenomenon in 1896, appeared to have exhausted his subject, for he published seven papers on uranium rays in that year, only two the following year, and none in 1898. On the other hand, research on x rays and on other forms of radiation was pursued at an enormous rate. To understand why radioactivity was not regarded as particularly significant, it is necessary to examine the work performed in the years from 1896 to 1898. The developments during this period point out how a proliferation of related phenomena can obscure the relevant facts, and how a sterile approach can effectively kill a newborn science.

I. URANIUM RAYS

The discovery of x rays in late 1895 led to a search for other penetrating radiations. Because the x rays originated in the fluorescent spot on the Crookes tube, where the cathode ray beam hit, Henri Poincaré suggested before the French Academy of Sciences that many luminescent bodies may emit similar invisible rays. Thus, Henri Becquerel and others sought for radiations which could pass through sheets of aluminum and affect photographic plates. And, Becquerel and others found such radiations.

These various rays were almost always considered to be an unusually long-lived form of phosphorescence. Certainly, this was Becquerel's view of radioactivity for quite a while. Other erroneous conclusions by the discoverer of uranium rays further offset his numerous correct interpretations of the phenomena and left the subject rather confused. Adding to the disorder were a great many other emitters and emissions, real and imagined, which fascinated physicists of the time.

Of interest to us here are not the discoveries of Becquerel, but the work of those who tried to build upon his findings. Suffice it to say that after his discovery of the rays from uranium salts, he localized the phenomenon in the metal itself and found that stimulation by sunlight was unnecessary. Although he persisted in the belief that some form of phosphorescence was involved, he showed that the nonphosphorescent uranous series of salts was also active. In addition, he examined the absorption of the rays by different materials and their ability to discharge the gold leaves of an electroscope. Most important, he felt (incorrectly) that he had shown that the rays could be refracted and polarized—which meant they were electromagnetic.

II. HYPERPHOSPHORESCENCE

In England, Silvanus P. Thompson, the Professor of Physics at the City and Guilds Technical College, Finsbury, London had, simultaneously and independently, observed the

* A report based upon this paper was presented to the annual meeting of the History of Science Society, 5 April 1963, at Bloomington, Indiana.

strange action of uranium rays.[1] He was among the many scientists throughout the world who, in January 1896, were repeating Röntgen's newly announced experiments. Like a number of the others, he conceived the idea of placing fluorescent substances in contact with the photographic plate in order to shorten the long exposures then needed to obtain x-ray shadow pictures. The fluorescence stimulated by the x rays was expected to hasten the action in the emulsion.[2] In conjunction with these experiments, Thompson also tested sunlight and arc light for invisible radiations but found none that would pass through a sheet of aluminum.

During the course of this work, Charles Henry, of the *Ecole Pratiqüe des Hautes Etudes* in Paris, published his findings that zinc sulphide apparently augmented the transparency of aluminum to x rays.[3] This paper seems to have stimulated Thompson into testing the luminous materials themselves for invisible radiation. He therefore placed a variety of them on a sheet of aluminum under which was a photographic plate. The arrangement was "left for several days upon the sill of a window facing south to receive so much sunlight (several hours as it happened) as penetrates in February into a back street in the heart of London."[4] Upon developing, he noted definite photographic action had occurred only beneath the uranium nitrate and uranium ammonium fluoride. This result was recognized as significant. News of it was sent to the Cambridge mathematician, Sir George Stokes, a former president of the Royal Society, who replied on 29 February 1896:

> Your discovery is extremely interesting; you will, I presume, publish it without delay, especially as so many are now working at the X-rays. For my own part I am not at all disposed to believe that the Röntgen Rays are due to normal vibrations, the hypothesis to which Röntgen himself leans. I think it far more probable that they are transversal vibrations of excessive frequency. That being the case, I think what you have discovered belongs to the same class of phenomena as Tyndall's calorescence . . . . I am in correspondence with Lord Kelvin about the Röntgen Rays, and I should like to refer to your discovery, but do not mean to do so till you have published your result. I should be glad, therefore, of a notice of the publication. Perhaps you may be writing to him yourself. Of course if you have done so I am free to say anything to him. He is very enthusiastic, and might let something slip out without thinking about it.[5]

Less than a week later, Stokes called Thompson's attention to the work being done across the Channel: "I fear you have already been anticipated. See Becquerel, *Comptes Rendus* for February 24th, p. 420, and some papers in two or three meetings preceding that."[6]

Having lost priority, Thompson gave no thought to the rapid publication of his results, and his paper did not appear until July. No further experiments seem to have been performed, for he then cited Becquerel's work in describing the properties of the rays. He found Becquerel's (erroneous) proof of their refraction and polarization very significant, since this meant they "consist of transverse waves of an exceedingly high ultra-violet order."[7] The persistent emission from uranium compounds is, he suggested, to the transient emission of x rays in the Crookes tube, as the persistent emission of visible light by phosphorescence is to the transient emission by fluorescence.

> "Hence the writer ventures to give to the new phenomenon thus independently observed by M. Becquerel and by himself the name of *hyper-phosphorescence*. A hyper-phosphorescent body is one which, after due stimulus, exhibits a persistent emission of invisible rays not included in the hitherto recognized spectrum."[8]

But his title never took hold, for the radiation,

---

[1] Phil. Mag. [5], **42**, 103 (1896).

[2] This method was successful, particularly in the hands of Michael Pupin. Thompson also was able to achieve a shortening of the exposure by designing a curved cathode which, by focusing the cathode rays, made the x-ray beam more intense.

[3] Compt. Rend. **122**, 312 (1896).

[4] Reference 1, p. 104.

[5] Jane Smeal Thompson and Helen G. Thompson, *Silvanus Phillips Thompson, his Life and Letters* (T. Fisher Unwin, London, 1920), pp. 185–86. It must be noted that the original letter from Thompson to Stokes, which is *not* printed in the former's biography, contains *no* mention of uranium compounds. This letter, dated 28 Feb. 1896, does appear in *Memoir and Scientific Correspondence of the Late Sir George Gabriel Stokes*, selected and arranged by Joseph Larmor (University Press, Cambridge, 1907, 2 vols.), vol. 2, p. 495. In a note on p. 496, Larmor writes that "in answer to an inquiry [Thompson] now states that he had been trying various substances, including uranium nitrate, and that he had found, conclusively on Feb. 26–7, that the latter was the only one to which the aluminum foil was not opaque. . . ." While the independent discovery of uranium rays by Thompson is not questioned, we may wonder why he did not mention this element to Stokes on 28 Feb., if he had seen its properties on 26–27 Feb. 1896.

[6] J. S. and H. G. Thompson, *ibid.*, p. 186.

[7] Reference 1, p. 105.

[8] Reference 1, p. 107.

while unnamed by Becquerel, was already being called Becquerel rays or uranium rays.

Thus Silvanus Thompson *did* discover radioactivity, but his contribution was negligible compared to that of Becquerel, and his paper was rarely cited in the literature. He never recognized, from his own work, that the emission of invisible radiation from uranium had nothing to do with luminescence. Nor did he localize the properties in the metal itself. He was, in fact, far more interested in x rays. As was his custom, he went to Paris during the 1896 Easter vacation to attend meetings of the *Société de Physique*—and to present papers on x rays. In letters home to his wife, he mentioned many of the French scientists he saw, but Becquerel's name was strangely absent.[9] This, perhaps, is most indicative of his attitude toward uranium rays.

III. OTHER RADIATIONS

Yet Thompson cannot honestly be criticized for his predilection for x rays. He was in the majority, and a very respectable majority it was, of scientists who found Röntgen's discovery most fascinating. Not only could these rays be produced in any laboratory having a Crookes tube and a high voltage coil, but they possessed the remarkable ability to make shadow photographs of the bones in a hand. This latter property was the main reason for the enormous public interest, which overnight made x rays the most publicized scientific discovery the world had ever known.

In contrast to this, relatively few laboratories possessed the various luminescent uranium salts, which were, in any case, too weak to make satisfactory bone pictures. The pure metal, which had only in 1896 been prepared by Henri Moissan, was virtually unobtainable, and it too was of little value for photographs, compared with x rays.

Interest in uranium rays was even further subordinated by the proliferation of other types of newly discovered rays. There were, of course, such authentic varieties as cathode rays, x rays, and canal rays, but there was also a confusing abundance of other, questionable emissions. In the summer of 1895, Eilhard Wiedemann, of the University of Erlangen, announced a new type of invisible radiation from sparks, which he called "discharge rays." Silvanus Thompson divided cathode rays into dia- and para-kathodic classes. Freshly cleaned metallic surfaces and numerous other substances were found to emit penetrating radiations. And similarly with glow worms, fireflies, and other luminous materials, which were tested because x rays had been traced to the fluorescent spot on the Crookes tube where the cathode rays hit.

Becquerel's papers in the *Comptes Rendus* (starting in February, 1896) were abstracted in French, English, and German periodicals, often within a month or two. While on occasion the review was fairly extensive,[10] it usually was merely a short paragraph among many other reviews.

Then Becquerel rays began to receive attention in papers by other scientists. Jean Perrin, of the *Ecole Normale Supérieure*, cited his countryman's work in an article entitled "Rayons cathodiques, rayons x et radiations analogues," and pointed out that these rays constitute an important theoretical transition between ultra-violet light and x rays.[11] In the spring of 1896, a detailed review of Becquerel's first five papers appeared in the *Journal de Physique*, written by Georges Sagnac, of the University of Paris, who was known for his own work with x rays.[12] And, in June, from Athens, Constantin Maltézos presented a one sentence discussion of uranium rays in a longer paper on x rays, suggesting that the former are hyper-ultra-violet transverse vibrations, since they are capable of refraction and polarization.[13]

IV. ORIGINAL RESEARCH ON URANIUM RAYS

The first original research performed, after the initial experiments of Becquerel and S. P. Thompson, was described before the Physical Society of Berlin on 23 October 1896. This in itself is a measure of the indifference with which the discovery was received, for it was over half a year before others saw fit to investigate the phenomenon. P. Spies then reported that fluorspar, placed between uranium and a piece

---

[9] J. S. and H. G. Thompson (note 5), pp. 187–89.

[10] For example, Electrician **36**, 680 (1896).

[11] Séances Société Française de Physique 121 (8 Apr. 1896).

[12] J. Phys. **5**, 193 (1896).

[13] Compt. Rend. **122**, 1474 (1896).

of bromide paper, caused the paper to be much more darkened than it would ordinarily be.[14] While this effect was perhaps due to some chemical action or secondary emission, it is interesting to see that efforts to concentrate uranium rays proceeded similarly to those already described for x rays. The following year, F. Maack put forth the claim that colophony or resin possessed the ability to concentrate these rays.[15]

The second research contribution came, oddly enough, from the United States. This was not to be expected in the context of nineteenth century physics, for this country usually lagged far behind the European developments. Yet, A. F. McKissick, professor of electrical engineering at the Alabama Polytechnic Institute (now Auburn University), had tested "all such [luminescent] substances that are known as available," hoping to discover other sources of Becquerel rays.[16] Nor did he stop with known luminescent materials; a list of the "active" emitters he discovered includes lithium chloride in solution, barium sulphide, calcium sulphate, quinine chloride, sugar, chalk, glucose, and uranium acetate. His procedure was to place the object to be photographed (e.g., a coin or key) directly upon the emulsion, cover it with the plate holder, and sprinkle on top of this the materials to be tested, all of which were previously exposed to the sun for two hours. Development 48–72 hours later revealed weak but distinct images of the objects. A complication was injected into the analysis of the results, however, for McKissick often obtained two to four images of each object. Granulated sugar was pronounced the best source, since with it he had gotten a fairly clear negative through two and a half inches of wood. A colleague, Professor B. B. Ross, suggested the phenomenon might be due to a high *molecular* weight, but McKissick found no satisfactory verification of this.

Except for the uranium compounds, McKissick was, of course, completely wrong. His images were more likely results of chemical action than penetrating radiations. Still, he is worthy of our attention, since he was very typical of the casual experimenters who added greatly to the confusing array of emitters and emissions. The suggestion of Professor Ross is also interesting, both because radioactivity was later shown to be found in elements of high *atomic* weight, and because it demonstrates that Becquerel's evidence that the rays come from *elemental* uranium (published in May, 1896) was not in November regarded as a relevant factor. This short paper by McKissick originally appeared in the *Electrical World* (New York), and was reprinted in its entirety in the *Scientific American Supplement* and *The Electrician* (London). All three journals were popular scientific weeklies, which assured wide dissemination of these false results.

Hanichi Muraoka, who received his training in Germany and was now professor in the Physical Institute of the Daisan Kotogakko, Kyoto, Japan, was another to be inspired by the discovery of Becquerel rays in phosphorescent uranium salts.[17] Could not the glow worms, which abounded during the summer, also be sources of something similar to x rays? He found not Becquerel rays, however, but something new. The glow worms produced a photographic impression *through* cardboard, but *not* where the cardboard was cut away. Muraoka concluded cardboard exerted a "suction effect" for his new rays, analogous to the permeability of iron for lines of force. There were others who examined glow worms and fireflies, and also detected types of radiations, but they, fortunately, did not compare them so closely to uranium rays. This line of research was discredited a year later when Muraoka and M. Kasuya published the results of more careful experiments on the next generation of the insects.[18] The photographic effect was due, at least in part, to vapors coming from the glow worms—it had been necessary to keep them moist. Damp brown paper gave the same effect.

On 1 March and 4 April 1897, the best executed and most serious investigations of uranium rays during this period were described before the Royal Society of Edinburgh.[19] Un-

[14] Verh. Physik. Ges. Berlin **15**, 101 (1896).

[15] Beibl. Ann. Physik **21**, 366 (1897).

[16] Elec. World **28**, 652 (1896); Sci. Amer. Suppl. **43**, 17542 (1897); Electrician **38**, 313 (1897).

[17] Ann. Physik [3], **59**, 773 (1896).

[18] Ann. Physik [3], **64**, 186 (1898).

[19] Proc. Roy. Soc. Edinburgh **22**, 131 (1897); **21**, 393 (1897).

fortunately, however, the excellent quantitative results were given no interpretation. The authors were the famous Lord Kelvin, John Carruthers Beattie, and Maryan Smoluchowski de Smolan, the latter two being research fellows at the University of Glasgow. Using a quadrant electrometer they measured the potential differences between an insulated disk of uranium metal (obtained from Moissan) and disks of other metals. They found that each metal acquired a distinct voltage when placed across the air gap from the uranium. Further, any charge given to the metal was dissipated quickly and its peculiar voltage reassumed.

Quantitative absorption experiments were next performed with screens of various materials and thicknesses. Then the size of the air gap was varied as well as the pressure and the type of gas. From these last experiments came the most significant information: the leakage (i.e., rate of activity) at ordinary pressure, measured in scale divisions per minute, increased with the applied voltage up to a point, where it effectively leveled off. The discharge of electricity by uranium was not proportional to voltage.[20] Throughout the papers the data are clearly presented. Yet, nowhere are they explained, not even here where it might be hoped that the authors would recognize a saturation current. One is left wondering what enables the conduction to occur, for there is no mention of ions, even though J. J. Thomson and Ernest Rutherford had almost half a year earlier shown that x rays produce ions and the current can become saturated.[21] It was, in fact, left to Rutherford to state explicitly that uranium rays produce ions in the gas.[22]

Another paper was read to the Royal Society of Edinburgh on 7 June 1897, this time by J. C. Beattie alone.[23] In it he described experiments in which he tested the electrification of the air surrounding uranium when the containing vessel and the uranium were held at different potentials. In all cases, the air acquired the same type of charge, positive or negative, as the uranium, and its electrification reached a maximum when the uranium was approximately 10 V. Beattie attempted no explanation for the decrease above 10 V, but Kelvin did in an appended note. He suggested that the lines of force around the small piece of uranium have a much greater voltage per centimeter than at the surface of the larger container. Above 10 V the rate of discharge of electricity into the air from the uranium ceases to increase much, while that from the container, of opposite charge, is still increasing. Hence, a maximum is reached, and the electrification must then diminish. It was an ingenious idea, showing that the seventy-three-year-old Kelvin had lost none of his inventiveness. But, in the light of modern understanding of the behavior of electricity in gases, the explanation seems to be that, with increased voltages, the samples of air Beattie drew off from the container were found lacking in charges because the ions formed by the uranium had been too quickly drawn to the electrodes.

V. SPECULATION AND MORE CONFUSION

In the Wilde Lecture, before the Manchester Literary and Philosophical Society, entitled "On the nature of the Röntgen rays," Sir George Stokes emphasized the refrangibility and polarizability of uranium rays and, with an apparent love of mechanical models, he explained the emission process:

> What takes place? My conjecture is that the molecule of uranium has a structure which may be roughly compared to a flexible chain with a small weight at the end of it. Suppose you have vibrations communicated to such a chain at the top; they travel gradually to the bottom, and near the bottom produce a disturbance which deviates more from a simple harmonic undulation. So, if a vibration is communicated to what I will call the tail of the molecule of uranium, it may give rise to a disturbance in the ether which is not of a regular periodic character. I conceive, then, that you have vibrations produced in the ether, not of such a permanently regular character as would constitute them vibrations of light, and yet not of so simple a character as in the Röntgen rays—something between. And accordingly, there is enough irregularity to allow the ethereal disturbance to pass through black paper, and enough regularity on the other hand to make possible a certain amount of refraction. You can also obtain evidence of the polarization, and, consequently, of the transverse character of these rays.[24]

[20] Just eight days later Becquerel presented a paper to the *Académie des Sciences* containing similar findings, and also suggesting a formula for the leakage–voltage relationship: Compt. Rend. **124**, 800 (1897).

[21] Phil. Mag. [5], **42**, 392 (1896).

[22] Phil. Mag. [5], **44**, 422 (1897).

[23] Proc. Roy. Soc. Edinburgh **21**, 466 (1897).

[24] Mem. Proc. Manchester Lit. Phil. Soc. **41**, 1 (1897). The lecture was delivered on 2 July 1897.

Throughout 1897 there also appeared other one-paragraph contributions to the study of uranium rays, contained in longer papers on related topics. In Paris, Gustave Le Bon found that many bodies, particularly shiny metal surfaces, when struck by light possess the property of discharging the gold leaves of an electroscope.[25] He thereby concluded that the properties of uranium are only a case of a very general law. In an extract from his Erlangen dissertation, W. Arnold fed the general confusion by declaring that Becquerel rays are emitted from zinc sulphide, mixtures of various sulphides and of tungstates, fluorspar, and even the organic compound retene.[26] From St. Petersburg, Ivan Ivanovitsch Borgman reported that Becquerel rays and Röntgen rays are capable of producing thermoluminescence in a mixture of calcium sulphate and manganese sulphate, an effect formerly attributed only to Wiedemann's discharge rays.[27] This paper is interesting in that it shows Borgman exposed the double sulphate of uranium and potassium to ultraviolet light before use. Thus, a year after Becquerel indicated that no external stimulation was necessary for the emission of these rays, another property lay unrecognized as relevant.

VI. CONFIRMING THE PROPERTIES

But there were also those who appeared to be on the right track. While offering no new properties, they at least confirmed Becquerel's work. Adolf Miethe, the scientific director of the Voigtländer optical works, verified that the rays from uranium salts and metal pass through aluminum foil.[28] At the *gymnasium* in Wolfenbüttel, the famous team of Julius Elster and Hans Geitel, who collaborated for almost forty years, including much later work on radioactivity, verified the physical properties of the rays and the fact that they persisted when the salts were kept in darkness for months.[29] They showed that zinc, aluminum, and other substances which present photoelectric effects, do *not* emit dark radiations of sufficient intensity to electrify the surrounding air in the manner of uranium. This enabled them to conclude that the source of the radiant energy was still unknown, and that true photoelectric phenomena cannot be attributed to hyperphosphorescence (a rare instance of the use of the name given by S. P. Thompson).

The experiments by the Italian, E. Villari, also confirmed Becquerel's results but are more interesting to us historically because of the way they came provokingly close to later, extremely significant, discoveries.[30] Using uraninite, he tested the ability of a current of air to discharge a gold leaf electroscope. Then he placed the sample in a paper envelope and noticed no electrical action. As was learned later by Rutherford and Mme. Curie, thorium and radium would have exhibited an effect, since they evolve a gaseous emanation, whereas uranium does not. Even more interesting, however, is his use of uraninite. This variable mixture, consisting mainly of UO_2 and UO_3, is none other than the ore pitchblende, from which the Curies were to extract the new elements polonium and radium. Had Villari performed *quantitative* work, it is possible that he would have discovered that uraninite was far more active than accounted for by its uranium content.

The final experiments to be cited in this period are those of C. T. R. Wilson, who was "Clerk Maxwell Student" in the University of Cambridge.[31] Like Rutherford, he was working in the Cavendish Laboratory under J. J. Thomson, and in the process of perfecting his cloud chamber. Wilson showed that uranium rays would produce nuclei for the condensation of water about them. Since the nuclei are swept away by an electric field, they must be charged particles, or ions. Whether the uranium compounds were placed within or without the expansion apparatus, they produced nuclei requiring the same degree of supersaturation as those produced by x rays. Prophetically, Wilson noted that "expansion experiments probably furnish one of the most delicate methods of detecting these rays."[32]

[25] Compt. Rend. **124**, 892 (1897).
[26] Ann. Physik [3], **61**, 313 (1897).
[27] Compt. Rend. **124**, 895 (1897).
[28] Beibl. Ann. Physik **21**, 606 (1897).
[29] Beibl. Ann. Physik **21**, 455 (1897).

[30] L'Éclairage Électrique **15**, 32 (1898); Sci. Abstr. **1**, 392 (1898).
[31] Proc. Cambridge Phil. Soc. **9**, 333 (1897); Proc. Roy. Soc. London **64**, 127 (1898).
[32] Proc. Roy. Soc. London **64**, 128 (1898).

VII. SUMMARY

What, then, can we say about these first two years of the development of radioactivity? In addition to the nine papers by Becquerel, there were perhaps a dozen original research contributions, including a number in which the discussion of uranium rays was a minor topic, and a number which presented misleading results. Becquerel's papers were widely abstracted but given no special prominence. News of the discovery does not appear to have been printed in the daily newspapers, but this is not strange; the reception accorded x rays was unique.

We are led to conclude that in the morass of various radiations uranium rays did not stand out as anything particularly significant. This subservient role is underscored by the difficulty in locating papers on the subject in the indexes of many scientific periodicals. They are to be found not under the headings of Becquerel rays, radiation (the name radioactivity was coined later), or even uranium, but often under Röntgen rays. As another measure of regard, we may compare these few publications on uranium rays with the approximately 50 books and pamphlets and 1000 papers on x rays *in 1896 alone*.[33]

As for the scientific content of these two years, Becquerel laid the foundation and built most of the edifice. But his mistaken belief that uranium rays could be refracted and polarized resulted in the misinterpretation of the nature of the phenomenon and the general satisfaction that these rays were even better understood than x rays (which were not known to possess these properties).[34] Certainly, they were a form of electromagnetic radiation, lying somewhere between x rays and ultraviolet light, and having some properties of each. It is perhaps ironic that the uranium rays were not long after shown to be corpuscular (at least those which produced the effects observed: the alpha and beta rays), while von Laue, Friedrich, and Knipping in 1912 proved x rays to be *identical* to light, though of a different wavelength.

It is also of note that the search for these rays was undertaken on the false premise that x rays were associated with the fluorescence of the Crookes tube. Thus, uranium rays were not only misinterpreted when found, but their discovery was caused by an incorrect assumption.

Two other points stand out as being crucial. (1) The deluge of other radiations, authentic and imagined, and especially those called Becquerel rays or allied to them, buried the significant findings to all but the "better" scientists. Becquerel, Elster and Geitel, and the Kelvin–Beattie–Smoluchowski de Smolan team appear not to have been led astray, but the relevant properties were often obscured to others. Even so, the "better" men seemingly exhausted the subject with their physical inquiries, and radioactivity lay moribund until analytical chemistry was applied by the Curies.

(2) While the experiments of this period cannot be characterized as completely qualitative and unsystematic, one is left with the impression that Marie Curie's systematic examination of the elements (a relevant characteristic) with a very accurate quadrant electrometer, and the obtaining of quantitative data of their relative emissive strengths, is a break with the past. The photographic method of testing, in which the active substance would merely cause a dark spot, or smudge, or the outline of an object to appear on the developed plate, was far too qualitative and subject to misleading interpretations. In more than a few cases, a strong photographic effect was later shown to be due to a chemical process. Other published reports of invisible radiations were so lacking in descriptions of the experiments that any explanation from inept technique to placement of the material directly on the emulsion (which is begging for a chemical reaction) is possible. Also, black paper was found to be no protection against chemical action. Finally, though no direct evidence has been uncovered to substantiate this belief, it is possible that nonuniform emulsions may have contributed to the confusion.

In summary, the work performed during this period by those other than Becquerel was largely unimportant. In early 1898, radioactivity was something of a "dead horse"—it was there,

[33] Otto Glasser, *Wilhelm Conrad Röntgen and the Early History of the Roentgen Rays* (Charles C. Thomas, Springfield, Illinois, 1934), Chap. 23.

[34] It must be noted that these properties were never among those confirmed by other investigators during this period.

but no one knew what to do with it. It took not only the discovery of thorium's activity, first by Gerhard C. Schmidt[35] and then by Marie Curie,[36] but the subsequent discoveries of polonium[37] and radium[38] by the Curies to produce a sustained renewal of interest. For then it became apparent that this was an atomic phenomenon of great significance.

[35] Verh. Physik. Ges. Berlin **17,** 14 (1898).
[36] Compt. Rend. **126,** 1101 (1898).
[37] Compt. Rend. **127,** 175 (1898).
[38] Compt. Rend. **127,** 1215 (1898).

ACKNOWLEDGMENTS

It is with pleasure that I express my appreciation to Professor Alfred Romer, of the St. Lawrence University, Dr. W. James King, of the American Institute of Physics, and Professor Derek Price, of Yale University, with whom I have had fruitful discussions of this subject.

The Discovery of the Beta Particle

MARJORIE MALLEY
541 Schwarz Road
Lawrence, Kansas 66044
(Received 12 February 1971; revised 26 April 1971)

Natural radioactivity was discovered by Henri Becquerel early in 1896. Only after three and a half years had passed was it determined that (at least) part of this radiation consisted of material particles. Towards the end of 1899 three different sets of investigators in as many countries found independently that radium rays were bent when they traversed a magnetic field. This paper delineates both the reasons for this delay and the events which converged in 1899 to make likely the simultaneous discovery of the deflection of the beta rays in a magnetic field. It also indicates the significance this discovery had for the future development of radioactivity.

I. INTRODUCTORY REMARKS

The first few years in the history of radioactivity were closely linked with the history of the x rays. The latter, discovered late in 1895 by Wilhelm Röntgen, created a major sensation in both scientific and popular circles. When Becquerel discovered radioactivity a few months later he was looking for new x-ray sources in phosphorescent substances.[1] At first Becquerel associated his rays with the visible light of phosphorescence and found that the invisible rays from uranium could be reflected, refracted, and polarized.[2] However, experiments by others contradicted Becquerel's findings, and it was eventually determined that, like their other properties, the optical behavior of the Becquerel rays approached that of the Röntgen rays. Thus, when Ernest Rutherford found in 1898 that the Becquerel rays consisted of a penetrating (beta) radiation and an easily absorbable component (alpha), he interpreted his findings in the light of the recent discovery of secondary x rays.[3] By the beginning of 1899 all physicists considered the Becquerel rays to be a mixture of primary and secondary x rays of unknown origin. Speculation placed the source of these rays in an external radiation or in the atom itself.[4] But initially there was little interest in the Becquerel rays themselves since they were considered as something of an epiphenomenon in the wake of the Röntgen rays.[5]

The general apathy towards radioactivity ended abruptly in 1898 when Marie and Pierre Curie succeeded in obtaining impure samples of two new Becquerel ray emitters which were many times more powerful than even pure uranium or thorium.[6] In the face of the discovery of two new chemical elements, polonium and radium, chemists as well as physicists decided that radioactivity was worth investigating. The vivid effects of these substances openly challenged physicists to find their source of energy. The availability of radium and polonium also made possible a number of experiments which could not be performed successfully with low-intensity uranium or thorium radiation. The consequent upsurge of interest in radioactivity and influx of new investigators made it probable that someone would soon test the effect of a magnet on radium and polonium rays.

II. THE EXPERIMENTS

The first such experiments were made in the spring of 1899 by Julius Elster (1854–1920) and Hans Geitel (1855–1923) in their laboratory in

Wolfenbüttel, Germany.[7] Elster and Geitel had come to study the Becquerel rays through their longstanding interest in atmospheric electricity. Their earlier researches had shown that the conductivity in air produced by heated electrodes and the photoelectric effect decreased in a magnetic field if the field was not parallel to the discharge. They had interpreted these results to indicate that the magnet deflected the gas ions out of the region of maximum electric field, thereby reducing the number which reached the far electrode in any given time.[8] Now they wished to determine whether the conductivity produced by Becquerel rays was affected in the same way.

Their preliminary tests with uranium were inconclusive, so Elster and Geitel obtained a sample of radium from Friedrich Giesel, a chemist at the nearby quinine factory in Brunswick. They placed the radium (contained in an aluminum bottle) inside a glass tube, on an electrode maintained at a potential of 500 V (Fig. 1). The other electrode was connected to a grounded electrometer. Elster and Geitel then evacuated the tube to a pressure of 1 mm (as they had in their earlier experiments) and noted qualitatively the effect of an electromagnet on the electrometer's rate of discharge. They found that the discharge produced by radium rays, like that in their earlier experiments, was noticeably delayed by the magnetic field, and inferred as before that the field had deflected the gas ions.

But the possibility remained that in this experiment the Becquerel rays themselves, as well as the gas ions, were deviated by the magnet, thereby reducing the ionization along the direct path between the electrodes. To test this explanation Elster and Geitel used a phosphorescent screen to determine the direction the rays traveled. In order to prevent particles of atomic dimensions from affecting the screen, they sealed the tube off about 15 mm from the radium by an airtight aluminum plate 1 mm thick. Elster and Geitel found, as they had expected, that the magnet did not affect the position of the bright spot on their screen produced by the radium rays. They concluded that

> the Becquerel rays experience no deflection by magnetic forces which might be comparable with that of the cathode rays. Thus, they agree in this relation too—as in all other characteristics known to date—with the Röntgen rays.[9]

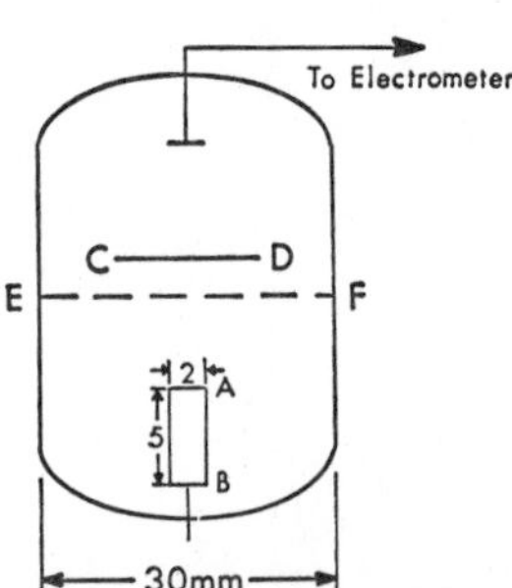

Fig. 1. Elster and Geitel's experimental arrangement to determine the effect of a magnetic field on the ionization produced by radium rays. The radium was contained in an aluminum tube AB and maintained at a potential of 500 V. CD is a phosphorescent screen. The airtight aluminum plate EF was used later to test the effect of the magnetic field on the radiation alone.

These results did not remain unchallenged for long. Stefan Meyer (1872–1949), a young physicist at the University of Vienna, became interested in radioactivity while watching a demonstration of the radioactive properties of radium and polonium by Friedrich Giesel at the Munich Naturforscherversammlung in September, 1899.[10] At that time Meyer had been studying the magnetic properties of the chemical elements, so he naturally wanted to test these new substances. For this purpose he obtained 2 g of barium–radium chloride and a little impure polonium from Giesel, as well as small quantities of radium and polonium preparations from the Curies.

Meyer performed his experiments on radium and polonium with a colleague at the University, Egon Ritter von Schweidler (1873–1948), who was familiar with the properties of electricity in gases from his own earlier work. Their first results were announced to the Viennese Academy of Sciences on 3 and 9 November 1899.[11] After they made the magnetization measurements, Meyer and von Schweidler investigated the phenomenon which Elster and Geitel had reported, using a horseshoe electromagnet which could produce fields as high as 17 000 G. They placed some radium, enclosed in paper, underneath an open, grounded brass tube which was connected to an electrometer (Fig. 2). When they applied the

Marjorie Malley

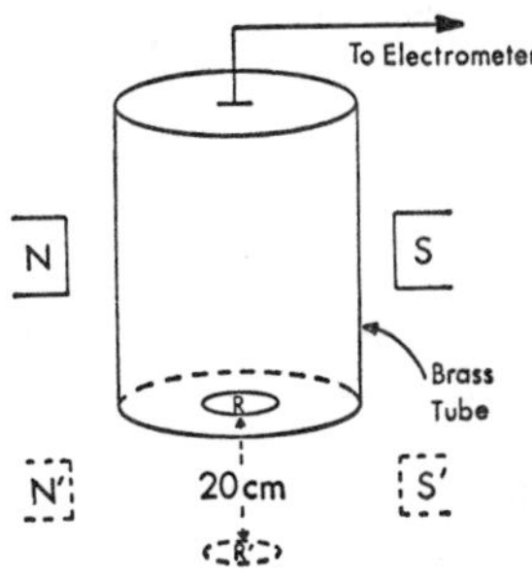

Fig. 2. Meyer and von Schweidler's experiment with radium rays. In the second part of the experiment the magnet *NS* and the radium *R* were moved away from the brass tube as shown.

magnetic field, they observed a decrease in the electrometer's rate of discharge, as Elster and Geitel had at low pressure.

Further investigation showed that when the radium was in the field but the rest of the apparatus was kept about 20 cm away, the discharge took place even more slowly. This result indicated that the magnet must have affected the radiation itself, either by deflecting the Becquerel rays or by retarding their emission. Meyer and von Schweidler eliminated the second alternative qualitatively by noting that the luminescence produced on the radium's paper covering did not appear to decrease when they applied the field. To test the former possibility they used the same type of phosphorescent screen Elster and Geitel had employed. But unlike their colleagues, they found that when they placed the screen a few centimeters from the radium and applied the field, the fluorescence on the screen disappeared completely. If they set the radium (still in its cover) directly on the screen without applying the field, the specimen cast a sharp image. When they turned the magnet on two bright streaks appeared a small distance from the original image, showing that the powerful magnetic field had bent the rays sent from both the top and the bottom of the radium specimen back onto the screen (Fig. 3). Meyer and von Schweidler determined that the direction of deflection corresponded to that of negatively charged particles.

But the Austrians' results had in part been anticipated. On 2 November 1899, the editors of the *Annalen der Physik und Chemie* received a paper from Giesel announcing that he had deflected both radium and polonium rays in a magnetic field.[12] Friedrich Giesel (1852–1927), an industrial chemist, had become involved with radioactivity as a sort of hobby through an earlier interest in phosphorescence and x rays.[13] He began isolating highly active substances from commercial ores at about the same time the Curies started working along these lines, and by 1899 he had advanced sufficiently to be able to supply others with radium and polonium preparations, as well as to use them in demonstrations.[14] When Giesel managed to find time to make a few experiments himself, he decided to check the work of Elster and Geitel on the behavior of Becquerel rays in a magnetic field.

In this experiment Giesel placed a phosphorescent screen on the poles of a horseshoe electromagnet, and 0.1 g of his freshly prepared (impure) polonium between the poles and about 1 cm underneath the screen (Fig. 4). When he turned the magnet on, the bright spot produced on the screen by the rays developed a cometlike tail, which showed that the field deflected the rays. He found that the direction of bending depended on the relative orientation of the source and poles but did not try to determine the sign of the charged particles as Meyer and von Schweidler did soon afterwards. Giesel observed similar but less marked effects with radium, including the sample he had lent Elster and Geitel earlier.

Since neither Elster and Geitel nor Giesel indicated the strength of their magnet, it is difficult to be sure how important experimental limitations were in producing the former's negative results. It may be that after Giesel had observed a deflection with his impure polonium he was predisposed to see one with radium also—which

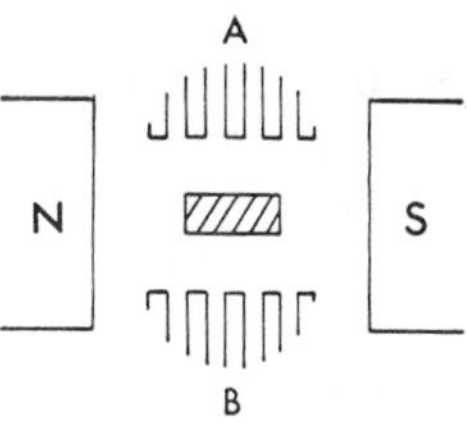

Fig. 3. The images *A* and *B* of the deflected rays obtained by Meyer and von Schweidler.

effect he noted was less clear. But after Elster and Geitel learned of the results of Giesel, Meyer, and von Schweidler, they succeeded in obtaining similar effects with their own equipment, indicating that their magnet, though small, sufficed for these experiments.[15] Giesel's apparatus apparently was not good enough to pursue more exact experiments with the rays.[16]

Meanwhile, Henri Becquerel (1852–1908) had resumed work on radiactivity after having spent a year and a half investigating the Zeeman effect. Becquerel had obtained samples of radium and polonium from the Curies and like all others who had the opportunity, was eager to test the properties of these new substances. First he repeated his earlier experiments on the reflection, refraction, and polarization of the rays. This time his results agreed with those of other observers, and he conceded that the Becquerel rays behaved optically like x rays.[17] He then compared the behavior of various phosphorescent and fluorescent minerals under the influence of radium and Röntgen rays.[18] Becquerel found that in a number of cases these radiations produced quite different effects, and at first surmised that the two kinds of rays had different wavelengths. But he also noticed that certain minerals reacted to radium rays very much as they had to cathode rays in some earlier studies made by his father: Both radium radiation and cathode rays lengthened the phosphorescence period of certain substances and both colored glass violet, an effect discovered by the Curies. Most likely these observations led Becquerel to test whether rays from radium or polonium were deviated in a magnetic field. Though he had just recently acknowledged that Becquerel rays behaved more like x rays than like visible light, it is significant that he had done so only reluctantly, and was never committed to the x-ray theory of radioactivity the way most other physicists were.[19]

Becquerel's papers on the behavior of radium and polonium rays in a magnetic field were read on 11 and 26 December 1899. He tested the preparations at various locations in the field of a variable electromagnet (4000–10 000 G), using both a phosphorescent screen and photographic plates, and obtained results similar to those of Giesel, Meyer, and von Schweidler.[20]

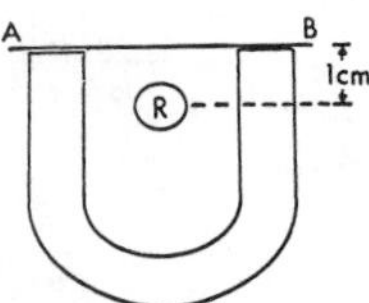

Fig. 4. Giesel's experiment with radium rays. The phosphorescent screen AB was placed on the poles of an electromagnet and the radium sample R was positioned approximately one centimeter below the screen.

III. ASSESSMENT

So, after a delay of three and a half years, scientists in Germany, Austria, and France reported the deflection of Becquerel rays by a magnetic field within a period of less than two months. Why did it take so long for this simple experiment to be performed successfully? What determined the timing and individual circumstances of the discovery?

In a sense, the answer to the first question is easy. The powerful radioactive substances polonium and radium were not discovered until 1898. In 1899 they were still very difficult to procure and could be obtained only from the Curies in Paris, Giesel in Brunswick, and the chemical firm of de Haen in Hanover.[21] The radiation from uranium and thorium is too low in intensity for the simple type of deflection experiments used in 1899; these rays will not activate a phosphorescent screen, and the long exposure required to yield satisfactory photographs greatly increases the chances of error.[22]

But granting the experimental difficulties, it is both curious and noteworthy that no one perservered in the attempt to bend Becquerel rays in a magnetic field before 1899. The technical problems could have been surmounted (and were eventually) with the resources available in 1896. The fundamental reason for the delay in the magnetic deflection experiments was that, before 1899, no one seriously considered that the Becquerel rays might contain material particles. Only in that very year, through the cathode ray researches of J. J. Thomson, had physicists in general come to accept the notion that certain particles could pass through metallic foil.[23] To most the idea seemed preposterous that any corpuscle could penetrate a millimeter or more of solid metal as the beta rays were known to do. The

Marjorie Malley

later Lord Rayleigh reported that both Thomson and Rutherford were skeptical of the first announcements that Becquerel rays had been deviated by a magnetic field for this reason.[24] The theory of the identity of Becquerel and x rays was held so firmly that no one appears to have explored possible relations between the Becquerel and the cathode rays before 1899, whereas after that year the journals of chemistry and physics abounded with reports of such experiments.[25]

However, by 1899 evidence was mounting rapidly for the existence of bodies smaller than atoms, particularly in studies of cathode rays and the Zeeman effect. Radioactivity was becoming more widely recognized as a *bona fide* new phenomenon, and more people were turning to serious study of the Becquerel rays.[26] These circumstances, coupled with the availability of two new radioactive elements, converged to produce the cluster of experiments on radium and polonium rays in 1899. It was no coincidence that of the four persons involved in the successful magnetic deflection experiments, three were newcomers to radioactivity, and one was a chemist. As such they could be expected to have harbored fewer preconceived notions than the few physicists who were established in that field. We have seen that the fourth, Becquerel himself, had never been very enamoured of the x-ray theory of radioactivity; Becquerel was led to the discovery through his familiar interest in phosphorescence.[27]

Nor was it a coincidence that all but one of the persons involved were from German-speaking nations, for personal contact was instrumental in precipitating the simultaneous discovery in Germany and Austria.[28] Giesel was in constant association with Elster and Geitel through their local scientific society in Brunswick. We have seen how Giesel supplied radioactive materials to both Elster and Geitel and Meyer and von Schweidler, and how he decided to try his own experiment because of his friends' results. Elster and Geitel's first researches on radioactivity had been inspired by the hypothesis that a relationship might exist between "hyperphosphorescence" and the puzzling photoelectric effect; it is possible that von Schweidler, who wrote several papers on the latter pheonomenon, was also led to radioactivity through this interest.[29] Meyer became interested in radioactivity through Giesel; he and von Schweidler began their joint research by repeating the experiments of Elster and Geitel. Although the two physicists from Wolfenbüttel were advocates of the x-ray theory, and obtained negative results in their magnetic deflection experiment with the Becquerel rays, they played a key role in the discovery.

IV. FINAL REMARKS

After the successful experiments were reported, work progressed rapidly. Pierre Curie showed that only the more penetrating or beta radiation appeared to be deflected.[30] He and Marie Curie found that the beta rays carried a negative charge.[31] Ernst Dorn in Halle succeeded in deflecting the beta rays in an electric field.[32] This last result made it possible to determine the ratio of charge to mass for the beta particles by sending them through crossed magnetic and electric fields. The first such measurements were made by Becquerel, who found a ratio of the same order of magnitude as that of the cathode ray particles.[33] By the middle of 1900 most scientists viewed the beta rays as high speed cathode ray particles.

It might seem that this identification of the beta ray particle with J. J. Thomson's "corpuscle" would immediately displace the x-ray theory of radioactivity. As it happened, the discovery at first strengthened this interpretation. Since the Becquerel rays contained corpuscles, it followed that secondary x radiation might also include them. This inference was confirmed in April 1900, when Pierre Curie and Georges Sagnac, one of the pioneers in the study of secondary x rays, announced that secondary x radiation indeed carried a negative charge.[34] This result accorded well with the view that Becquerel rays were a mixture of primary and secondary x rays.

Nevertheless, the discovery that part of the Becquerel radiation was material did encourage more radical speculation. The beta rays, the highest velocity particles then available, were valuable for studies of the theoretically important effect of velocity on the e/m ratio. The researches of Walter Kaufmann, which showed that this ratio decreased approximately as required by the current electrodynamics, led a number of physicists to wonder whether all mass might not be of electrodynamic origin.[35] More pertinent to radioactivity *per se* was the theorizing which pointed to the disintegration of the atom and the transmutation of the elements. For instance, in 1900 the Curies as usual maintained that a number of general explanations of radioactivity were possible, but they gave pre-eminence to a "ballistic hypothesis," whereby

> Radium would continuously emit small negatively charged particles. The utilizable energy, stored as potential energy, would disappear little by little, and this viewpoint would lead to the supposition of a mutable atom.[36]

The next year two other French physicists postulated that a subatomic change was the cause of radioactivity. Jean Perrin, whose researches had helped to verify the existence of Thomson's corpuscles, proposed an atomic model in which these negatively charged bodies orbited around positively charged "suns" of greater mass. Perrin supposed that the outermost "planets" in the heaviest atoms could be detached from their suns so easily that these elements would appear to be spontaneously radioactive.[37] Later in 1901 Becquerel developed a theory of atomic disintegration to explain radioactivity, in which he conceived the atom as an aggegrate of Thomson's corpuscles and positively charged particles of atomic dimensions.[38]

But by then many new and startling discoveries had been made which supported the theory of atomic transmutation more directly than the speculations inspired by the identification of the beta particle.[39] As it turned out the British, who had lagged with respect to that finding, put forth the definitive theory of subatomic change. The discovery of the beta particle, however, was an essential turning point in the new science of radioactivity which culminated in the acceptance of the Rutherford–Soddy theory of 1902–1903.[40]

ACKNOWLEDGMENTS

I wish to express my appreciation to Professor John L. Heilbron of the University of California, Berkeley, for valuable assistance with the manuscript, and to Mme. Monique Bordry for access to and information about documents at the Archives de l'Institut du Radium, Paris.

[1] For bibliography, see E. Whittaker, *A History of the Theories of Aether and Electricity* (Harper, New York, 1960), Vol. I, p. 357 and Vol. II, p. 1; O. Glassner, *Wilhelm Conrad Röntgen and the Early History of the Roentgen Rays* (Thomas, Springfield, Ill., 1934). The discovery was made independently by S. P. Thompson [Phil. Mag. **42,** 103 (1896)] also in the course of studies of x rays.

[2] H. Becquerel, Compt. Rend. **122,** 559 (1896) (received 9 March 1896). Most likely Becquerel had been misled by secondary radiations. See H. Becquerel, *Rapports présentés au Congrès International de Physique* III (1900), pp. 47–78.

[3] E. Rutherford, Phil. Mag. **47,** 109 (1899) (but dated 1 September 1898); G. C. Schmidt, Ann. Physik Chem. **65,** 141 (1898) (received 24 March 1898); J. L. Heilbron, Arch. Hist. Exact Sci. **4,** 247 (1968). At that time x rays had not yet been shown to be reflected (specularily), refracted, or polarized. The gamma rays were not discovered until 1900.

[4] The first suggestion was favored by the Curies, the second by J. J. Thomson, Rutherford, and Elster and Geitel. A third idea, that the energy was obtained from atmospheric molecular motion in violation of the second law of thermodynamics, did not receive much support. See M. Curie, Compt. Rend. **126,** 1101 (1898) (received 12 April 1898); E. Rutherford, Ref. 3; J. Elster and H. Geitel, Ann. Physik Chem. **69,** 83 (1899) (received 5 August 1899); J. J. Thomson, Proc. Cambridge Phil. Soc. **9,** 393 (1898) (received 24 January 1898); Sir Wm. Crookes, Presidential Address, Rept. Brit. Assoc. Advan. Sci. **68,** 26 (1898).

[5] Compare L. Badash, Amer. J. Phys. **33,** 128 (1965). See also F. Giesel's letter to the Curies, 22 December 1899 (Archives de l'Institut du Radium, Paris):

> Ihre schöne Entdeckung ist anfangs bei uns leider kaum beachtet worden, da mann in Folge der le Bon'schen Arbeiten [French psychologist whose work on "black light" had been discredited] etwas misstrauisch geworden war. Selbst auf der Münchener Naturforscher-Versammlg. war sehr vielen Physikern die Sache noch vollständig neu. Erwähnen möchte ich hierbei, dass sogar Röntgen nicht an die Existenz der Becquerelstrahlen glaubte und erst durch Aufnahmen mit Urankaliumsulfat auf Bromsilberplatten von Elster & Geitel überzeugt werden musste.

[6] P. Curie and M. Curie, Compt. Rend. **127,** 175, 1215 (1898) (received 18 July and 26 December 1898).

[7] J. Elster and H. Geitel, Verhandl. Deut. Physik. Ges. **1,** 136 (1899) (meeting on 5 May 1899).

[8] See, e.g., J. Elster and H. Geitel, Ann. Physik Chem. **37,** 315 (1889) (signed March 1889) and **39,** 167 (1890) (signed July 1890).

[9] Reference 7, p. 138. The shape of the aluminum bottle containing the radium probably caused the illumination to be concentrated in a central spot. The polarity of the terminals was not indicated.

[10] S. Meyer, Naturwiss. **3,** 129 (1949); F. Giesel, Physik. Z. **1,** 16 (1899) (submitted 24 August 1899).

Marjorie Malley

[11] S. Meyer and E. von Schweidler, Anz. Kaiserl. Akad. Wiss. Wien **36,** 309, 323 (1899) (received 3 and 9 November 1899); Physik. Z. **1,** 90, 113 (1899) (published 25 November and 2 December 1899).

[12] F. Giesel, Ann. Physik Chem. **69,** 834 (1899) (received 2 November 1899). Giesel observed the magnetic deflection on October 21; see S. Meyer, Ref. 10, p. 129. The reason for Giesel's anomalous results with polonium, which emits only alpha rays in this case (alpha rays were not shown to be deviable until 1902), long remained obscure, but it was soon observed [first by Meyer and von Schweidler, Anz. kaiserl. Akad. Wiss. Wien **36,** 309 (9 Nov. 1899)] that polonium prepared by the Curies did not emit deviable rays. Moreover, Giesel found that his polonium lost its power of emitting deflectable rays rapidly. In 1906 he finally determined that the beta ray emitter in his polonium was identical with Rutherford's radium *E* (^{210}Bi, half-life five days). See F. Giesel, Ber. Deuts. Chem. Ges. **39,** 780, 1014 (1906).

[13] Giesel to the Curies, 6 December 1899 (Archives de l'Institut du Radium, Paris): "Indirekte Veranlassung [of my interest in radioactivity] war mein schon seit mehreren Jahren den Phosphorescenzerscheinungen dargebrachtes Interesse." Compare Giesel's reports on the Röntgen rays and related phosphorescence in Jahresber. Ver. Naturwis. zu Braunschweig **10,** 73, 99, 113 (1896) (received 6 February, 19 March, and 22 October 1896).

[14] Giesel to the Curies, Ref. 5:

> Ich konnte bereits im vorigen Jahre, als Ihre Arbeit über Radium erschien, dem hiesigen Naturwissenschaftlichen Vereins an einem kleines Präparat die Phosphoresi. des Schirms zeigen, und hatte ich viel früher schon gefunden, dass die Hauptmenge der in meinem Rohstoff befindl. aktiv Subst. garnicht die chemischen Eigenschaften Ihres Polonium besass, vielmehr ein neuer aktiver Stoff sein musse.

There is no record of this demonstration in the reports of the Brunswick Society. Compare F. Giesel, Ann. Physik Chem. **69,** 91 (1899) (received 5 August 1899).

[15] Compare L. Badash, Centaurus **11,** 236 (1966).

[16] On 12 January 1900, apparently in reply to a request from the Curies, Giesel wrote:

> Ob es mir unter diesen Umstanden gelingen wird mich meinen eigenen kleinen Mitteln das magnetische Feld so lange zu erhalten, um auf photogr. Wege noch die von Ihnen gewünschte Auskunft über die noch vorhanden Ablenkbarkeit, constatiren zu können, bezweifle ich. Ich werde mich aber bemühen.

Nothing further about these experiments occurs in the correspondence preserved in the Archives de l'Institut du Radium. Meyer, however, attributed Geisel's timidity to his not being competent as a physicist: see Meyer, Ref. 10, pp. 130, 132.

[17] H. Becquerel, Compt. Rend. **128,** 771 (1899) (received 27 March 1899).

[18] H. Becquerel, Compt. Rend. **129,** 912 (1899) (received 4 December 1899).

[19] Although Becquerel did not indicate any direct connection between the experiments on phosphorescence and the magnetic deviability, he noted the relationship in a later publication: See his *Recherches sur une propriété nouvelle de la Matière* (Firmin Didot et Cie, Paris, 1903), pp. 126, 129. Becquerel may also have been influenced by a suggestion made by Mme. Curie early in 1899 that the Becquerel rays might be material [M. Curie, Rev. Gén. Sci. **10,** 41 (1899) (published 30 January 1899)]; he did know that the Curies had already tried to deflect uranium rays in a magnetic field [H. Becquerel, Compt. Rend. **129,** 1205 (1899) (received 26 December 1899)]. See M. Curie's notebook, Archives Pierre et Marie Curie, Bibliotheque Nationale, Section 1-II, 20 March 1898.

[20] H. Becquerel, Compt. Rend. **129,** 996, 1205 (1899) (received 11 and 26 December 1899).

[21] E. de Haën, Ann. Physik Chem. **68,** 902 (1899) (received 14 July 1899).

[22] Beta rays from uranium were first deflected in a magnetic field by Becquerel in 1900 [Compt. Rend. **130,** 1583 (1900) (received 11 June 1900)]. By devising a special but simple experimental arrangement, Becquerel was able to succeed with a field of only 1500 G. Rutherford and A. G. Grier showed that thorium emitted deviable (beta) rays in 1902 [Phil. Mag. **4,** 315 (1902)] using a field of 2200 G.

[23] E. Whittaker, Ref. 1, Vol. I, pp. 360–65.

[24] Lord Rayleigh, *The Life of Sir J. J. Thomson* (Cambridge U. P., Cambridge, Eng., 1943), p. 133; J. L. Heilbron, Ref. 3, p. 253. Compare E. Rutherford to Elster and Geitel, 12 June 1899 (Berlin, Staatsbibliothek Preussischer Kulturbesitz, früher Preussische Staatsbibliothek): "I was pleased to see you had investigated the effect of a magnetic field at low pressures on the rate of discharge. Your results are exactly what one would expect on the theory of ions."

[25] Georges Sagnac had intended to try to deflect the secondary x rays with a magnet as early as 12 March 1898 [L'Éclairage Électrique **14,** 466 (1898)], but he did not link this to the Becquerel rays at that time. Compare Ref. 34.

[26] Compare Ref. 5. A check of the bibliography in Jahrbuch der Radioaktivität und Electronik **1,** 413 (1904) for these years shows eight names listed for 1896, 16 for 1897, 14 for 1898, 23 for 1899, and 43 for 1900.

[27] Becquerel's grandfather, Antoine César Becquerel (1788–1878) and especially his father, Alexandre Edmond Becquerel (1820–1891) had studied phosphorescence. Becquerel discovered radioactivity while testing his father's collection of phosphorescent minerals.

[28] This is documented by Meyer, who published some correspondence relating to the magnetic field experiments (Ref. 10).

[29] J. Elster and H. Geitel, Ann. Physik Chem. **66,** 735 (1898) (received 26 September 1898); J. Elster, Jahresber Ver. Naturwiss. Braunschweig. **10,** 149 (1896) (meeting on 10 December 1896).

[30] P. Curie, Compt. Rend. **130,** 73 (1900) (received 8 January 1900).

December 1971

[31] P. Curie and M. Curie, Compt. Rend. **130,** 647 (1900) (received 5 March 1900).

[32] E. Dorn, Abhandl. Naturforsch. Ges. Halle **22,** 47 (1900) (signed 11 March 1900); Physik. Z. **1,** 337 (1900) (recieved 5 May 1900).

[33] H. Becquerel, Compt. Rend. **130,** 206, 809 (1900) (received 29 January and 26 March 1900).

[34] P. Curie and G. Sagnac, Compt. Rend. **130,** 1013 (1900) (received 9 April 1900). Compare Heilbron, Ref. 3, pp. 251–52.

[35] M. Jammer, *Concepts of Mass* (Harper and Row, New York, 1964), Chap. 11; Whittaker, Ref. 1, Vol. II, p. 53.

[36] P. Curie and M. Curie, in *Rapports presentes au Congres International de Physique* (Gauthier-Villars, Paris, 1900), Vol. III, pp. 79–114.

[37] J. Perrin, Rev. Sci. **15,** 46 (1901) (published 13 April 1901).

[38] H. Becquerel, Compt. Rend. **133,** 977 (1901) (received 9 December 1901). Becquerel's speculations were prompted by his experiments with uranium and uranium x.

[39] Such as the discovery of the so-called induced or excited radioactivity, radioactive emanations (gases), and radioactive deposits left by these gases; and the separation of active and inactive components of radioactive elements, together with the finding that these lost and regained their activity, respectively.

[40] E. Rutherford and F. Soddy, Trans. Chem. Soc. **81,** 321, 837 (1902); Phil. Mag. **4,** 370, 569 (1902); **5,** 576 (1903).

Reprinted from AMERICAN JOURNAL OF PHYSICS, Vol. 32, No. 9, 681–686, September 1964
Printed in U. S. A.

Rutherford and his Alpha Particles

THOMAS H. OSGOOD
Michigan State University, East Lansing, Michigan
AND
H. SIM HIRST
Culmer, Golf Links Road, Ferndown, Dorset, England

(Received 24 March 1964)

The authors give some personal recollections of their participation in the experiments of Rutherford and Chadwick on the disintegration of light elements by impact of alpha particles in the 1920's.

AS Blackett[1] and others have pointed out, Rutherford's scientific life can be divided into four fairly distinct periods according to the nature of the problems he worked on and the discoveries he made, rather than according to the laboratories over which he presided. Of these, the third period covered the years 1914–19 in Manchester plus 1919–30 at the Cavendish Laboratory in Cambridge. During these sixteen years, Rutherford and his students and colleagues refined their earlier measurements of the radiations from naturally occurring radioactive substances and discovered the disintegration of light elements by alpha particles. The alphas were peculiarly Rutherford's particles; he almost regarded them as playthings, an attitude thoroughly justified by his masterly handling of them with remarkably simple equipment.

The timing of the end of the third period of Rutherford's scientific work was determined, not by accident, but by the initial development of electronic techniques for detecting and counting ionizing particles.[2] Prior to that time there were, indeed, three methods of recording the arrival or passage of alpha particles. First, though not so in point of time, C. T. R. Wilson showed that they could be recorded photographically from their tracks in a cloud chamber. This method, though capable of beautiful precision (as in the hands of Blackett[3]) could deal with only a few particles at a time—far too few to be useful in the broad survey of the disintegration of the elements which occupied Rutherford and Chadwick during the early 1920's. Second, during his early Manchester days in 1908, Rutherford had collaborated with Geiger in constructing a workable point counter which, however, could handle only one particle at a time, and lacked the resolution and speed that were desired. Third, the ability of an alpha particle to cause a scintillation on a zinc sulphide screen, first observed in 1903 by Crookes, and by Elster and Geitel, gave early promise of developing into the most versatile and flexible of the three methods, considering the state of the art at that time. There was some doubt, at first, if every alpha particle would cause one scintillation; but while working on the point counter, Geiger and Rutherford checked this matter, using techniques of preparing screens perfected by others. They found to their pleasure that within the limits of experimental error each alpha particle produced one scintillation on a properly prepared screen of zinc sulphide. Later, methods were worked out for evaluating the efficiency of a particular screen, and the stage was then set for Rutherford to employ the scintillation method for a decade and half in a magnificent series of simple experiments culminating in the artificial disintegration of light elements under impact of alpha particles.

The late 1920's were devoted to developing electrical methods of counting (which were used exclusively thereafter) so we may take the year 1925 to mark the end of an era, an era during which the nuclear model of the atom was firmly established by simple experiments on alpha

[1] *Rutherford at Manchester*, edited by J. B. Birks, (Heywood and Company, London, 1962), p. 102.
[2] E. Rutherford, F. A. B. Ward, and C. E. Wynn-Williams, Proc. Roy. Soc. (London) **A129**, 211 (1930).
[3] P. M. S. Blackett, Proc. Roy. Soc. (London) **A103**, 62 (1923).

particles recorded by counting scintillations. The same year also marked the beginning of present-day experimental nuclear physics, in which electronics, engineering, and digital recorders took over the job of making observations whose reliability was formerly dependent on the skill and visual acuity of the physicist himself. That time also marks the beginning of physics-with-big-machines, especially accelerators for nuclear particles.

Contrary to the impression given in some modern textbooks, the visual counting of alpha particles in a research setting was not a simple operation *if quantitative results were desired.* Feather[4] described it as "a most fatiguing occupation, the eye very quickly losing its sensitivity to those very faint flashes of light unless it is frequently rested." Eve[5] wrote of many physicists spending "weary but exciting hours trying their eyes in counting single alpha particles." Rutherford himself knew it was not a game for old eyes, and admired[6] the enduring freshness which his young colleague Geiger could bring to the tedious task. Scintillation counting was plagued by a frustrating but never satisfied demand for a capability to handle greater numbers of particles; and it involved a continual strain on the part of the observer to distinguish with certainty between objective and subjective flickers of light, particularly when alpha particles were being counted whose energies were failing near the ends of their ranges.

In his early Cambridge days, Rutherford, then in his fifties, seldom or never relied entirely on his own visual counting of alpha particles, although he did a fair share of observing. It was obvious that he did not like to feel that he was, in this type of experiment, dependent upon the observations of others, but the numbers of scintillations he counted were often much higher than those of the other observers. However, he cheerfully accepted this limitation of his activity, and spent most of his time during the counting sessions encouraging the other observers and examining the results obtained. With his brilliant young colleague Chadwick, who had come with him from Manchester, he designed the experiments that were to be done, then enlisted the help of specially trained research students who assisted in the actual counting. When the results were written up and published, it was the generous custom of Rutherford and Chadwick, at the ends of their papers, to acknowledge their young helpers, who would, of course, have been honored to do the work with no recognition whatever, merely for the sake of the close association it gave them with the focus of nuclear research at that time.

We two, the authors of this account, appear to have been the last of these privileged assistants in the Cambridge experiments,[7] and it occurred to us that we could contribute something to the intimate history of a great scientific personality by setting down our recollections of those exciting days at the Cavendish Laboratory. We had both graduated from the ancient University of St. Andrews in Scotland where we had been good friends as undergraduates. Being both bent upon scientific careers there was only one place to go for further study—Cambridge, which for scientists in those days was like Broadway for an actor. Considering the small number of students the Cavendish was able to accept from British universities other than Cambridge (Rutherford, being of New Zealand origin himself, always kept a considerable number of places for students from the colonies, and a few from the U. S.) we think ourselves, in retrospect, to have been extraordinarily lucky. Technically, Hirst was one of J. J. Thomson's students, Osgood one of Rutherford's.

Apparently we passed the six-weeks' training course of alpha-particle experiments (supervised by Chadwick) in acceptable fashion, but why we, in preference to a number of other new research students, should have been chosen at all as prospective alpha-particle counters for Rutherford's experiments was never made clear. We suspect, however, that Rutherford noted that we came from St. Andrews, the traditional cradle of golfers, and had heard that we both enjoyed there at least modest reputations as

[4] N. Feather, *Lord Rutherford* (Blackie and Son, London, 1940), p. 150.

[5] A. S. Eve, *Rutherford* (The Macmillan Company, New York, 1939), p. 177.

[6] Reference 5, p. 180. Letter from Rutherford to Bumstead.

[7] E. Rutherford and J. Chadwick, Phil. Mag. **48**, 509 (1924); **50**, 889 (1925).

players. Hence, Rutherford argued (we think), they must have good eyes, possess considerable powers of concentration and calmness under stress, and should therefore be good and accurate observers.

The experiments which formed the training course made us familiar on a first-hand basis with the simple properties of alpha particles. We observed the rates of decay of pure radioactive substances, measured the ranges of alpha particles in air and in mica sheets, verified the statistical character of the emission of the particles from a weak source, plotted the growth of a daughter substance from its parent, all with equipment of which the most complex items were a gold-leaf electroscope, a microscope, a reciprocating vacuum pump, and a stopclock. We were not, ourselves, permitted to prepare the radioactive sources (this was before the days of artificial radioactivity) which was done by Chadwick and Rutherford's technician, Crowe, on the forbidden top floor of the laboratory. Our only visit there was a guided preliminary tour of inspection. We remember distinctly, however, seeing Chadwick come running down the long flight of stairs, carrying a quick-decaying radioactive sample he had just prepared, and hearing him call "Out of the way! This won't wait! Is everything ready?" or words to that effect.

At this point special reference must be made to George Crowe, in view of the great service he rendered, not only to Rutherford, but to his successors during fifty years at the Cavendish Laboratory. In the regular handling of radioactive materials, he faced an ever present but usually dormant danger to fingers and general health. The period with which we are concerned was, of course, long before the now recognized subject, health physics, had developed as a result of the dangers of handling massive quantities of artificially produced radioactive materials. Rutherford, however, encouraged Crowe to lead as much an outdoor life as possible. After a weekend he would regale us with an account of his fishing on the Norfolk Broads, and one day even brought for our inspection the desiccated head of a very large pike he had caught, so that we might see and feel the creature's powerful teeth. Crowe's mode of life kept him in good health until his retirement at the appointed time in 1959. He was then honored by a brief but accurate biography[8] in *New Scientist*.

During the years 1923–25, Rutherford and Chadwick carried out their experimental observations according to a fixed schedule, two sessions per week, from 4 to about 6 or 6:30 p.m. Nothing was allowed to displace or postpone these sessions, and in spite of the multitude of demands made on Rutherford's time, he missed very few of them,—an indication that they were really his life, and that what else came along was just incidental. By the end of a session, the usual source of alphas, RaC,C′ (half-life about 20 min) had decayed to such an extent that the intensity was no longer adequate for making reliable observations, and the afternoon's work came to a natural end.

Promptly at 4 o'clock, then, on those days, leaving our own researches, we went along the stone-flagged ground floor of the Cavendish, through G. I. Taylor's two laboratories into Rutherford's own research room. Always it was closely shuttered with not the tiniest pencil of light coming in at the edges of the windows, and was pleasantly cool even on the few hot days of July. Its dimensions were perhaps 30 ft by 20 ft. In the room was a brick pillar which had supported the spinning coil used in Rayleigh's determination of the ohm, and, some years later, the apparatus by which J. J. Thomson identified the electron.[9] When we arrived, Crowe had already set up the equipment necessary for the day's experiment, adjusted the current in the electromagnet (to deflect unwanted beta rays) and we waited the arrival of Chadwick and the source of alpha particles. Meanwhile Rolfe, the laboratory assistant (first appointed to his job by Lord Rayleigh, certainly not later than 1884) put in a momentary appearance, bringing a metal tea tray with a brown pot full of strong tea, some "rock" buns of which we were all very fond, and four or five cups and saucers. The brief tea party which followed compensated us for missing the daily assembly of the Cavendish research students for tea in the library, where we used to munch buns (which we paid for) drank tea (provided by the professor) and discussed

[8] New Scientist, 6, No. 149 (1959). A Profile.

[9] A. Wood. *The Cavendish Laboratory* (Cambridge University Press, Cambridge, 1946), p. 20.

university sporting achievements and failures. At our more intimate teas in Rutherford's room we spent the first few minutes exchanging news. Rutherford, naturally, occupied the center of the stage, and although, as Ellis[10] put it, he enjoyed "talking of all things under the sun" in his customary lively, trenchant, and boisterous manner he knew also how to be a good listener. Quite often, R. H. Fowler would come in for tea, and then we sat quiet and listened while the three of them discussed the results of previous experiments.

Rutherford enjoyed a game of golf nearly every Sunday morning with Aston, Taylor, and Fowler, and because of his interest in the game used to follow the fortunes of the Cambridge University golfers. From time to time both of us (the authors) joined the university golf team in their week-end matches (there were no inconvenient eligibility rules which prevented a graduate student from playing) often played against well-known London golf clubs. Now amateur golf, in those days, was something of a society sport, and it was usual for *The Times* and *The Telegraph* on Mondays to devote a column to the week-end matches played by both the Cambridge and Oxford teams. This we used to read with eagerness and amusement to see what the golf correspondents said about our persons and our playing, and we had to be ready to endure some teasing from Rutherford and some dry but never malicious humor from Chadwick, both of whom took a paternal interest in our sporting activities. It took us a long time, for example, to live down newspaper descriptions of either of us as a "big man with enormous hands," or one who "made a floundering sort of effort," or who gave "a lamentable display of putting."

When the whole crew had assembled in Rutherford's room, all the electric lights were put out and Chadwick lit an old-fashioned fish-tail gas burner near the door, adjusting the flame to be precisely $\frac{1}{4}$ in. high. This faint illumination was just enough, after 10 or 15 min, to let our eyes focus on prominent objects in the room, but not bright enough to make out details. Then while for 10 or 15 minutes we mingled tea and conversation, our eyes became adjusted to the faintness of the light and we were ready to begin counting.

Seldom were we, the student assistants, fully informed about the nature or specific purpose of the experiment that was to be done, though we naturally knew in general what was happening. Our job was merely to count alpha particles with a microscope, and to count them reliably, a minute at a time, with Chadwick as time caller. To avoid bias, the counts of other observers were not disclosed until the end of a short series. It was easy, with dark adapted eyes, in a dark room, to see scintillations caused by alpha particles or protons, but counting them (up to 90 or 100 per min) without missing any, and without imagining scintillations when there were none was another matter, particularly if the particles were near the ends of their ranges. Intense concentration was necessary, and one was quite useless if out of sorts or overtired.

Matters like the choice and preparation of the zinc sulphide screen were taken care of by Chadwick and Crowe. We remember one occasion when a new screen was being tried out, coated with zinc sulphide sent by a friend from Germany, and recommended by him as excellent for counting. It turned out to be a dreadful flop and was immediately discarded. Only after some correspondence and inquiry did the reason for failure come to light. Seeing that the zinc sulphide crystals were considerably coarser than he had been in the habit of using, Crowe had ground them in a mortar, and this mechanical breakage apparently diminished enormously the quality of the resulting screen.

To keep the focus of the eye from wandering during the occasional few seconds which could elapse between the appearances of successive scintillations, a very faint adjustable illumination was maintained on the screen by a lamp and rheostat, just enough to make the field of view faintly distinguishable as a large gray circle on a black background. Without this precaution, scintillations could easily be missed while the eye lost its focus from one event to the next.

For experiments whose interpretation depended on measurements lying near the limit of observational precision, which at the best in scintillation experiments is of the order of a few percent, there was always a drive toward

[10] Reference 5, p. 274.

greater intensities of the source, or greater collecting area of the screen. After these had been exploited to the utmost under the conditions of technique then available, an attempt was made to improve[11] the experiment by using a special microscope designed by Mr. Perry of Adam Hilger and Company. As low-power microscopes go, it was a brute. The numerical aperture was very large; the field of view was so extensive that an observer had to rely continuously on peripheral vision; the objective lens was noticeably convex outwards, and the focal surface was the exterior of the objective lens itself, upon which the zinc sulphide layer was stuck. Not unnaturally, such a microscope suffered from some remarkable optical defects which did not, however, affect its use. For example, the image of a scintillation occurring near the edge of the field of view was not a point, but a line of appreciable length. This microscope, however, turned out to be so tiring for the eye that it was used on only a few occasions.

From time to time the counting was interrupted while Crowe, at Chadwick's or Rutherford's request, made some change in the experimental set up. Ordinarily, such changes took only two or three minutes, but they required that lights be turned on. To protect the investment of time that had already been made in getting our eyes adapted to the dark the two or three of us who were doing the actual counting retired to a small wooden closet built on one side of the room. This closet contained three or four straight chairs in a row, with very little space to spare even for one's knees and feet. We, (and Rutherford too) being frequent pipe smokers, used to take advantage of the few minutes of incarceration to fill and light our pipes. When the call went up inside the closet "Close your eyes!" we knew exactly what to do. We closed our eyes, even the man who was lighting his pipe, until the match was put out. On occasion, the atmosphere inside the closet was so thick with smoke that we would keep our eyes closed all the time until we were permitted to emerge into the fresher air of the laboratory. Nobody ever thought, apparently, of putting a ventilating fan in the closet roof!

Chadwick and Rutherford always had a good idea of the order of magnitude of the counts to be expected in any particular experiment. Occasionally, the numbers of scintillations far exceeded expectations, sometimes increasing until the screen looked like the sky during a display of fireworks, and the ominous word went round: "Contamination!" The cause of the difficulty could sometimes be spotted, more often not. It might be caused by radioactive gas emerging from the source as the surrounding pressure was reduced; a fragment of the source might break off and fall near the counting screen; sometimes the whole interior of the experiment box seemed to be contaminated, as indeed occurred (on only one occasion that we remember) because Crowe's hands were radioactive. He was accidentally carrying one or two milligrams of radioactive material under his finger nails.

When contamination set in to any significant extent, it was usually necessary to abandon the planned experiment or make a drastic modification. There is recorded[12] one occasion when an unexpected effect was at first mistaken for contamination. The experiment consisted in bombarding various elements with alphas, and looking for disintegration particles. A specially pure piece of Swedish iron had been obtained for the iron sample, and when it was put in place in the apparatus, the count was far larger than anticipated. Contamination was at once suspected, but contrary to the usual behavior of contamination, the counts did not go up, but remained proportional to the intensity of the source. It turned out that we were counting disintegration particles not from iron, but from nitrogen adsorbed on the iron. Before the next counting session the Swedish iron was given a prolonged baking in vacuum, after which it behaved much more normally.

We have attempted here to convey something of the background of the apparently simple experiments at which we were privileged to assist as scintillation counters, but at the time we certainly had no real idea of the significance of the results obtained. Indeed, it is doubtful if the designers of the experiments, Rutherford and Chadwick, had in the early 1920's any inkling of

[11] J. Chadwick, Phil. Mag. **2**, 1056 (1926), especially p. 1061.

[12] E. Rutherford and J. Chadwick, Proc. Phys. Soc. London **36**, 417 (1924), especially p. 418; Nature, **113**, 457 (1924).

the ultimate consequences of their work, or of the immensity of the scientific field they were opening up. For us, the assistants, the abiding memory is of the honor we felt in being asked to take part, and of the chance it gave us during those intimate counting sessions to observe the great at work, and to get to know something of the outlook and personalities of both Rutherford and Chadwick. They made a wonderful team, both full of enthusiasm and new ideas,—Rutherford always cheerful and optimistic and not easily put out when things did not go well, Chadwick more impatient and chafing at anything which delayed the production of useful results. Their example has undoubtedly assisted us in our respective spheres since we left the Cavendish. That we proved reasonably satisfactory at the job was shown by the fact that we were kept on, contrary to our expectations, for the full years of our stay in Cambridge, even though one of us (H. S. H.) moved to another department at an early stage in our association.

We wish to thank Professor Norman Feather of the University of Edinburgh, himself a former worker with Rutherford's alpha particles, for reading the manuscript, for verifying some of the facts, and confirming several of our recollections.

Reprinted from American Journal of Physics, Vol. 37, No. 6, 577–584, June 1969
Printed in U. S. A.

The Richtmyer Memorial Lecture—Some Historical Notes

S. Chandrasekhar
Laboratory for Astrophysics, University of Chicago, Chicago, Illinois 60637

(Received 4 February 1969)

Response of the Richtmyer Memorial Lecturer to the American Association of Physics Teachers, 4 February 1969.

I

I am afraid that the topic of my lecture today does not fit into the pattern of the earlier lectures in this series: The previous Richtmyer lecturers have generally, if not invariably, chosen topics that were centered in their own active scientific interests. But it did not seem to me that any of the areas in which I am currently working is suitable for this occasion. I have therefore chosen to recall certain incidents of long ago. These incidents are related to matters that are currently under discussion; and it is possible that they may provide some historical perspective.

II

The incidents I wish to narrate have their origin in two of Einstein's three famous tests for his general theory of relativity. Let me first state what these tests are.

Einstein's third test, which is really a consequence of the principle of equivalence (and therefore takes precedence over the first two tests), is that the rate maintained by a clock, or any time-keeping device for that matter, will be affected by the gravitational field in which it may be located. Thus

$$\frac{\nu(U)-\nu(0)}{\nu(0)}=\frac{\Delta\nu}{\nu(0)}=\frac{U}{c^2}+O\left(\frac{U^2}{c^4}\right),$$

where $\nu(U)$ is the frequency, in terms of the local proper time, maintained by the device in a location where the Newtonian gravitational potential is U, and $\nu(0)$ is the frequency as it would be measured at zero potential.

S. Chandrasekhar, Richtmyer Memorial Lecturer.

Einstein's first and second tests, are a consequence of the explicit form of his field equations. The first test relates to the precession of the perihelion of Mercury; and of this test I shall not have anything to say. Einstein's second test requires that a light ray passing a body is deflected, and in particular a ray passing at a distance R from a spherical body of mass M is deflected by the amount

$$\theta = 4GM/Rc^2.$$

For a light ray grazing the sun, the deflection amounts to 1.75 in.; and it decreases inversely as R as we proceed outwards.

It is curious that a test that verifies the chosen form of Einstein's field equation should have been given precedence over the prediction which is a consequence of the principle of equivalence, and does not depend on the chosen form of the field equation. I suppose the reason for this is that fifty years ago the verification of the deflection of light during a solar eclipse appeared more practicable than the measurement of the gravitational redshift. In any event it was the second test which was attempted by Sir Frank Dyson and Eddington. Let me tell this story in Eddington's words.[1,2]

> The bending affects stars seen near the sun, and accordingly the only chance of making the observation is during a total eclipse when the moon cuts off the dazzling light. Even then there is a great deal of light from the sun's corona which stretches far above the disc. It is thus necessary to have rather bright stars near the sun, which will not be lost in the glare of the corona. Further, the displacements of these stars can only be measured relatively to other stars, preferably more distant from the sun and less displaced; we need therefore a reasonable number of outer bright stars to serve as reference points.
>
> In a superstitious age a natural philosopher wishing to perform an important experiment would consult an astrologer to ascertain an auspicious moment for the trial. With better reason, an astronomer today consulting the stars would announce that the most favourable day of the year for weighing light is May 29. The reason is that the sun in its annual journey round the ecliptic goes through fields of stars of varying richness, but on May 29 it is in the midst of a quite exceptional patch of bright stars—part of Hyades—by far the best star-field encountered. Now if this problem had been put forward at some other period of history, it might have been necessary to wait some thousands of years for a total eclipse of the sun to happen on the lucky date. But by strange good fortune an eclipse did happen on May 29, 1919 . . . .
>
> Attention was called to this remarkable opportunity by the Astronomer Royal (Sir Frank Dyson) in March 1917; and preparations were begun by a committee of the Royal Society and Royal Astronomical Society for making the observations.
>
> Plans were begun in 1918 during the war, and it was doubtful until the eleventh hour whether there would be any possibility of the expeditions starting . . .. Two expeditions were organized at Greenwich by Sir Frank Dyson, the one going to Sobral in Brazil and the other to the Isle of Principe in West Africa. Dr. A. C. D. Crommelin and Mr. C. Davidson went to Sobral; and Mr. E. T. Cottingham and the writer went to Principe.
>
> It was impossible to get any work done by instrument-makers until after the armistice; and, as the expeditions had to sail in February, there was a tremendous rush of preparation. The Brazil party had perfect weather for the eclipse; through incidental circumstances, their observations could not be reduced until some months later, but in the end they provided the most conclusive confirmation. I was at Principe. There the eclipse day came with rain and cloud-covered sky, which almost took away all hope. Near totality the sun began to show dimly; and we carried through the programme hoping that the conditions might not be so bad as they seemed. The cloud must have thinned before the end of totality, because amid many failures we obtained two plates showing the desired star-images. These were compared with plates already taken of the same star-field at a time when the sun was elsewhere, so that the difference indicated the apparent displacement of the stars due to the bending of the light-rays in passing near the sun.
>
> As the problem then presented itself to us, there were three possibilities. There might be no deflection at all; that is to say, light might not be subject to gravitation. There might be a "half-deflection," signifying that light was subject to gravitation, as Newton had suggested, and obeyed the simple Newtonian law. Or there might be a "full deflection," confirming Einstein's instead of Newton's law. I remember Dyson explaining all this to my companion Cottingham, who gathered the main idea that the bigger the result, the more exciting it would be. "What will it mean if we get double the deflection?" "Then," said Dyson, "Eddington will go mad, and you will have to come home alone."
>
> Arrangements had been made to measure the plates on the spot, not entirely from impatience, but as a precaution against mishap on the way home, so one of the successful plates was examined immediately. The quantity to be looked for was large as astronomical measures go, so that one plate would virtually decide the question, though, of course, confirmation from others would be sought. Three days after the eclipse, as the last lines of the

---

[1] A. S. Eddington, *Space, Time and Gravitation* (Cambridge University Press, Cambridge, England, 1953), pp. 113–114.

[2] Joseph Needham and Walter Pagel, Eds., *Background to Modern Science* (Cambridge University Press, Cambridge, England, 1938), pp. 140–142.

calculation were reached, I knew that Einstein's theory had stood the test and the new outlook of scientific thought must prevail. Cottingham did not have to go home alone.

It is perhaps hard for us to realize today what a great impact these measurements made at the time. Thus the meeting of the Royal Society on 6 November 1919 at which Dyson gave an account of the results of the Sobral and the Principe expeditions was described by A. N. Whitehead, the distinguished philosopher-mathematician, in the following terms[3]:

> The whole atmosphere of tense interest was exactly that of the Greek drama: we were the chorus commenting on the decree of destiny as disclosed in the development of a supreme incident. There was dramatic quality in the very staging—the traditional ceremonial, and in the background the picture of Newton to remind us that the greatest of scientific generalizations was now, after more than two centuries, to receive its first modification. Nor was the personal interest wanting: a great adventure in thought had at length come safe to shore.

As you may have gathered from what I have said, Dyson's role in arranging and organizing the eclipse expeditions was a major and an essential one. This is not generally recognized. So it is pertinent to recall that the part Dyson played in them was one of the grounds for which he was awarded the Gold Medal of the Royal Astronomical Society in 1924. And Jeans in presenting the Medal said[4]:

> In 1918, in the darkest days of the war, two expeditions were planned, one by Greenwich Observatory and one by Cambridge, to observe, if the state of civilization should permit when the time came, the eclipse of May 1919, with a view to a crucial test of Einstein's generalized relativity. The armistice was signed in November 1918; the expeditions went, and returned bringing back news which changed, and that irrevocably, the astronomer's conception of the nature of gravitation and the ordinary man's conception of the nature of the universe in which he lives. If the credit of this achievement had to be divided between Sir Frank Dyson and Professor Eddington I frankly do not know in what proportion the division should be made. To my mind, however, it is not so much an occasion for sharing out credit as for attributing the whole credit to each, for if either had failed to play his part, either from want of vision, of enthusiasm, or of capacity of seizing the right moment, I doubt if the expeditions would have gone at all, and the great credit of first determining observationally what sort of things space and time really are would probably have gone elsewhere.

Having in mind these remarks of Jeans, I once expressed to Eddington my admiration for his scientific sensibility in planning the expedition under circumstances when the future must have looked very bleak indeed. To my surprise Eddington disclaimed any credit on that account and told me the following story. It has to my knowledge never been told before; and it is a remarkable testimony of the times.

In 1917, after more than two years of war, England enacted conscription for all able-bodied men; and Eddington, who was then thirty-four, was eligible for draft. But as a practising and devout Quaker, he was a conscientious objector; and it was known and expected that he would claim deferment from military service on that account. Now the climate of opinion in England during World War I was very adverse with respect to conscientious objectors: it was in fact a social disgrace to be associated with one. And the stalwarts of the Cambridge of those days, Sir Joseph Larmor (of the Larmor precession), Professor H. F. Newall, and others, felt that Cambridge would be disgraced by having on its faculty a declared conscientious objector. They therefore tried through the Home Office to have Eddington deferred on the grounds that he was a most distinguished scientist and that it was not in Britain's long-range interests to have Eddington serve in the army. The case of Moseley killed in action at Gallipoli was very much in the minds of British scientists. And Larmor and others very nearly succeeded in their efforts. A letter from the Home Office was sent to Eddington, and all he had to do was to sign his name and return it. But Eddington added a postscript to the effect that if he were not deferred on the stated grounds, he would claim it on grounds of conscientious objection anyway. This postscript naturally placed the Home Office in a logical quandary since a confessed conscientious objector must be sent to a camp; and Larmor and others were very much piqued. But

[3] A. N. Whitehead, *Science and the Modern World* (The Macmillan Company, New York, 1926), p. 43.

[4] J. H. Jeans, Monthly Notices Roy. Astron. Soc. **85**, 672 (1924–25).

as Eddington told me, he could see no reason for their pique. As he expressed himself, many of his Quaker friends found themselves in camps in Northern England pealing potatoes, and he saw no reason why he should not join them. In any event, apparently at Dyson's intervention—as the Astronomer Royal he had close connections with the Admirality—Eddington was deferred with the express stipulation that if the war should end by May 1919, then Eddington should undertake to lead an expedition for the purpose of verifying Einstein's prediction! Well, that was how, in Jeans's words, "The great credit of first determining observationally what sort of things space and time really are," did not go "elsewhere."

III

I should like to turn now to the second part of my story which starts with the attempts to verify Einstein's third test, the gravitational redshift of spectral lines. Einstein, of course, was aware of its logical antecedence to his theory. Thus in reply to a letter from Eddington in which he was informed of the results on the deflection of light, Einstein wrote[5] on 15 December 1919:

> Above all I should like to congratulate you on the success of your difficult expedition. Considering the great interest you have taken in the theory of relativity even in earlier days, I think I can assume that we are indebted primarily to your initiative for the fact that these expeditions could take place. I am amazed at the interest which my English colleagues have taken in the theory in spite of its difficulty.

Einstein then stressed the importance of the third test, saying,

> If it were proved that this effect does not exist in nature, then the whole theory would have to be abandoned.

The first efforts to detect the redshifts of spectral lines in a gravitational field were naturally directed to a careful measurement of the solar lines; and the early efforts of St. John and Evershed were conflicting and inconclusive. But astronomical evidence from an entirely different source came from a study of the companion of Sirius.

The mass of the companion of Sirius is derived from the double star orbit and is trustworthy; a value of about one solar mass is generally accepted. From its distance, spectral type, and luminosity one can deduce that its radius is about 20 000 km. With the known mass and estimated radius it was concluded that M/R for the companion of Sirius is about 30 times that for the sun; and this value of M/R should lead to a spectral redshift of some 20 km. Also, the redshift as determined by Walter Adams in 1925 from a measurement of the H_γ line seemed to agree with the predicted value. Eddington concluded,[6] "Professor Adams has killed two birds with one stone: he has carried out a new test of Einstein's general theory of relativity and he has confirmed the suspicion that matter 2000 times denser than platinum is not only possible but actually present in the universe."

While there appeared to be good agreement between the expected and the observed values of the redshift, the observational evidence is no longer felt to be as secure. In any event, with the discovery of this new class of objects, interesting theoretical developments in astrophysics followed, and to these I now turn.

Considering the structure of white dwarfs with the knowledge available in 1925, Eddington formulated a paradox concerning these stars. He stated, "*The stars will need energy in order to cool.*" In less cryptic terms the paradox amounts to the following.

Let E_V denote the negative electrostatic energy per gram; at the high pressures we are concerned with, this energy per atom is essentially the sum of all the ionization energies required to strip the atom successively of all the electrons to the bare nucleus. And let E_K denote the kinetic energy per gram of completely ionized matter. If such matter were released of the pressure to which it is subject, then it could resume the state of ordinary un-ionized matter only if

$$E_K > E_V.$$

Eddington's paradox was this (though Eddington, at a later time disclaimed to this particular

[5] A. Vibert Douglas, *The Life of Arthur Stanley Eddington* (Thomas Nelson and Sons Ltd., London, 1957), pp. 41, 43.

[6] A. S. Eddington, *Internal Constitution of the Stars* (Cambridge University Press, Cambridge, England, 1926), p. 173.

formulation): If the quantities E_V and E_K are calculated for the densities and temperatures expected in the interiors of the white dwarfs, then one finds that

$$E_K \text{ (perfect gas)} < E_V.$$

In other words, on these premises, the matter will not be able to resume its ordinary state—an extraordinary situation if true. R. H. Fowler's resolution of this paradox in 1927 in terms of the then very new statistical mechanics of Fermi and Dirac is, in my opinion, one of the great landmarks in the development of our ideas on stellar structure and stellar evolution. May I spend a few minutes on it.

Dirac's paper, which contains the derivation of what has since come to be called the "Fermi-Dirac distribution," was communicated by Fowler to the Royal Society on 26 August 1926. On 3 November Fowler communicated a paper of his own in which the application of the laws of the "new quantum theory" to the statistical mechanics of assemblies consisting of similar particles is systematically developed and incorporated into the general scheme of the Darwin–Fowler method. And by 10 December (i.e., before a month had elapsed) his paper entitled "Dense Matter" was read before the Royal Astronomical Society. In this paper Fowler drew attention to the fact that the electron gas in matter as dense as in the companion of Sirius, must be degenerate in the sense of the Fermi–Dirac statistics. Thus, to Fowler belongs the credit for first recognizing a field of application for the then "very new" statistics of Fermi and Dirac, though among physicists this credit is generally given to Pauli for his explanation of the paramagnetic susceptibility of the alkali metals.

Fowler's resolution of Eddington's paradox is simply this: since at the temperature and densities in the white dwarfs the electron assembly will be highly degenerate, E_K should be evaluated by using the formulae appropriate in this limit. He showed that when E_K is so evaluated, it is indeed much greater than E_V. And Fowler concluded his paper[7] with the following statement.

> The black dwarf material is best likened to a single gigantic molecule in its lowest quantum state. On the Fermi–Dirac statistics, its high density can be achieved in one and only one way, in virtue of a correspondingly great energy content. But this energy can no more be expended in radiation than the energy of a normal atom or molecule. The only difference between black dwarf matter and a normal molecule is that the molecule can exist in a free state while the black dwarf matter can only so exist under very high external pressure.

It was at this stage that I became interested in the theory of white dwarfs. I hope that you will forgive and tolerate my bringing myself into this account. But most of what I shall have to say happened more than 38 years ago; and as far as I feel now, it could equally well relate to someone else.

My knowledge of physics, astronomy, and mathematics was rudimentary in the extreme at the time, the spring and early summer of 1930 to be precise. But I had read Fowler's paper, and I was familiar with the Fermi–Dirac statistics from Sommerfeld's exposition of it in his marvelous papers on the electron theory of metals. And, on the astrophysical side, I was familiar with the theory of polytropes, i.e., the theory of the equilibrium under their own gravitation of gaseous masses in which the relation between the pressure and the density is of the form $p = K\rho^{1+1/n}$, where n, called the polytropic index, and K are constants.

As I said, Fowler had shown that the matter in the white dwarfs must be highly degenerate. Under these circumstances the pressure is proportional to $\rho^{5/3}$. Accordingly white dwarfs must be polytropes of index $n = \frac{3}{2}$, and the conclusions followed.[8]

> (i) The radius of a white dwarf is inversely proportional to the cube root of the mass; (ii) the density is proportional to the square of the mass; (iii) the central density would be six times the mean density $\bar{\rho}$.

A further examination based on these deductions showed that the momenta of the electrons near the Fermi threshold at the densities at the centers of white dwarfs of solar mass are already comparable to mc. So it appeared that the special relativistic mass-variation with velocity should be taken into account in obtaining the equation

[7] R. H. Fowler, Monthly Notices Roy. Astron. Soc. **87**, 114 (1926).

[8] S. Chandrasekhar, Phil. Mag. **11**, 592, 594 (1931).

of state; and it was not difficult to show that in the extreme relativistic limit, the relation between the pressure and the number of electrons per cubic centimeter is given by[9]

$$p=(n^{4/3}hc/8)(3/\pi)^{1/3}.$$

In this limit, then, the white dwarf is a polytrope of index $n=3$; and the mass of the configuration becomes uniquely determined. And I was quite puzzled by this result.

For an assumed mean molecular weight $\mu=2.5$ (the "canonical value" in 1930) this mass is[9]

$$0.918\odot \text{ (and quite generally } 5.76\mu^{-2}\odot).$$

All these results were obtained in the early summer of 1930, and some of them, including the one on the critical mass, were obtained on the long voyage from India to England in July 1930.

Soon after arriving in England, I showed these results to R. H. Fowler. Fowler drew my attention to two papers[10,11] by Stoner, one of which had appeared earlier that summer. In these two papers Stoner had considered the energetics of homogeneous spheres on the assumption that the Fermi–Dirac distribution prevailed in them. While Stoner's results gave some valid inequalities for the problem, he had not derived the structure of the equilibrium configurations in which all the governing equations are satisfied. Fowler, of course, appreciated this difference, and he was satisfied with my detailed results pertaining to the nonrelativistic configurations. But he appeared skeptical of my result on the critical mass, and so was E. A. Milne to whom he had communicated it.

As I said I was puzzled by the emergence of the critical mass when I first obtained it. But by October of that year it was clear that what was happening was that the relation $R\propto M^{-1/3}$ given by the nonrelativistic theory was modified by the inclusion of the relativistic effects in the following way. If for purposes of approximation we consider a white dwarf as consisting of a "$(n=\frac{3}{2})$-envelope" (in which $p\propto\rho^{5/3}$), and a "$(n=3)$-core" (in which $p\propto\rho^{4/3}$), then "the completely relativistic model, considered as a limit of this composite series is a point mass with $\rho_c=\infty$!".[12] On this account, I tried the modification of putting cores of constant density (at the density of nuclear matter) as M approached the critical mass and increased beyond it.

But the existence of the critical mass continued to puzzle me. What does it mean for the evolution of the stars? And how is one to relate it to the ordinary stars in which, according to Eddington's theory, radiation pressure plays a dominant role. Soon it became clear that *if the ratio* $(1-\beta)$ *of the radiation pressure to the total pressure exceeds the value* 0.092, *then matter cannot be degenerate.* This result follows from a comparison of the two equations

$$\begin{aligned} p&=p_{\text{gas}}+p_{\text{rad}},\\ &=(k/\mu H)\rho T+\tfrac{1}{3}aT^4,\\ &=\beta p+(1-\beta)p,\\ &=\{(k/\mu H)^4(3/a)[(1-\beta)/\beta^4]\}^{1/3}\rho^{4/3}, \end{aligned}$$

and

$$p=[hc/8(\mu H)^{4/3}](3/\pi)^{1/3}\rho^{4/3},$$

valid for the perfect gas and the relativistically degenerate gas, respectively. (In the foregoing equations k is the Boltzmann constant, a is the Stefan constant, H is the mass of the hydrogen atom, and μ is the mean molecular weight.)

Since the relative magnitude of the radiation pressure increases with increasing mass, the result

$$1-\beta>0.092$$

means that massive stars during their evolution can never develop degenerate cores. And on Eddington's standard model (in which $1-\beta$ is constant and depends only on the mass) the result as I stated in 1932 is the following[13]:

> *For all stars of mass greater than* $\mathfrak{M}(=6{\cdot}6\mu^{-2}\odot)$, *the perfect gas equation of state does not break down, however high the density may become, and the matter does not become degenerate. An appeal to the Fermi–Dirac statistics to avoid the central singularity cannot be made.*

---

[9] S. Chandrasekhar, Astrophys. J. **74,** 81, 82 (1931).

[10] E. C. Stoner, Phil. Mag. **7,** 63 (1929).

[11] E. C. Stoner, Phil. Mag. **9,** 944 (1930).

[12] S. Chandrasekhar, Monthly Notices Roy. Astron. Soc. **91,** 456, 463 (1931).

[13] S. Z. Chandrasekhar, Z. Astrophys. **5,** 321, 342, 327, (1932).

And I concluded

> Great progress in the analysis of stellar structure is not possible before we can answer the following fundamental question: *Given an enclosure containing electrons and atomic nuclei (total charge zero) what happens if we go on compressing the material indefinitely?*

I should state at this point that in a short paper[14] published early in 1933, Landau isolated the critical mass apparently without knowledge of my results published two years earlier. Since Landau's is a most distinguished name in physics, the tendency in some current literature is to give him priority. But questions of priority are not of any great interest; of far greater interest is the fact that Landau was concerned with astrophysical problems already at that time.

While the constitution of the white dwarfs was fully understood by early 1931, the structure of the white dwarfs based on the exact equation of state for degenerate matter was derived only in 1934. In this theory[15] the approximation used in the earlier work of representing the equation of state by either of its two limiting forms is not made, and the exact equation was used instead. In a preliminary announcement of the results of these exact calculations I concluded with the following statement.[16]

> Finally, it is necessary to emphasize one major result of the whole investigation, namely, that it must be taken as well established that the life-history of a star of small mass must be essentially different from the life-history of a star of large mass. For a star of small mass the natural white-dwarf stage is an initial step towards complete extinction. A star of large mass ($>\mathfrak{M}$) cannot pass into the white-dwarf stage, and one is left speculating on other possibilities.

This rather bold and emphatic statement—I am surprised at my own confidence of 35 years ago—produced an adverse reaction on the part of Eddington, Milne, and others. Thus, in a paper[17] with which Eddington followed my account of these results at the January 1935 meeting of the Royal Astronomical Society he claimed that my results had no validity since the underlying formula for relativistic degeneracy is wrong. Let me quote from the published account of this meeting.[18]

> I do not know whether I shall escape from this meeting alive, but the point of my paper is that there is no such thing as relativistic degeneracy!...
>
> Chandrasekhar, using the relativistic formula which has been accepted for the last five years, shows that a star of mass greater than a certain limit $\mathfrak{M}$ remains a perfect gas and can never cool down. The star has to go on radiating and radiating, and contracting and contracting until, I suppose, it gets down to a few km radius, when gravity becomes strong enough to hold in the radiation, and the star can at last find peace.
>
> Dr. Chandrasekhar had got this result before, but he has rubbed it in in his last paper; and, when discussing it with him, I felt driven to the conclusion that this was almost a *reductio ad absurdum* of the relativistic degeneracy formula. Various accidents may intervene to save the star, but I want more protection than that. I think there should be a law of Nature to prevent a star from behaving in this absurd way!

And E. A. Milne who believed that *all* stars *must* have degenerate cores—a conclusion untenable if one accepts relativistic degeneracy—climbed on Eddington's bandwagon which immediately started rolling. Two days after the meeting of the Royal Astronomical Society to which I have referred, Milne wrote a Letter to *The Observatory* in which he stated[19]:

> In view of the fundamental character of the paper read by Sir Arthur Eddington at the meeting of the Royal Astronomical Society on 11 January 1935, perhaps I may be allowed to state that the basis of the calculation just completed by Mr. Norman Fairclough (referred to in my remarks at the meeting) is the equation of state $p=K\rho^{5/3}$ for a degenerate gas.... Sir Arthur Eddington's investigations may now confer on our work a justification to which it is only accidentally entitled.

At about this time, I wrote to Milne that I had discussed the physical basis of the relativistic equation of state with several physicists and that none of them, to the extent I was able to gather, found Eddington's arguments either convincing

---

[14] L. Landau, Physik. Z. Sowjetunion **1**, 285 (1933).

[15] S. Chandrasekhar, Monthly Notices Roy. Astron. Soc. **95**, 207 (1935).

[16] S. Chandrasekhar, Observatory **57**, 373, 377 (1934).

[17] A. S. Eddington, Monthly Notices Roy. Astron. Soc. **95**, 197 (1935).

[18] Proceeding of the meeting of the Royal Astronomical Society, January 1935, Observatory **58**, 37 (1935).

[19] E. A. Milne, Observatory **58**, 52 (1935).

or satisfactory. To this letter Milne replied

> Your marshalling of authorities such as Bohr, Pauli, Fowler, Wilson, etc., very impressive as it is, leaves me cold. If the consequences of quantum mechanics contradict very obvious, much more immediate, considerations, then something must be wrong with the principles underlying the equation of state derivation. Kelvin's gravitational age-of-the-sun calculation was perfectly sound; but it contradicted other considerations which had not then been realized. To me it is clear that matter cannot behave as you predict.... A theory must not be used as a doctrine, to *compel* belief....
>
> Eddington is nearly always wrong in his work in the long run, and I am quite prepared to believe that he is wrong here, in his details. But I hold by my general considerations.

With the prestige of Eddington and Milne arrayed against me, I am afraid that I appeared to most astronomers of the thirties as Don Quixote.

IV

Let me conclude with two observations with respect to subsequent developments. In 1939 Oppenheimer and Volkoff[20] pointed out that as we approach the limiting mass along the sequence of the completely degenerate configurations, the central density will become high enough for the electrons at the Fermi threshold to have energies that exceed the maximum energy of the β-ray spectrum emitted by the neutrons when they decay into protons. When this happens, the neutron becomes the stable nucleon instead of the proton. Neutron stars thus become possible. And Oppenheimer and Volkoff studied their structure with the aid of the general relativistic equations for hydrostatic equilibrium. Based on this work it was possible to say[21]

> If the degenerate cores attain sufficiently high densities (as is possible for these stars) the protons and electrons will combine to form neutrons. This would cause a sudden diminution of pressure resulting in the collapse of the star onto a neutron core giving rise to an enormous liberation of gravitational energy. This may be the origin of the supernova phenomenon.

This was stated thirty years ago. Today with the discovery of a pulsar at the center of the Crab Nebula, it does appear that there is a neutron star in it; and the Crab Nebula is, of course, the remnant of a supernova.

[20] J. R. Oppenheimer and G. Volkoff, Phys. Rev. **55**, 374 (1939).

[21] S. Chandrasekhar, *Colloque International d'Astrophysics XIII*, 17–23 July 1939; A. J. Shaler, *Novae and White Dwarfs* (Hermann and Cie., Editeurs, Paris, 1941), p. 245.

The Scattering of X Rays as Particles*

A. H. Compton
Washington University, St. Louis, Missouri
(Received March 22, 1961)

The experimental evidence and the theoretical considerations that led to the discovery and interpretation of the modification of the wavelength of x rays as a result of scattering by electrons are reviewed, as is the controversy between Duane and the author that took place in 1923–24. The confirmatory evidence obtained by Bothe, Geiger, Simon, and Compton is summarized.

I HAVE been asked to say something about how the study of the scattering of x rays has led to the concept of x rays acting as particles.

In the interest of conserving time I shall summarize the first part of the story by noting that, beginning in 1917, I spent five years in an unsuccessful attempt to reconcile certain experiments on the intensity and distribution of scattered x rays with the electron theory of the phenomenon that had been developed by Sir J. J. Thomson. Then a series of experiments that I performed at Washington University, beginning in 1922, confirmed an observation by J. A. Gray[1] of Queen's University of Kingston, Ontario, that the secondary rays produced when x rays pass through matter are in fact of the nature of scattered rays, showing the same polarization and approximately the intensity predicted by Thomson's electron theory and, further, that in the process of scattering, these rays are in some way altered to increase their absorbability. From my absorption measurements I was able to estimate that over a wide range of wavelengths of the primary rays the increase in the absorbability of scattered rays was what it should be if their wavelength was increased by about 0.03 A over the wavelength of the primary ray. This result I checked with an x-ray spectrometer, measuring an increase in the wavelength of approximately 0.02 A.

At this point I found myself engaged, as a member of a committee of which William Duane of Harvard was the chairman, in preparing a report for the National Research Council on secondary radiations produced by x rays. When it came to publication of the report, Duane objected to including my revolutionary conclusion that the wavelength of the rays was increased in the scattering process just described because he felt that the evidence was inconclusive. At the insistence of A. W. Hull, however, this portion of my report was included in the publication.[2]

At this point I paused in my experiments in order to concentrate on their theoretical interpertation. I found at once that the change of wavelength that I observed for scattering at 90° was what should be expected if the scattering electrons were moving in the direction of the primary beam at about half the speed of light, which would mean that each electron had a momentum equal to that of a quantum of energy of the frequency of the primary x rays. It was obvious, however, that not all of the electrons in the scattering material, which was fixed in my apparatus, could be moving forward at such a velocity; yet according to the theory all of the electrons should participate in the scattering

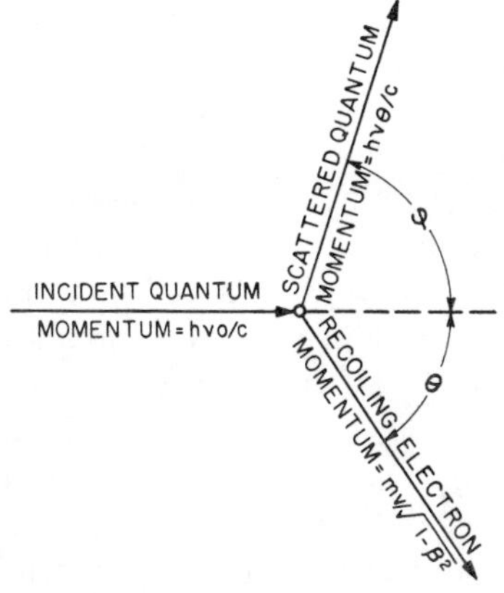

Fig. 1. Recoil of an electron upon scattering of an x-ray photon.

* Paper delivered as part of a program on "Topics in the history of modern physics" on February 3, 1961, at a joint session of the American Physical Society and the American Association of Physics Teachers during their annual meetings in New York City.

[1] J. A. Gray, J. Franklin Inst. **189**, 643 (1920).

[2] A. H. Compton, Bull. Natl. Research Council, No. 20, 19 (1922).

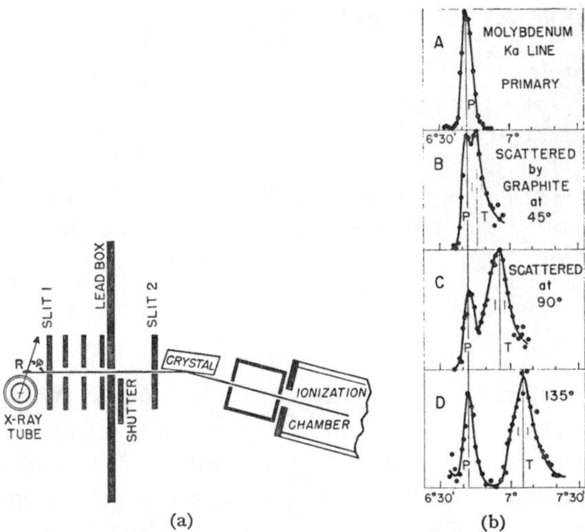

FIG. 2. (a) Schematic arrangement of the apparatus for determination of the spectrum of the scattered x rays. (b) Experimental results.

process. This led me to examine what would happen if each quantum of x-ray energy were concentrated in a single particle and would act as a unit on a single electron. Thus I was led to the now familiar hypothesis, illustrated in Fig. 1, of an x-ray particle colliding with an electron and bounding elastically from it with reduced energy, the lost energy appearing as the recoil energy of the electron. This idea, of an x-ray quantum losing energy by collision with an electron, must have been already in the mind of Peter Debye, then working at Zürich, for immediately upon the appearance of my report in the Bulletin of the National Research Council, he published a paper[3] in the Physikalische Zeitschrift in which he presented an explanation of the change in wavelength of the scattered rays identical in principle with my own hypothesis, and appearing in print only a few days after my first full publication.[4]

Assuming that the energy and the momentum of the incident quantum and of the electron are conserved in this collision process, one is led to a group of three expressions representing a change of wavelength, the energy of the recoil electron, and the relation between the angle of recoil and the angle of scattering of the photon. Each of these formulas, expressed in Eqs. (1)–(3), is subject to precise experimental test.

$$\delta\lambda = \lambda' - \lambda = (h/mc)(1-\cos\phi), \tag{1}$$

$$E_{kin} = 2\alpha \cos^2\theta/(1+\alpha)^2 = \alpha^2 \cos\theta, \tag{2}$$

$$\cot\theta = (1-\alpha)\tan(\phi/2), \tag{3}$$

where

$$\alpha = h\nu/mc^2.$$

The change in wavelength I measured repeatedly at Washington University. Figure 2 shows the results of one series of these experiments.

The results, confirming accurately the theoretical predictions, immediately became a subject of the most lively scientific controversy that I have ever known. I reported the results shown in Fig. 2 before the American Physical Society in April, 1923. At the meeting of the American Physical Society during the Christmas holidays of that year there was arranged a rather formal debate between Duane and myself on the validity of the results. Having frequently repeated the experiments I entered the debate with confidence, but was nevertheless pleased to find that I had support from P. A. Ross of Stanford and M. de Broglie of Paris, who had obtained photographic spectra showing results similar to my own. Duane at Harvard with his graduate students had been able to find not the same spectrum of the scattered rays, but one which they attributed to tertiary x rays excited by photoelectrons in the scattering material. I might have criticized his interpretation of his results on rather obvious grounds, but thought it would be wiser to let Duane himself find the answer. Duane followed up this debate by visiting my laboratory (at that time in Chicago) and invited me to his laboratory at Harvard, a courtesy that I should like to think is characteristic of the true spirit of science. The result was that neither of us could find the reason for the difference in the results at the two laboratories, but it turned out that the equipment that I was using was more sensitive and better adapted than was Duane's to a study of the phenomenon in question.

During the following summer at Toronto there occurred a meeting of the British Association for the Advancement of Science, with Sir William Bragg presiding over the physics section. In the previous decade Sir William, as also Ernest Rutherford, had been greatly impressed by the

[3] P. Debye, Physik. Z. **24**, 161 (1923).
[4] A. H. Compton, Phys. Rev. **21**, 484 (1923).

forward momentum of the secondary electrons ejected from matter by both x rays and gamma rays and had been led thereby to defend a corpuscular theory of the scattering of the rays. This interpretation, however, he had abandoned following the experiments by von Laue and by himself on the reflection of x rays by crystals, which had given him confidence in the wave interpretation of x rays. At this Toronto meeting a full afternoon was set aside for a continuation of the debate. The result was inconclusive. It was summarized by Sir C. V. Raman by this statement to me privately after the meeting. "Compton," he said, "you are a good debater, but the truth is not in you." Nevertheless, it seems to have been this discussion that stimulated Raman to the discovery of the effect which now bears his name. Duane followed up this meeting by a new interpretation of the change in wavelength which he attributed to what he called a "box" effect, explaining that surrounding the scattering apparatus with a lead box had in some way altered the character of the radiation. This interpretation I answered by repeating the experiment out of doors with essentially the same results, and at the same time Duane and his collaborators in a repetition of their own experiments began to find the spectrum line of the changed wavelength in accord with my collision theory. At the next meeting of the American Physical Society they reported a very good measurement of this change in wavelength.

In the meantime other experimenters had not

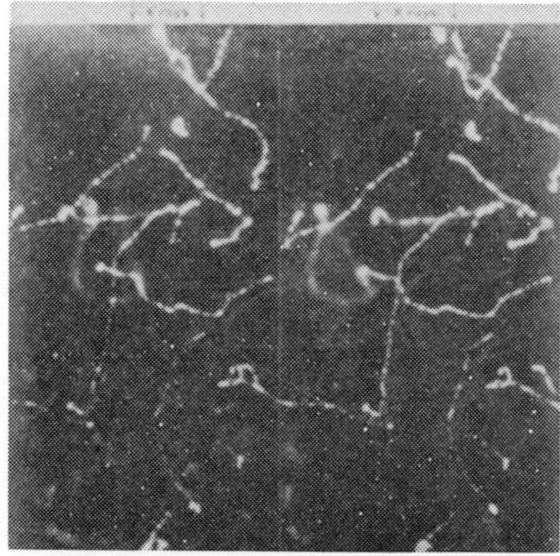

FIG. 3. Cloud chamber tracks produced by recoil electrons (after C. T. R. Wilson).

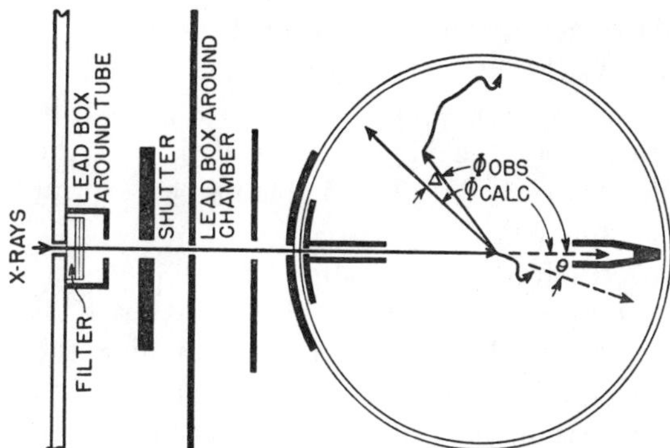

FIG. 4. Apparatus for measurement of scattering angle.

been idle. Within a few months after my first paper C. T. R. Wilson at Cambridge and W. Bothe in Germany had found the recoil electrons predicted by the corpuscular theory. Figure 3 shows one of C. T. R. Wilson's photographs of the cloud tracks left by these electrons in air traversed by x rays. The appearance of the trails led Wilson to call them "fish" tracks, with their tail toward the x-ray tube and their head pointed in the direction of the beam. A. W. Simon and I, repeating these experiments, showed that the number of the tracks and their ranges were just what should be expected according to the theory that each recoiling electron was the result of the impact of one photon of x rays that it scattered.[5] I had the opportunity to show some of these fish tracks in a cloud chamber to S. K. Allison, who was at that time working in Duane's laboratory. It is possible that it was these tracks, rather than the evidence of the x-ray spectra, that convinced Duane of the validity of the corpuscular theory. In any case, since that time no one seems to have questioned the correctness of our experimental results.

Immediately following their observation of the recoil electrons, Bothe and Geiger reported an observation of coincidences of recoil electrons and associated scattered photons as observed in a pair of counters. Simon and I were engaged in checking the angles at which the recoil electron and the associated scattered photon would occur. The apparatus that we used is shown diagrammatically in Fig. 4. According to the theory, associated with an electron recoiling at an angle θ, any effect of the associated scattered photon

[5] A. H. Compton and A. W. Simon, Phys. Rev. **25**, 309 (1925).

should occur in the direction of ϕ, as given by Eq. (3). With a specially designed cloud chamber, out of 850 photographs 38 showed a β particle resulting from the photon associated with the recoil electron.[6] Figure 5 shows a typical photograph.

This result is of especial interest because it shows that it is possible to follow the path of an x-ray particle or photon by examining the secondary electrons that it ejects along its way. It is clear that the x rays thus scattered proceed in direct quanta of radiant energy; in other words, that they act as photon particles. This test of the relation between the angles θ and ϕ is a crucial test of the conservation of energy and momentum as related to the process of the scattering of photons by electrons. The results of Simon and myself have accordingly been re-examined and refined by a number of experimenters, as summarized recently by Robert Shankland in his *Atomic and Nuclear Physics*.[7] The net result is a full confirmation of the angular relation given by Eq. (3).

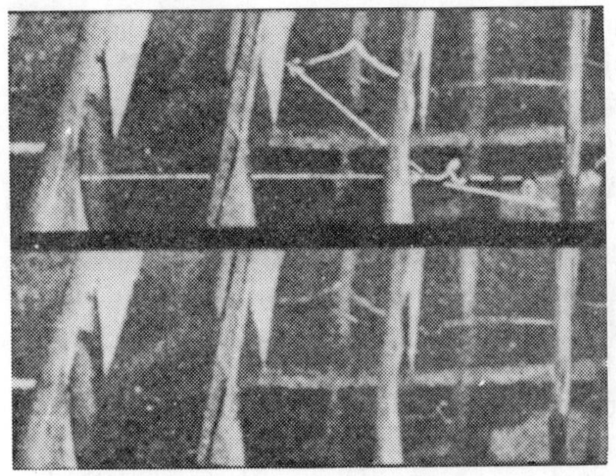

FIG. 5. Typical cloud chamber photograph of recoiling electron track.

Time does not permit me to review the evidence that was accumulating in the meantime that gave full support likewise to the electromagnetic wave character of the x rays: complete polarization of x rays scattered at 90°, the diffraction of x rays from ruled gratings, as well as from crystals, interference phenomena, and refraction phenomena, precisely analogous to results obtained with light. It became evident that though x rays moved and did things as particles, they nevertheless have also the characteristic optical qualities that identify them as waves. Thus we were introduced to the concept of light as having the nature of waves and particles as having a kind of reality, a difficult concept to which L. de Broglie was, however, at the same time giving a theoretical meaning.

It may be fair to say that these experiments were first to give, at least to physicists in the United States, a conviction of the fundamental validity of the quantum theory.

[6] A. H. Compton and A. W. Simon, Phys. Rev. **26**, 289 (1925).

[7] Robert Shankland, *Atomic and Nuclear Physics* (The Macmillan Company, New York, 1960), 2nd ed., p. 204.

AMERICAN JOURNAL of PHYSICS

A Journal Devoted to the Instructional and Cultural Aspects of Physical Science

VOLUME 15, NUMBER 5 SEPTEMBER–OCTOBER 1947

The Franck-Condon Principle and Related Topics*

E. U. CONDON
National Bureau of Standards, Washington 25, District of Columbia

A REVIEW of the historical development and present status of certain topics in molecular physics may serve as a reminder of the progress made in the past two decades. Let us recall that in 1925 there was no subject of nuclear physics, and high voltage equipment was almost non-existent except in the x-ray laboratories. About all that was known about cosmic rays was that the ionization increases in high balloon flights in the atmosphere and that its cause penetrates deeply into lakes. Nobody had heard of closed electron bands in a solid, and the word semi-conductor was essentially unknown.

"What did physicists find to be interested in?," young physicists may wonder. The answer is, of course, that they worried about a lot of things that are nowadays taught to and learned by beginning graduate students with such glibness that it is hard to realize there ever was a time when these things were not known.

In 1925 there was no quantum mechanics. Though de Broglie's thesis was published in 1924, I have never met anyone who read it seriously until later. We worked with quantization by the $\int p dq = nh$ method of Bohr and Sommerfeld and tried to get at radiative transition probabilities by approximate and unclearly formulated procedure based on the Bohr correspondence principle, according to which quantum radiative jumps were associated with, or set in correspondence with certain terms in the Fourier analysis of the quasiperiodic motions in the mechanical system.[1]

The Bohr theory had given a beautiful account of the spectrum of atomic hydrogen and of the arc spectra of the alkalis. However, why the D lines of sodium were double was a great mystery that was only cleared up by the electron-spin hypothesis of Goudsmit and Uhlenbeck. Similarly, the old quantum theory had thrown a good deal of light on the infra-red and electronic spectra of diatomic molecules. Pure rotation, rotation-vibration, and electronic band systems were recognized, analyzed and utilized to get interesting quantitative data on important molecules. Nevertheless, here too there were puzzling things; for example, the isotope shift between HCl^{35} and HCl^{37}, and in other cases, required that the vibrational levels be assigned half-integral quantum numbers.

First Ideas on the Principle

At that time I was a graduate student in the University of California in Berkeley. Ernest Lawrence was a graduate student at Yale, and

* Address of the Retiring President, American Physical Society, New York, January 31, 1947. My apology for including my own name in the title of this paper is that it conforms to a well-established usage. In the text I shall refer simply to the Principle. Professor Rabi once said to me that the Principle was a great boon to lecturers in courses on atomic physics, for it is so easy to understand that they are always assured of one or two lectures which they do not have to prepare.

[1] W. Lenz, *Zeits. f. Physik* **25**, 299 (1924). This paper represents an attempt to deal with the problem of nuclear transitions associated with electronic jumps by means of the correspondence principle.

there were no cyclotrons in the world—not one, not even in Berkeley. Professors R. T. Birge and L. B. Loeb were the stimulating research leaders of the department, the former on molecular spectra, the latter on every phase of electric conduction in gases.

During the year 1925–26, Birge conducted a seminar on molecular spectra which it is impossible for me to praise too highly. Then as always his work was distinguished by the most exact and painstaking scrutiny of the data in their relation to the theory. I still remember how exciting it was when he first showed, with precision, that the swelling of molecules by rotation, as inferred from their pure rotation spectrum, agreed with the values found from the vibration-rotation spectrum and from electronic systems. It was he who had carefully compiled all the existing data on analyzed electronic band systems—then about a dozen in number—and who recognized the basic empirical facts of the intensity distribution, later to be explained by the Principle. I mention this point so explicitly because without his stimulating guidance I would never have been aware of the problem of intensity distribution in band systems.

In Göttingen, Professor James Franck was very much interested in photochemical reactions and, in particular, the dissociation of iodine vapor by absorption of light. He gave a paper before the Faraday Society in London[2] in which the basic idea of the Principle was first presented in connection with this problem.

Proof sheets of this paper were sent by Franck to his student, Dr. Hertha Sponer, who was in Berkeley that year on an International Education Board Fellowship. She let me read them, and, because I had just learned the empirical problem about intensity distribution in band systems from Birge's seminar, it was immediately evident how to generalize Franck's ideas a little more in order to get the full story. What is more, from Birge's seminar I had at hand a good critical compilation of the existing data which made possible a quick quantitative test of the ideas.

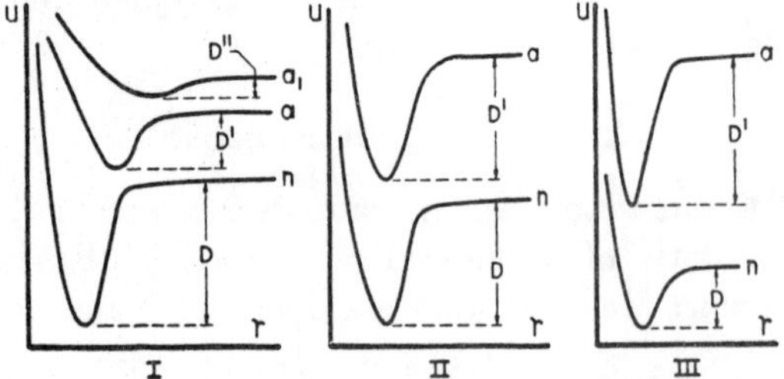

Fig. 1. Potential-energy curves from Franck's Faraday Society paper (reference 2).

This work was all done in a few days. Doctor Sponer showed me Franck's paper one afternoon, and a week later all the quantitative work for my 1926 paper[3] was done. But let us see just what was the situation.

Figure 1 is from Franck's paper.[2] He pointed out that, owing to the large masses of the nuclei in a molecule, their relative momentum cannot be directly affected by an electronic transition, so that those transitions will be most likely that conform most closely to the Principle. Therefore, *if* in the iodine molecule the curves are related as in set *I* of Fig. 1, a molecule initially not vibrating will most readily absorb light that carries it into states of high vibration, or to states higher than the energy needed for dissociation, resulting in photochemical dissociation of the molecule.

It was, of course, not a very difficult step to recognize that if the molecule is vibrating initially, then the Principle asserts that transitions are favored to those states which require least instantaneous adjustment of the relative position *and momentum*. Moreover, it was intuitively felt that the electronic transition was sufficiently independent of the nuclear vibration that it was equally likely to occur at any phase of the nuclear vibratory motion. This is not self-evident and perhaps not exactly true. It might be, for example, that the electron jump is stimulated by vibratory motion in the molecule, perhaps in such a way that it is more likely to occur at a phase of maximum relative velocity. However, this is not the case. With all times of electron jump equally likely, the most favored vibrational transitions will be those associated with the turning points of the nuclear vibration; for, as the nuclei move slowly here, a larger fraction of the time is spent in such regions.

[2] J. Franck, *Trans. Faraday Soc.* **21**, 536 (1925). The paper in which the Principle was first advanced to explain the photochemical dissociation of iodine vapor.

[3] E. U. Condon, *Physical Rev.* **28**, 1182 (1926). First application of the Principle to intensity distribution in band systems.

This idea led to Fig. 2, taken from my 1926 paper,[3] which indicates that there are two most favored vibrational quantum-number changes associated with nonvanishing values of the initial vibrational quantum number.

The proof of the pudding lay in the fact that a number of band systems were well analyzed, and therefore it was possible to put down the potential-energy curves in their approximately correct form and relative location. Near the minimum a curve is parabolic:

$$V(r)=\tfrac{1}{2}k(r-r_0)^2+\cdots.$$

One can get the value of k from the vibration frequency ν, since, if μ is the reduced mass, $2\pi\nu=(k/\mu)^{\frac{1}{2}}$; and one gets r_0 from the moment of inertia, which in turn is given quantitatively from the observed rotational energy levels. Evidently in Fig. 2, when there is little change in ν or r_0, the curves are similar and lie directly over each other, so the Principle requires small or zero change in the vibrational quantum number. But if there is a big change in r_0, the curves are widely displaced and the Principle requires large changes in the vibrational quantum number. Both cases were found in the data available in 1926. The Principle was triumphant in that those with large changes in vibrational quantum number were correctly correlated with large changes in the equilibrium internuclear distance, and *vice versa*.

It must be remembered that the intensity data available were simply rough estimates of plate blackening, uncorrected for variations in plate sensitivity over the rather great range of wavelength involved in some band systems. Nevertheless, there was no doubt of the essential correctness of the Principle.

Curiously enough, my calculations indicated a bad disagreement with the facts for iodine, the very molecule which led Franck to his qualitative discovery of the idea. This caused me much worry, and I searched for an error in calculation a long time before sending in the 1926 paper with such a bad discrepancy for iodine. About a year later the error was discovered by Professor Wheeler Loomis, then of New York University. I had used the value of the moment of inertia for the 26th vibrational state in the electronically excited level, which is involved in the fluorescence spectrum of iodine, thinking it was the moment of inertia of the nonvibrating molecules. When this error was corrected, iodine agreed with the Principle as well as the others.[4]

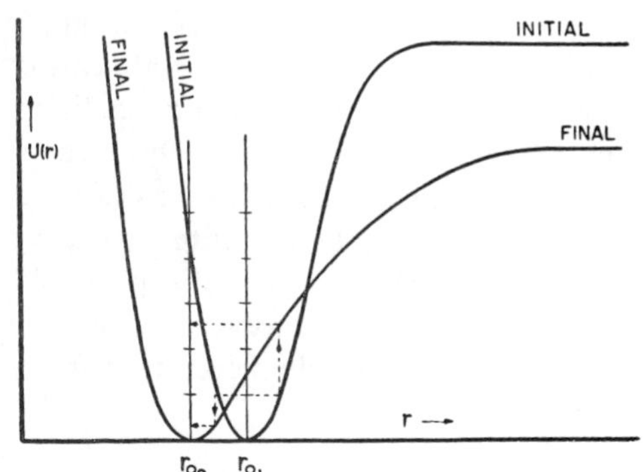

FIG. 2. Graphical construction of favored transitions from my 1926 paper (reference 3).

The faculty of the University of California was broad-minded enough to accept the paper as a doctor's thesis. However, it is interesting to note that orthodox theoretical physics was so tied to Bohr's correspondence principle at the time that the referee for the *Physical Review* was reluctant to recommend the paper for publication. He felt it could not be correct because it did not go at the problem in terms of Fourier amplitudes of the classical motion.

But the whole story was not given by the simple classical picture of 1926. The main puzzle might be stated thus: How exact is the Principle? or, what determines the extent of its inexactness?

The potential-energy curves lie in a definite location, so the Principle leads to a rather definite indication of most favored transition. Actually, although the predictions of the most favored transition agreed well with the facts, there was no indication of why other transitions could occur or how to calculate their intensities. This was a serious shortcoming of the theory, which Professor Birge did not hesitate to point out to me at the time.

A year later, in 1927, we all became familiar with Heisenberg's uncertainty principle as one of the broad implications of quantum mechanics. But in 1926 it did not occur to anyone that there was anything basically wrong with talking about

[4] E. U. Condon, *Physical Rev.* **32**, 858 (1928). Development of quantum-mechanical formulation of the principle, including basic idea of internal diffraction.

definite values of both nuclear position and momentum being carried over without alteration in an electronic transition. Looking back on it later one can see that it might have been thought of, because statistical mechanics was dealing in terms of elementary cells of finite extension in phase space and the uncertainty principle is closely related to that. However, as elsewhere, hindsight is often better than foresight in theoretical physics.

Quantum-Mechanical Formulation

All that I just described happened in the spring of 1926. Matrix mechanics had been developed, but the matrix calculus was so difficult mathematically that it was hard to put any physical insight into it.* Schrödinger's famous series of papers on wave mechanics was just appearing; but it was not until the fall of 1926 that Born first recognized the probability interpretation of $|\psi|^2$ which collision problems forces on us, as contrasted with the more hydrodynamic views which are possible alternatives for closed systems.

I had the good fortune to be sent by the International Education Board from Berkeley to Göttingen in the fall of 1926 and thus was able to plunge into the study of the new quantum mechanics at one of the few points of high concentration of original thinking in this field. Great ideas were coming out so fast that period (1926–27) that one got an altogether wrong impression of the normal rate of progress in theoretical physics. One had intellectual indigestion most of the time that year, and it was most discouraging.

Besides studying the current papers I tried to solve the basic problem in quantum mechanics that underlies all molecular dynamics—the basic justification for the method of first working out the electronic states for fixed positions of the nuclei, so that the electronic energy levels depend parametrically on the nuclear coordinates and serve as the potential-energy functions for determination of the nuclear motions.

The justification is, of course, connected with the smallness of electron mass relative to nuclear mass, but I never could see how to work it out. Later the problem was handled in a basic paper by Born and Oppenheimer[5] which, however, I have never felt that I properly understood. But Born is such a great master at seeing all physics in terms of "Entwicklungen nach einem kleinen Parameter Kappa" that I suppose it must be all right. The paper of Born and Oppenheimer is among those difficult ones that are more often cited than read.

These studies did, however, serve to make clear the place of the Principle in relation to the general ideas of quantum mechanics. Although this topic was treated in a short paper written from Göttingen, it was not properly handled until the fall of 1928 in a paper[4] written in Princeton. The essence of the argument is that the wave function for a diatomic molecule is approximately of the form of a product of an electronic wave function, in which the nuclear coordinates appear as parameters, and of a wave function for the nuclear motion, say

$$\Psi = u(x_e, x_n)\cdot v(x_n),$$

where x_e stands symbolically for the electron coordinates and x_n for the nuclear coordinates. Therefore, the matrix element for a quantity like the dipole moment, which determines the radiation transition probabilities, is of the form

$$M_{12} = \int_e\int_n \Psi_1 M(x_e, x_n)\Psi_2 dx_e dx_n$$

$$= \int_n \bar{v}_1(x_n) M_{12}(x_n) v_2(x_n) dx_n,$$

where

$$M_{12}(x_n) = \int_e \bar{u}_1(x_e, x_n) M(x_e, x_n) u_2(x_e, x_n) dx_e.$$

The quantity $M_{12}(x_n)$ is the matrix element of the dipole moment calculated by regarding the nuclear coordinates as parameters of the electronic problem rather than as dynamical coordinates. It is characteristic of the electronic jump in question and so is the same for all the vibrational transitions of a band system.

* In fact, I remember very well, in the fall of 1926 at Göttingen, that Professor David Hilbert said to his class in this connection, "die Physik wird zu schwer für die Physiker!"

[5] M. Born and J. R. Oppenheimer, *Ann. Physik* **84**, 457 (1927). Basic quantum-mechanical justification of use of electronic potential-energy curves to determine nuclear motion in molecules.

The nuclear wave functions $v(x_n)$ may be quite complicated for a polyatomic molecule, but for a diatomic molecule they are of the form of a function of the radial coordinate $R(r)$ multiplied by a spherical harmonic for the simple motion under conservation of angular momentum.

It is hard to say much in general about $M_{12}(x_n)$. In fact, no explicit calculation of an example has been made even yet. But it is natural to suppose that it will be a slowly varying or smooth function of r over the small range of r in which the radial wave functions have appreciable value.

The radial functions $R(r)$ are rather closely related to the corresponding classical vibratory motion, according to the general correlation provided by the Wentzel-Brillouin-Kramers approximation. According to this, the wave function is of the form

$$R(r)=\frac{1}{4(p)^{\frac{1}{2}}}\cos\left[(2\pi/h)\int^{r} p\mathrm{d}r+\alpha\right]$$

within the range of the classical motion, and falls off rapidly to zero outside the range of the classical motion, being dominated by a factor of the form

$$\exp\pm\left[(2\pi/h)\int^{r}|p|\mathrm{d}r\right],$$

where p is given by

$$(1/2\mu)p^2+V(r)=W,$$

where $V(r)$ is the effective potential energy of the radial motion, including the effects of the rotational energy;

$$V(r)=V_0(r)+\frac{h^2}{2\mu}\frac{J(J+1)}{r^2},$$

when J is the rotational quantum number and $V_0(r)$ is the potential-energy function applicable in the case of a nonrotating molecule.

It is easy to see qualitatively that the value of an integral of the form

$$\int R_1(r)R_2(r)\mathrm{d}r,$$

with wave functions given by the Wentzel-Brillouin-Kramers approximation, is in the main determined by the conditions set up in the classical formulation of the Principle as given in my 1926 paper:

(1) The integral will be small unless the wave functions overlap, that is, unless there is little sudden change in the internuclear distance called for in the transition;

(2) The integral will be small if the wave functions are related in such a way that a nonoscillatory part of one wave function overlaps a rapidly oscillatory part of the other, for this means that the electron jump would have to be accompanied by a considerable change in the radial momentum.

This formulation goes considerably further than that of the 1926 paper in three main respects:

(*i*) It follows in a definitely deductive way from a well-founded theory whose other successes have given it a definite standing as an adequate basis for atomic mechanics;

(*ii*) It gives, in principle, definite results for the relative strength of each possible quantum transition; it therefore goes beyond the simple classical picture based on the potential-energy curves and provides a basis for calculating the extent to which the less favored transitions actually do occur;

(*iii*) It leads to the prediction of some specific effects arising from the wave nature of matter for which I want to introduce the term "internal diffraction;" these will be dealt with at length in the next section.

When it comes to setting up explicit formulas for the integrals $\int R_1(r)R_2(r)\mathrm{d}r$ it is not possible to get results of very general applicability. The natural thing to do is to start with the approximation that the initial and final potential-energy functions are Hooke's law parabolas, differing as to force constant and as to equilibrium separation. The formulas on this supposition were set up and evaluated by Hutchisson.[6] However, the results are rather complicated.

But a more severe limitation is that the cases which are most interesting physically are those in which there is a fairly large change in the equilibrium separation occasioned by the electronic transition, so that large changes in the vibrational quantum number occur. In such cases the

[6] E. Hutchisson, *Physical Rev.* **36**, 410 (1930); **37**, 45 (1931). Explicit calculation of integrals for transitions using harmonic-oscillator wave functions.

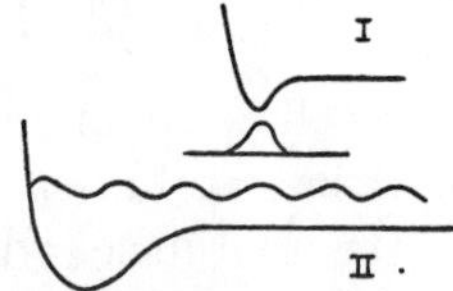

FIG. 3. Wave functions as related for internal diffraction in continuous spectra, from my 1928 paper (reference 4).

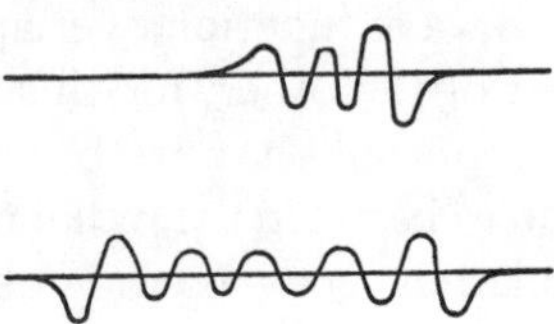

FIG. 4. Rough sketch of wave functions showing sensitivity of integral to relative location of nodes.

wave functions have to be known with fair accuracy at some distance away from the equilibrium separation. The actual force law is not parabolic, so that the harmonic-oscillator wave functions are no longer good approximations over an important part of the range of the coordinates.

Since the actual departure from the harmonic law is different for each molecule, general formulas are of small value. The only proper test of the theory is therefore explicit calculation based on the specific facts in particular cases.

Internal Diffraction

I particularly want to stress the specifically nonclassical, or wave-mechanical, features of the theory, for it seems to me they should be more widely known. These features seem to me to be more than mere quantum-mechanical refinements of detail, as they have sometimes been presented in books. On the contrary, internal diffraction is to me just as real and forceful a proof of the wave nature of nuclear motion as any of the basic external diffraction experiments, such as those in which a beam of electrons or neutral hydrogen atoms is diffractively scattered by a crystal.

Figure 3, from my 1928 paper,[4] presents one particular situation in which internal diffraction might rise. It does not matter whether curve *I* is above or below curve *II*: if *I* is above, the phenomenon will appear in emission, while if *I* is below, then it will appear in absorption. The wave function of the lowest vibrational level in state *I* will be approximately a Gauss error function as shown. A radial wave function for a typical energy value in the continuum above the dissociation limit of *II* will look like the one sketched in the figure. As the energy of the final state is increased the radial wave function will vary in this way: its first loop will have a fairly constant relation to the turning point for the classical motion; but, as the de Broglie wavelength slowly decreases, the nodes become more closely spaced, so that the phase of the quasisine wave located under the center of the Gauss error wave function gradually changes. Clearly, when a loop of the sinoidal wave function is under the center of the Gaussian wave function we shall get a large value of the integral which governs radiative transition. But when a node is under the Gaussian wave function, the integral is very small.

In this way the initial-state wave function is able to "see" the wave nature of the final-state wave function. Transitions can occur to the energies of the continuum for which a loop is under the initial wave function and are much weaker when they occur to the energies having a node there. The result is a rippling variation in the intensity of the continuous spectrum. Such a manifestation of the wave detail of the ψ function I call *internal diffraction*, in analogy with external diffraction, which is also determined by a phase relation between initial and final wave functions.

The term internal diffraction is a convenient one to use more generally to describe specific

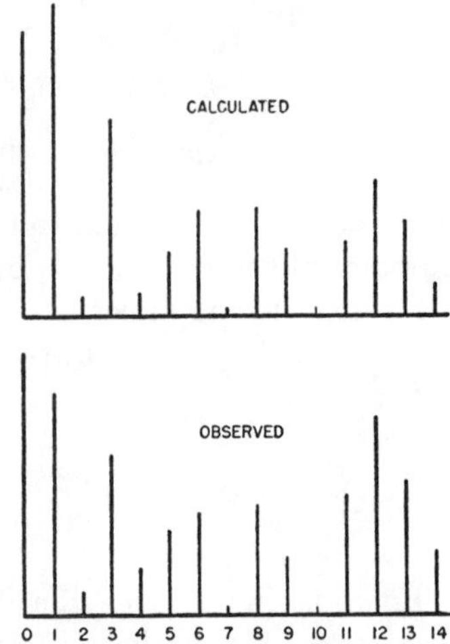

FIG. 5. Brown's calculations on sodium fluorescence bands (reference 7).

variations of the transition probabilities away from those indicated by the Principle after it is fuzzed out by uncertainty principle requirements. In the more general case we may suppose that two wave functions for initial and final states are related approximately as in Fig. 4. Evidently, the exact value of the integral of the product of two such functions is quite sensitive to the exact relative positions of the nodes and loops. The integrand is itself a roughly oscillatory function, being positive where the factors are of the same sign and negative where they are of opposite signs. Hence the value of the integral is quite sensitive to the average relative phase of the oscillations in the wave functions of initial and final states. This is a specifically quantum-mechanical effect, and definite results calculated from it afford a sensitive test of the reality of de Broglie waves associated with the nuclear motions in the molecules.

Although it did not seem feasible to make sufficiently accurate calculations to check the point at the time, this view was put forward in 1928 to explain the intensity fluctuations observed by Wood in the fluorescence spectrum of iodine vapor. On excitation by the green line of mercury, diatomic iodine molecules are raised to the 26th vibrational level of an excited electronic state from which they emit a long series of fluorescence

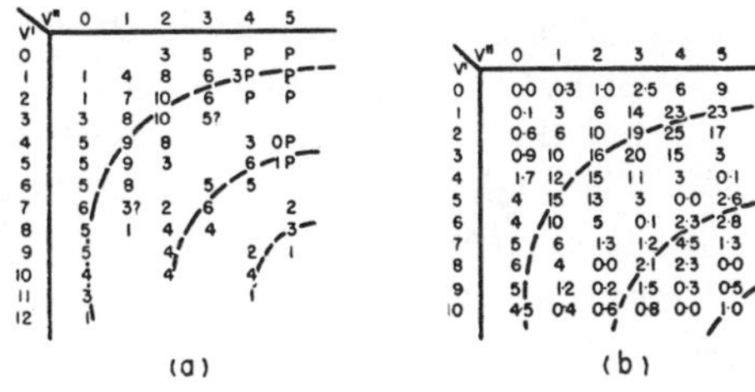

(a)

v′ \ v″	0	1	2	3	4	5
0			3	5	P	P
1	1	4	8	6	3P	P
2	1	7	10	6	P	P
3	3	8	10	5?		
4	5	9	8		3	0P
5	5	9	3		6	1P
6	5	8		5	5	
7	6	3?	2	6		2
8	5	1	4	4		3
9	5		4		2	1
10	4		4		4	
11	3				1	
12	1					

(b)

v′ \ v″	0	1	2	3	4	5
0	0·0	0·3	1·0	2·5	6	9
1	0·1	3	6	14	23	23
2	0·6	6	10	19	25	17
3	0·9	10	16	20	15	3
4	1·7	12	15	11	3	0·1
5	4	15	13	3	0·0	2·6
6	4	10	5	0·1	2·3	2·8
7	5	6	1·3	1·2	4·5	1·3
8	6	4	0·0	2·1	2·3	0·0
9	5	1·2	0·2	1·5	0·3	0·5
10	4·5	0·4	0·6	0·8	0·0	1·0

FIG. 7. Intensities for rubidium hydride as given by Gaydon and Pearse: (*a*) observed visual estimates on a scale of 10; (*b*) calculated, reduced to a scale of 25.

doublets on jumping to various vibrational levels of the normal electronic state. These doublets form a series the intensity of which varies in quite an irregular way, namely: 10, 9, 1, 9, 3, 8, 8, 2, 9, 0, 8, 3, 2, 7, 0, 7, 0, 2, 1, 0, $\cdots$. All of these transitions are permitted by the approximate form of the Principle. Undoubtedly the fluctuations are due to the particular phase relations of the nodes in the initial and final states, although because of the high quantum numbers involved an explicit calculation to test this statement would be very laborious and has not been made.

But the lack has been pretty well supplied by two explicit calculations which will now be mentioned. A similar irregular variation of intensity occurs in the fluorescence bands of Na_2. W. G. Brown[7] in 1933 calculated the approximate integrals and obtained the comparison of observed and calculated intensities that is shown in Fig. 5.

In 1939 similar calculations were made by Gaydon and Pearse[8] the band system of rubidium hydride. Figure 6, from their paper, shows the potential-energy functions and wave functions that they used. The dashed parabolas show the Hooke law approximation to the potential energy, and the full curves show the more accurate potential-energy functions inferred from the levels. The wave functions used were obtained by a reasonable approximate transformation from harmonic-oscillator wave functions rather than by numerical integration of the wave equation.

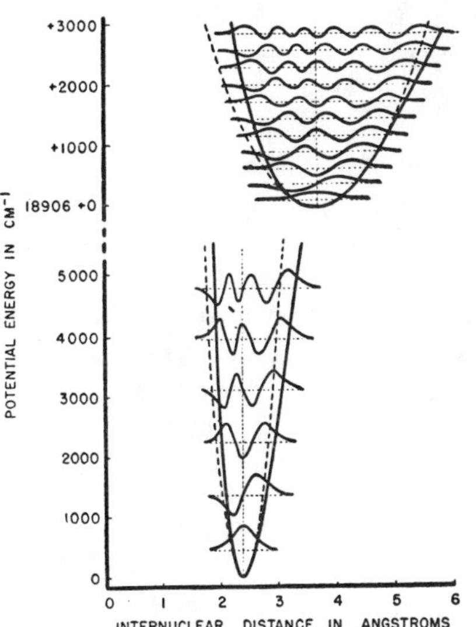

FIG. 6. Potential curves and wave functions used by Gaydon and Pearse in calculations on rubidium hydride (reference 8).

[7] W. G. Brown, *Zeits. f. Physik* **82**, 768 (1933). Explicit calculation of internal diffraction intensity variations in fluorescence bands of diatomic sodium molecule.

[8] A. G. Gaydon and R. W. B. Pearse, *Proc. Roy. Soc.* **A173**, 37 (1939). Careful calculation of transition probabilities in band spectrum of rubidium hydride, showing internal diffraction effects quantitatively.

The result of their calculations is shown in Fig. 7. The table on the left gives observed values (estimates) and that on the right the results of calculation. The main parabolic locus of strong bands is the familiar result of the classical Principle. The secondary weaker loci are the result of special phase relations between initial and final wave functions, that is, internal diffraction.

Molecular Hydrogen

Diatomic hydrogen is a particularly interesting molecule since it has the simplest structure, one that is simple enough to permit of making some fairly accurate theoretical calculations. These calculations, initially made by Heitler and London,[9] provided a host of new viewpoints applicable in general to chemistry. In particular, they cleared up the basic nature of the electron-pair or homopolar valence bond. There has to be some limit to the scope of this review, so my remarks will be confined to some interesting points that grew out of the application of the Principle to the theoretical potential-energy curves of Heitler and London, and of Burrau[10] for H_2^+.

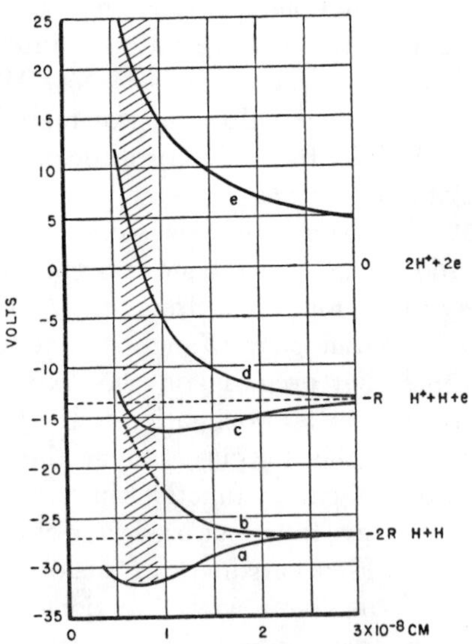

FIG. 8. Potential-energy curves for molecular hydrogen, from Bleakney's 1930 paper (reference 13).

[9] W. Heitler and F. London, *Zeits. f. Physik* **44**, 455 (1927). First quantum-mechanical calculation of potential-energy curves of molecular hydrogen.

[10] O. Burrau, *Kgl. Danske Videnskab. Selskab. Math-fysiske Medd.* **7**, 14 (1927). First calculation of potential-energy curves for normal state of ionized hydrogen molecule.

The application of the Principle to molecular hydrogen was first made by Winans and Stueckelberg[11] in 1928. They recognized that radiative transitions from the excited triplet levels on to the repulsive-force triplet sigma ($^3\Sigma$) state of Heitler and London could account for the very extensive ultraviolet continuous spectrum of molecular hydrogen. At about the same time a paper written with Smyth[12] gave some interpretations of observed critical potentials in terms of the repulsive curves. These two papers may be said to have provided the first evidence of the physical reality of the repulsive-force curves given by quantum mechanics.

A much more convincing proof of the physical reality of such electronic states in molecular hydrogen was given in 1930 by Bleakney[13] when we were together at the University of Minnesota.

In the older data on critical potentials in hydrogen a value close to 31.5 v had usually been reported. As this agrees quite well with the minimum value of the energy needed to strip the two electrons off a hydrogen molecule, it was assumed to correspond to the process of total dissociation of the hydrogen molecule into two protons and two electrons.

But how could this be? According to the Principle, the colliding electron would have to strip off the two electrons, leaving the two protons at essentially the same distance apart as they are in the normal molecule. At this distance, the two protons have a Coulomb interaction energy of about 20 ev; therefore the process of total dissociation would require about 51.5 v instead of 31.5 v. Subsequently, of course, the two protons would fly apart under their mutual repulsion, each of them getting 10 ev of energy.

The situation is shown in Fig. 8, taken from Bleakney's 1930 paper.[13] Curves (a) and (b) are derived from the basic Heitler and London solution for the mutual interaction of two normal hydrogen atoms. Curve (c) is from Burrau's

[11] J. G. Winans and E. C. G. Stueckelberg, *Proc. Nat. Acad. Sci.* **14**, 867 (1928). Interpretation of the continuous ultraviolet emission spectrum of molecular hydrogen.

[12] E. U. Condon and H. D. Smyth, *Proc. Nat. Acad. Sci.* **14**, 871 (1928). Interpretation of the critical potentials of molecular hydrogen.

[13] W. Bleakney, *Physical Rev.* **35**, 1180 (1930). Experimental proof that some ions formed on electronic impact with molecular hydrogen have kinetic energy.

calculation for the normal state of H_2^+. Curve (*d*) is the analogous repulsive-force state in H_2^+ as calculated by Morse and Stueckelberg.[14] Finally, curve (*e*) is simply e^2/r, the potential-energy curve of the molecule H_2^{++}.

As soon as these curves were drawn some interesting conclusions were evident. In the first place it is clear that the Principle does not allow us to dissociate the molecule simply by striking it with an electron of the minimum necessary energy, in this case about 4.4 v, for that would require a large change in position or momentum of the nuclei to accompany the electronic transition.

However, at about 11 v it ought to be possible to dissociate the molecule by electron impact by inducing transitions from state (*a*) to state (*b*). This transition is forbidden for light absorption but allowed for electron impact. The excited H_2 molecules would immediately fly apart, giving two normal hydrogen atoms each with about 3.5 ev of kinetic energy. As both products are neutral they could not be observed in a mass spectrometer. Observations have been made that indicate a rapid clean-up—that is, adsorption on the walls—of hydrogen, which could be due to formation of atomic hydrogen at about this energy, but not at lower voltages.

Considering state (*c*), the change in equilibrium separation indicates that the transition from H_2 to normal H_2^+ not vibrating is quite unlikely, and that it is likely that there will be a good yield of $H+H^+$ by direct transition from the normal state to the part of (*c*) lying above the dissociation limit. This was observed to be the case.

Coming now to curve (*d*), we notice the explanation of the 31.5-v critical potential. It has nothing to do with total dissociation; that would require transitions to curve (*e*). The 31.5-v potential is in reality the transition, indicated by the Principle, to curve (*d*). If this is the correct explanation, then the H^+ ions formed in this process should have about 6.5 ev of kinetic energy. By an appropriate use of retarding fields in his mass spectrometer, Bleakney could show that this was so. It constituted the first instance of observation of molecular-ion fragments formed with kinetic energy and the most direct and unambiguous proof yet found of the physical reality of the repulsive-force molecular states predicted by quantum mechanics.

Later more careful data were taken by Tate and Lozier,[15] establishing that molecular-ion fragments can be formed with kinetic energy in other molecules as well, although here the quantum-mechanical calculations are too difficult to permit detailed predictions to be made.

There is another point in this connection, which was studied briefly by Hipple in 1936, that provides another interesting example of a quantum-mechanical phase of the Principle. Consider again the transitions from curve (*a*) to curve (*c*), induced by electron impact. The minimum of curve (*a*) is so related to (*c*) that the most favored transitions are those leading to H_2^+ molecules in a rather high vibrational state. If we bombard H_2 with, say, 18-v electrons, we will get some H^+ but transitions to H_2^+ are more favored.

What happens if we use deuterium instead of hydrogen? Theory tells us that the potential-energy curves are quite accurately the same in the two isotopic molecules. Theory also tells us that the wave function of the normal zero-vibrational state in D_2 will be narrower than in H_2; because of its greater mass it behaves more classically. Hence we can predict a lower yield of D^+ relative to D_2^+ at 18 v in D_2 than of H^+ relative to H_2^+ in H_2. Hipple tried this and found a considerable effect. It would be interesting to get accurate data on this and to attempt a precise calculation.

In 1931 Finkelnberg and Weizel tried to get definite information on the shape of the basic Heitler-London repulsive-force curve by an interpretation of data on the potentials needed to excite different parts of the continuous spectrum of molecular hydrogen. Such a procedure involved using the Principle in a more precise way than was ever intended, as was pointed out in 1936 by Coolidge, James and Present.[16] They did a very careful job, first on an improved calculation from theory of the repulsive-force curve, then on numerical integrations of the radial wave-function products occurring in the theory,

[14] P. M. Morse and E. C. G. Stueckelberg, *Physical Rev.* **33**, 932 (1929). First calculation of repulsive-force potential-energy curve for ionized hydrogen molecule.

[15] J. T. Tate and W. Lozier, *Physical Rev.* **39**, 254 (1932).

[16] A. S. Coolidge, H. M. James and R. D. Present, *J. Chem. Physics* **4**, 193 (1936). Careful discussion of application of the Principle to continuous spectrum of molecular hydrogen.

in order to get a proper treatment instead of the usual rough graphical construction from the potential-energy curves.

The work of Coolidge, James and Present brought out some fairly good evidence that in this case one cannot treat the electronic dipole-moment matrix component as constant. This meant that the rough assumption I had made earlier in order to get the main idea straightened out needed to be improved. Because of this point they were led to say in the abstract that "It is concluded that the Franck-Condon principle leads to results definitely incompatible with observations." I cannot let that pass, even ten years later, without a word of protest. They can say that I misused the Principle if they like, but not that the Principle is incompatible; for the Principle is properly to be judged by its correct quantum-mechanical formulation, which they so beautifully worked out, rather than by the rough criterions I gave as approximate rules.

The whole situation with regard to these special properties of molecular hydrogen seems to me to be highly satisfactory.

Conclusion

In the past decade, despite both the competitive attraction of nuclear physics and the wasteful interruptions of the war, many more band systems have been analyzed so that today there are a large number of instances of electronic transitions in a wide variety of molecules, all of which conform nicely to the semiquantitative intensity relations given by the approximate form of the Principle. Nevertheless, the fact remains that the Principle has never been subjected to a really severe test in which intensities are carefully measured by good photographic photometry and compared with those calculated from highly accurate wave functions.

I will conclude this review by merely mentioning one other application. The Principle has also contributed a good deal to the elucidation of many points concerned with predissociation of molecules.[17] Here, except for a few cases in which feebly bound molecules can dissociate by rotational instability aided by potential barrier leakage, we are concerned with radiationless transition between two electronic states of equal energy. The Principle acts restrictively here also, in that such transitions cannot occur if they would require too much readjustment of the nuclear motions.

The Principle also applies "in principle" to polyatomic molecules,[18] although here the situation is much more complicated than in the atomic case. Not only are the band systems much more complicated, so that thus far little progress has been made in analyzing them, but also the potential-energy curves have to be replaced by potential-energy surfaces, which are not well known either empirically or theoretically.

---

[17] L. A. Turner, *Zeits. f. Physik* **68**, 178 (1931). Application of the Principle to predissociation.

[18] J. Franck, H. Sponer and E. Teller, *Zeits. f. physik. Chemie* **18**, 88 (1932). Application to predissociation in polyatomic molecules.

Recollections of physics and of physicists during the 1920's

David M. Dennison
Department of Physics
University of Michigan
Ann Arbor, Michigan 48104
(Received 26 February 1974)

These recollections of physicists and of the development of the quantum theory in the 1920's are taken, essentially unchanged, from a talk given at the joint Annual Meeting of the American Physical Society and the American Association of Physics Teachers in Chicago, 6 February 1974.

The decade of the 1920's was a period of confusion, of excitement, of optimism and of discovery for physics. I had the good fortune to have been born at just the right time—in 1900 to be exact—so that I experienced this era as a young man in my 20's. It made an indelible impression on me and I would like to give some of my recollections of this period of fifty years ago. I might warn you right at the outset however that I won't attempt any detailed or proper analysis of all that was going on in physics at that time. I am glad to leave all that to the specialists in the history of science who also can assess the correctness of the recollections of people like myself in the light of written records such as letters and published papers. I will try instead to give some feeling for the excitement and the flavor of those times. Unfortunately this means that a lot of what I have to say will sound like personal history—which I suppose it is.

I will begin with my last year of graduate work at the University of Michigan in the year 1923–24. Fortunately for me, the young Swedish physicist, Oskar Klein, had just joined the Michigan staff and he supervised my doctoral research which was on the vibrations and rotations of the methane molecule. Klein had been Bohr's assistant in Copenhagen during the previous year or two and through him I learned about the current physics then going on in Europe and, incidentally, acquired a burning desire to go to Copenhagen for further study. Among other things, Oskar Klein taught me a great deal about the art of scientific writing. This was during the latter part of the year while I was writing up my thesis work for publication. Actually, I had a pretty high opinion of myself as a writer but when I would take sections of the manuscript for Klein to read, he would object to nearly every sentence and make me revise it. Finally, I could stand it no longer and I said, "Professor Klein, *you* would write it one way but *I* would write it a different way." I remember his answer. "No," he said, "there is only one way to write a scientific paper. Each sentence must be rigorous and correct and the thought must follow smoothly from one sentence to the next!" Good advice—but not always easy to follow!

When Niels Bohr visited this country in the spring of 1924, Klein persuaded him to come to Ann Arbor for one or two lectures and I got my chance to talk with him. The upshot was that Bohr kindly said that I would be welcome to come to Copenhagen and study at the *Institut for Teoretisk Fysik*. So in the late summer of 1924, newly married, my wife and I arrived in Copenhagen aboard the old Scandinavian-American steamship, the *Hellig Olav*. For the next two years I remained in Copenhagen and was supported by one of the first General Education Board Fellowships which was financed by the Rockefeller Foundation. As I recall it the stipend was about $2000 with no travel allowance. You did not get very rich on that stipend and it was not what you would call plushy living but it was adequate and I was very grateful for it.

Copenhagen in those years, 1924 to 1926, as in so many other years, was one of the great centers for physics. Nearly everyone who was doing anything at that time came. Heisenberg, Kramers, and Nishina were there most of the time. Hund, Pauli, Fowler, Dirac, Goudsmit,

Fig. 1. Oskar Klein.

Fig. 2. Frederick Hund.

Uhlenbeck and many others all came at one time or another. It was indeed a very exciting time.

During the autumn of 1925 a photographer came and took pictures of a number of us. These were taken in the library of the Institute where many of us worked. It may be of interest to reproduce some of these old photographs.

The first (Fig. 1) is of Oskar Klein who had returned to Copenhagen from Ann Arbor and who was to remain there for the next two or three years until he took a position in Sweden.

The second (Fig. 2) shows Frederick Hund who specialized in molecular structure and whose name is certainly known to all molecular spectroscopists.

Next (Fig. 3) we have the young Heisenberg who was just reaching the height of his imaginative creativity. He does look young, doesn't he? He was only 23 when this picture was taken.

The next picture (Fig. 4) shows Charles Darwin, the English mathematical physicist. Incidentally he was the grandson of the first famous Charles Darwin.

There was a group of Japanese physicists at the Institute at that time (Fig. 5). You will recognize Nishina on the left side of the picture. Part of their work had to do with experimentally determining the values of x-ray energy levels. When these were plotted against atomic number one could see small kinks in the curves at the points where the outer shells of electrons were being completed. We usually think of Nishina as being a theorist from his later work with Klein resulting in the Klein–Nishina formula. However, when I first knew him he was doing experimental work.

Figure 6 shows our editor of the *Physical Review* , Sam Goudsmit and myself.

I must say that we all look pretty young—which indeed we were. At the time these pictures were taken Bohr, at the age of 39 was the old man among us.

Now to return to the story. The old quantum theory had had many successes. This theory, you recall, used the Sommerfeld–Wilson rules of quantization where the integral of *pdq* was set equal to a multiple of Planck's constant. Looking back now, it is astonishing to see how much could be explained with this inconsistent and inadequate theory—even the fine structure of the hydrogen spectrum as far as it was known experimentally at that time. In addition to the formal rules of quantization we had Bohr's Correspondence Principle. This principle, largely unknown and unused today was often a powerful tool at that time.

However, in 1924, the old quantum theory, in spite of its success, was dying. Its inconsistencies, the incompatibility of the wave and particle properties of radiation, the problem of the stability of orbits, and so on which had earlier been swept under the rug, were now being reexamined. The new quantum theory was emerging. It was a struggle. It was the struggle of trying to formulate new ideas, new images, new pictures of something that had not existed before. It is all quite different now. Every schoolboy knows something about the Uncertainty Principle and wave mechanics. It is hard to appreciate how difficult it was then to formulate something which was so radical and so new. I suppose that we were all under the spell of classical mechanics and classical electrodynamics which had had so many striking successes and which—in their areas of application—are as successful today.

Early in 1925, while I was in Copenhagen, Heisenberg brought out his first version of wave mechanics. This was essentially the first introduction to matrices for many physicists. Of course, the mathematicians knew all about matrices but very few physicists had ever used them. Possibly Heisenberg himself had never really worked with them since, actually, in this initial paper he did not use the word "matrix" although he correctly derived the rules of matrix multiplication and found the commutation relation between momentum and coordinate for a one dimensional system. It was not until some months later in a paper by Born and Jordan that matrices were formally introduced and a more comprehensive picture given.

During the winter of 1925–26, Kramers suggested that I try to apply Heisenberg's theory to a problem involving not one particle but rather a group of particles which were rigidly fastened together and which then constituted a symmetrical top. This had not been in Heisenberg's original program and it required a certain modification of the theory. Fortunately this modification worked and it was possible to give the correct energies and matrix elements for a symmetric top molecule. The subsequent paper was published in the Physical Review and, to the best of my knowledge, it was the first paper on matrix mechanics to appear in that journal.

Well, as you can see these were tremendously exciting days—the old concepts and inconsistencies were being swept away and the new quantum theory was emerging.

In 1926 Schrödinger wrote the three famous papers in which he introduced the wave equation and later his theory was shown to be equivalent to Heisenberg's matrix mechanics. Incidentally, I have sometimes wondered if Schrödinger ever quite understood the theory he had created. For example, somewhat later he wrote a paper in which he showed that, for an harmonic oscillator he could construct a wave packet which permanently oscillated back and forth just like a real particle. Schrödinger evidently thought that this was really fundamental, although it clearly will not work for any other system. He appeared to want to retain the concepts of classical physics—and he seemed to have rather little sympathy with the philosophy of the new quantum theory.

My two years had come to an end in 1926, but there was an opportunity for me to stay a third year. Professor Randall got me a very small fellowship from the University of Michigan and I decided that I'd like to go and work with Schrödinger.

I wrote to Schrödinger and asked if I could come and among the few records which I have from those days is this old letter. I thought I might read some of it to you just to show you the kind of old-world phraseology which he used. He starts in,

> Dear Mister Dennison:
>
> I was greatly enchanted with your kind letter from June 15, though I must, of course, refuse giving you *permission* of coming zu Zürich for everybody is allowed to come here and study here. I am glad you have given me the opportunity of stating, that it will be a great pleasure to me, if you do so, and so more, if you take interest in the new theory, which I am putting forward at present and, which, as you probably know, has several points of connection with Heisenberg's quantum mechanics on which you have been working on.

He goes on to say:

> I am very much interested in the work about which you wrote me, namely, the treatment of a more than two-atomic molecule by Heisenberg's method. We have not yet succeeded here in treating this problem by the wave-method though I must owe, that I have not earnestly tried it myself. Would you be kind enough to write me where your paper is going to be published? It seems to be the case that some problems are easier treated by Heisenberg's theory and some by mine and since *mathematically* to a far extent the results are identical, the two points of view can help each other very much. It is always possible to construct Heisenberg's matrices from the "Eigenfunktionen" by simple quadratures, though the reverse problem it seems so far as I can tell till now, more complicated. I am, dear Mr. Dennison, with kindest greetings, yours sincerely. Please give my kindest regards to Mr. Bohr and Mr. Heisenberg.

Fig. 4. Charles Darwin.

Fig. 3. Werner Heisenberg.

Well, it's interesting to see what the letters were like then.

I was very much puzzled about the problem of homopolar diatomic molecules when I went to Zürich in the fall of 1926. Actually there were a lot of things which were known as, for example, the formula for the wave functions and the rotational energy levels. Also, there had been a good many measurements of the rotation spectra. Of course, this had to be through electronic bands because a homopolar molecule has no permanent dipole moment. It was known that the rotation lines of these molecules often showed alternating intensities, and sometimes it looked as though certain bands had alternate missing lines.

We were all trying to understand things of this sort, and as a matter of fact during that fall I wrote a paper which, I am afraid, showed my state of confusion more than anything else. Very little of it is right, but it illustrated how one was trying to work one's way into the truth. Part of it used the following idea.

We put various demands on the wave function, one of which is that it should be single valued. Now actually following out the Heisenberg philosophy one could say that the real observable quantity is not the wave function itself, but rather the density function, $\bar{\psi}\psi$. It was therefore

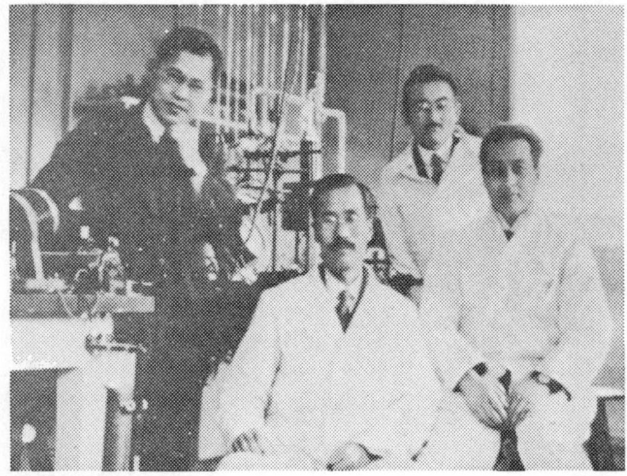

Fig. 5. Nishina and a group of Japanese physicists.

perfectly logical to say that the density should be the quantity which has to be single valued rather than the wave function. Well, if you do this for a homopolar molecule you immediately come to the possibility that instead of having integers for the angular momentum number J, O, 1, 2, etc., you could as well have the half-integers, 1/2, 3/2, etc. Of course, these exist all right but now we say that they are due to a different mechanism, namely, a spin 1/2 that is added on. Formally, though, the idea of making the density function unique gives these, and only these, two possibilities.

Meanwhile Heisenberg in Copenhagen had been working on somewhat similar problems and was beginning to get the right idea of what was up. He wrote a very fine paper during the winter of 1926–27 in which he examined the symmetry properties of wave functions for systems containing identical particles—for example, electrons. He applied these ideas to the spectrum of helium—which had been a great puzzle up to that time—and obtained the correct explanation of the ortho and para helium terms. As you know the helium spectrum can be divided into two classes of lines each with their own set of energy levels—the ortho and para levels. Transitions between ortho and para levels are possible but are very weak since they correspond to a reversal of the spin of one of the electrons. All of this was correctly explained by Heisenberg.

He also speculated about homopolar molecules in this paper. In fact, he suggested that possibly the origin of the alternating intensities found in band spectra was connected somehow or other with the existence of nuclear spins. But that's really about as far as he went and he confesses that he was very unsure about what it all meant.

Early in the spring of 1927 I returned to Copenhagen. This year, incidentally, was a very fruitful period for Heisenberg. It was in the spring of that year that he gave the first colloquium on the Uncertainty Principle and I had the good luck to be there. To all of us who heard him, there was never any question but that this was a really wonderful development. We felt that, suddenly, we could see very far ahead. . . .

Late in the spring of 1927 I went to Cambridge for a stay of about six weeks. This was at the invitation of the brilliant young theoretical physicist, Ralph Fowler, whose principal interest was in the field of statistical mechanics. Cambridge was also an exciting center for physics with people like Rutherford, Mott, Kapitza, Blackett and Cockcroft, some of whom were still graduate students. While there I met and talked with Sir J. J. Thomson who had discovered the electron some thirty years earlier. My recollection is of someone who seemed quite old, a little on the gruff side, and not altogether sympathetic with the new physics—or with the new generation of younger physicists.

While I was in Cambridge, Fowler asked me to give three lectures to his graduate class. Incidentally, it was a real thrill to be lecturing in the same lecture room, with its high ceiling and steeply sloping rows of benches, that had been used by Maxwell so many years earlier. Well it turned out that the work I had been doing on molecular structure and spectra supplied enough material for the first two lectures and these I gave. However, I could see that I was running out of material for the third lecture. I think that I realized this difficulty a week or so before I was actually to give the third talk and it occurred to me to take another crack at the problem of the specific heat of hydrogen.

Let me remind you briefly of the history of the problem.

In the first place the specific heat of hydrogen had been measured by a number of experimenters. They all agreed that the rotational part of the specific heat was represented by a simple monotonic curve.

Now, we all believed that the method for calculating this curve from the new quantum thermodynamics was known and was quite straightforward. The resulting formula contained only one adjustable parameter, namely the moment of inertia of the hydrogen molecule. This was really not adjustable either since the young Japanese physicist Taceo Hori had just completed measurements in Copenhagen on the ultraviolet spectrum of hydrogen which showed that moment of inertia in the ground state was around 4.67×10^{-41} g cm^2.

It should indeed have been all very straightforward but the trouble was that the calculated curve had a bump in it that was surely not there in the observed curve. This was a real puzzle since in every other case the predictions of

Fig. 6. S.A. Goudsmit and D.M. Dennison.

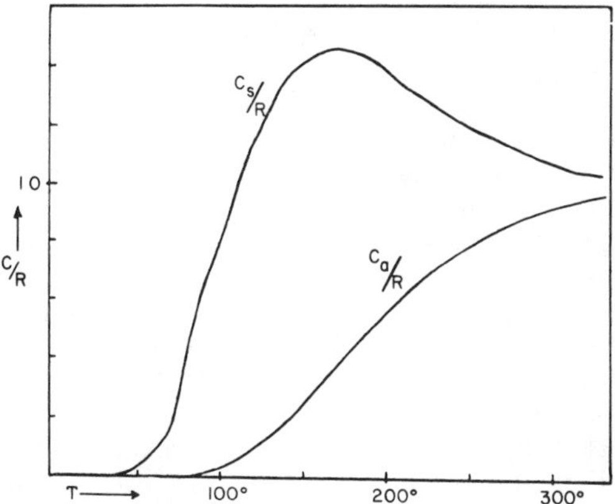

Fig. 7. Rotational specific heat curves for the separate gases of ortho and para hydrogen.

the new quantum theory had been verified by experiment. Hund, in Copenhagen, had just written a paper in which he had divided the energy levels into those for even J and those for odd J corresponding to the symmetric and antisymmetric rotational wave functions. He then gave these different weights as might be caused by a possible spin of the proton. He tried all kinds of combinations of weights but just couldn't fit the experimental curve or obtain the right value for the moment of inertia parameter.

Well, the idea that occurred to me, while I was sweating over material for the third lecture, was very simple. Since Heisenberg had shown that the ortho and para terms in the helium spectrum combined with each other so very infrequently, I began to wonder if perhaps the hydrogen rotational levels might not also divide into ortho and para terms and that since transitions between these would depend upon interactions with the very small magnetic moment of the proton, they might not occur at all during the time that the experimenters were making their measurements. This would mean that in effect we would have a mixture of two independent gases each with its own specific heat which would then be additive in the usual way.

Of course this was the right answer. Figure 7 shows the curves for the gas with even J—this is the one with the bump in it and the other rather featureless curve for the odd J gas. The only thing I had to do was to combine them. It turned out that the best fit with experiment was found by combining them in the ratio of 3:1 for the odd and even J's respectively. This is, of course, just what one would expect if the proton had a spin of 1/2. The result is shown in Fig. 8 taken from Fowler's book. The agreement, as you see, was very good. The final clincher for me was that the parameter for best fit gave a moment of inertia for the hydrogen molecule of 4.64×10^{-41} which was in almost perfect agreement with Hori's measured value of 4.67×10^{-41} g cm^2.

I can well remember that when I had finished my talk in that old lecture room of Maxwell's, Fowler was very much excited about the new theory. He had, in a way, a special reason for his interest. And that was that he was just finishing his book on statistical thermodynamics. He had written about the hydrogen specific heat with all of the discussion that the theory did not work and all the unhappiness about it and now here was the very best example that he could have of the wedding of quantum theory and of statistics as well as the first definite proof of the spin and statistics of the proton. He asked me, as I recall, would I let him have the graph which I had plotted out and of course I gave it to him. As far as I can tell, this is the identical graph—just inked in—which I have taken from his book to make Fig. 8. Then he said, "When are you going to publish this?" The question

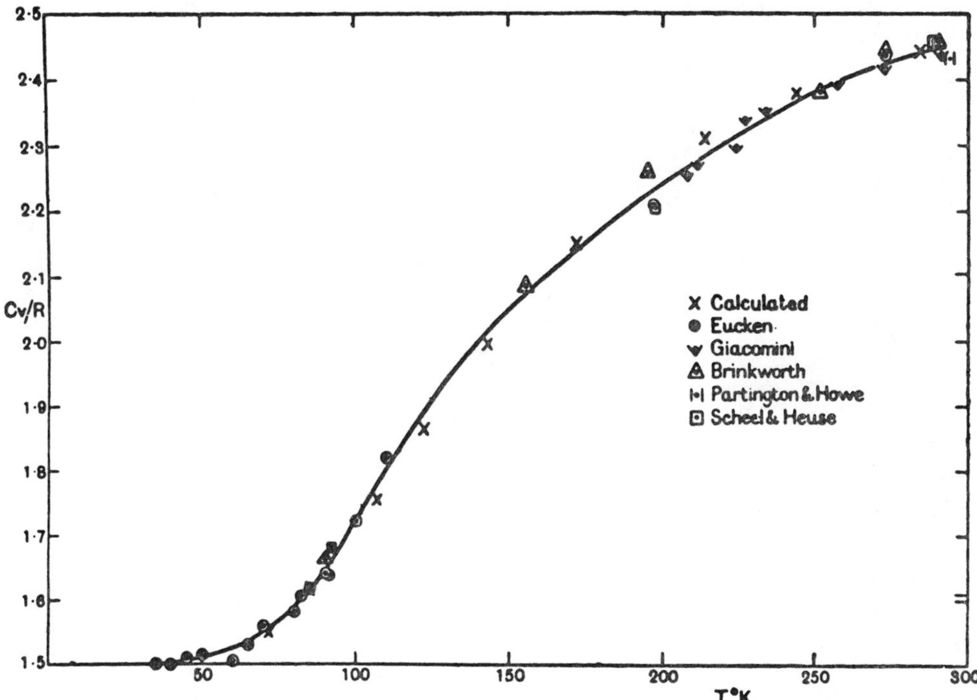

Fig. 8. The experimental and calculated rotational specific heat of Hydrogen gas.

hadn't really occurred to me because actually it was the lecture that I had concentrated on, to get out of the messy business of not having anything to say. I said, "Well, yes, all right, sure, I'll publish it." He wanted me to write it up right away. So I did; I think within a day or two, and he sent it in to The Proceedings of the Royal Society since he was a member.

I sent a copy of the manuscript to Bohr and within a few days received his reply. He said that he was convinced that this was the correct solution to the riddle of the specific heat of hydrogen. I was, of course, very much pleased to read that he also felt that it was an important contribution and was in conformity with the principles which Heisenberg had used in accounting so successfully for the fine structure of atomic spectra.

Bohr went on to make the suggestion that I be more explicit in stating that the 3:1 ratio of the anti-symmetric and symmetric modifications of hydrogen was just what was to be expected if the nuclear spin was equal to that of an electron and only the complete anti-symmetrical wave function was to be allowed in nature. Somehow this point had seemed so completely obvious to me that it had not seemed worth belaboring. However, I was glad to take Bohr's advice and add the necessary sentences in proof.

During this last minute or so I might just finish by describing the rest of my European tour. The last two or three weeks were spent in Leiden where Ehrenfest was of course the central figure. As you know, he was a very great teacher and a great physicist. Ehrenfest was a little skeptical about my theory of the specific heat. He thought it was pretty weird but finally he said, "Well, it *could* be right."

In spite of his healthy skepticism, Ehrenfest was very kind to my wife and myself. I was appreciative of the fact that he treated me in the same way as, for example, he did his students Goudsmit and Uhlenbeck who were about to join me on the staff at the University of Michigan. To further my education he suggested that I make an appointment to talk with H.A. Lorentz. Lorentz, who was by that time retired, lived in the neighboring town of Haarlem and I was invited to go there one afternoon for tea. I recall that Ehrenfest had been very explicit about how I was to behave. He said, "Now, when you get there don't tell him about your work and what you have been doing. Instead, listen to him and try to find out what it was like in the earlier days of his career." And so I did, and it was very delightful. Lorentz told me of his doubts of quantum theory and, in particular, his doubts about light quanta. He was, of course, so very familiar with electro-magnetic theory and he told me of one situation in which he was interested where light and presumably light quanta were going through various layers of dielectric. The classical calculation using Maxwell stresses showed that the net force on certain layers of the dielectric was in the opposite direction to that in which the light quanta were moving. He could not understand how this could happen and in fact it made him somewhat doubt the whole idea of light quanta.

Well, as you can see, this was indeed a wonderful period in the development of physics and I was very fortunate to have had a part in it. It was the period in which modern quantum theory was being born. I hope that I have somehow given you a feeling of the excitement of those days.

Development of concepts in the history of quantum theory*

Werner Heisenberg
Institut für Physik
Max-Planck-Institut für Physik und Astrophysik
Munich, Germany
(Received 22 May 1973)

Editor's note: *The Editor heard Professor Heisenberg deliver the substance of this article as a lecture at Harvard University. We are grateful that permission has been granted to reprint this presentation from its original source.*

In one of his lectures on the development of physics Max Planck said: "In the history of science a new concept never springs up in its complete and final form as in the ancient Greek myth, Pallas Athene sprang up from the head of Zeus." The history of physics is not only a sequence of experimental discoveries and observations, followed by their mathematical description; it is also a history of concepts. For an understanding of the phenomena the first condition is the introduction of adequate concepts. Only with the help of correct concepts can we really know what has been observed. When we enter a new field, very often new concepts are needed. As a rule, new concepts come up in a rather unclear and undeveloped form. Later they are modified, sometimes they are almost completely abandoned and are replaced by some better concepts which then, finally, are clear and well-defined. I would like to describe this development in three cases which have been important for my own work. First, the concept of the discrete stationary state, which obviously is a fundamental concept in quantum theory. Then, the concept of state, not necessarily stationary or discrete, which only could be understood after quantum mechanics and wave mechanics had been developed. And finally, closely connected with the first two, the concept of the elementary particle which is under discussion until now.

The concept of the discrete stationary state was introduced by Niels Bohr in 1913. It was the central concept in his theory of the atom, the intention of which was described by Bohr in the sentence: "It should be made clear that this theory is not intended to explain phenomena in the sense in which the word *explanation* has been used in earlier physics. It is intended to combine various phenomena, which seem not to be connected and to show that they are connected." Bohr stated that only after this connection had been established one could hope to give an explanation in the sense as explanations were meant in earlier physics. There were mainly three phenomena which had to be connected. The first was the strange fact of the stability of the atom. An atom can be perturbed by chemical processes, by collisions, by radiation, or by anything else and still it always returns to its original state—its normal state. This was one fact which could not be explained satisfactorily in earlier physics. Then there were the spectral laws, especially the famous law of Ritz that the frequency of the lines in a spectrum could be written as a difference between terms and these terms had to be considered as characteristic properties of the atoms. And finally there were the experiments of Rutherford which had led him to his model of the atom.

So these three groups of facts had to be combined, and as you know the idea of the discrete stationary state was the starting point for their combination. First of all, one had to believe that the behavior of the atom in the discrete stationary state could be explained by mechanics. This was necessary, because otherwise there would be no connection with Rutherford's model, since Rutherford's experiments were based on classical mechanics. Then also one had to combine the discrete stationary states with the frequencies of the spectrum. There one had to apply the law found by Ritz which now was written in the form that h times the frequency of the line was equal to the difference between the energies of the initial and the final state. This law, however, could best be explained by an assumption which Bohr did not accept, namely by Einstein's idea of the light quantum. Bohr was for a long time not inclined to believe in light quanta, and he therefore considered his stationary states as stations during the motion of the electron which in its orbit around the nucleus loses energy by radiation. The assumption was that during this process of radiation the electron stops radiating at some stations called discrete stationary states. For some unknown reason it does not radiate in these stations, and the final station is the normal state of the atom. When the radiation takes place, the electron goes from one of the stationary states to the next.

According to this picture the time in the stationary state was much longer than the time required for going from one state to another. But of course this ratio of time was never well defined.

What could be said about radiation itself? One could use the general ideas of Maxwell's theory. From this point of view the interaction between atom and radiation seemed to be the source of all trouble. In the stationary state there was no such interaction and therefore one could, so it seemed, apply classical mechanics. But could one use Maxwell's theory for radiation? I might mention that it was perhaps not necessary to take this point of view. One could have taken the light quanta more seriously. One could have said that the interference patterns which we see in light come about by some extra conditions on the motions of the light quanta. I remember vaguely a discussion with Wentzel in the old times where he explained to me the possibility that the motion of light quanta could be quantized and thereby possibly the interference patterns could be explained. But anyway this was not the point of view which Bohr took. Wherever one started one came into a lot of difficulties, and I would like to describe these problems in some detail.

First of all, there were strong arguments in favor of the mechanical model of the stationary state. I have mentioned the experiments of Rutherford. Then the periodic orbits of the electrons in the atom could easily be connected with quantum conditions. So the idea of the stationary state could be combined with the idea of a specified elliptic orbit of the electron. In his earlier lectures, Bohr very frequently showed pictures of electrons moving in their orbits around the nucleus.

This model worked perfectly well in a number of interesting cases. First of all in the hydrogen spectrum. Then in the theory of the relativistic fine structure of the hydrogen lines by Sommerfeld, and in the so-called Stark effect, the splitting of lines in an electric field. So there was an enormous amount of material which seemed to show that this connection of quantized electronic orbits with the discrete stationary states was correct.

On the other hand, there were other reasons for arguing that such a picture cannot be correct. I remember a conversation with Stern who told me that in 1913 when Bohr's first paper had appeared, he had said to a friend, "If that nonsense is correct which Bohr has just published, then I will give up to be a physicist."

I shall now point out the difficulties and errors of this model. The worst difficulty was perhaps the following: The electron made a periodic motion in the model defined by quantum conditions and therefore it moved around the nucleus with a certain frequency. However, this frequency never turned up in the observations. You could never see it. What you saw were different frequencies which were determined by the energy differences in the transitions from one stationary state to another. Then there was a difficulty with the degeneracy. Sommerfeld had introduced the magnetic quantum number. When we have a magnetic field in some direction, then the angular momentum of the atom around this field should be one or zero or minus one according to this quantum condition. But then if you take a different field with a different direction, quantization has to be carried out with respect to this different direction. But one may have an extremely weak field first in one direction and after a short time in another direction. The field is too weak to turn the atom around. Hence the contradiction with the quantum conditions seems unavoidable.

My first discussion with Niels Bohr—which was just fifty years ago—circled around one of these difficult points. Bohr had given a lecture in Göttingen and had stated that in a constant electric field one can calculate the energy of the stationary states according to the quantum conditions, and that a recent calculation of Kramers on the quadratic Stark effect should probably give correct results because in other cases the method had worked so well. On the other hand, there is very little difference between a constant electric field and a slowly varying electric field. When we have an electric field varying not very slowly but say with a frequency which comes near to the orbital frequency, then we know that of course resonance takes place not when the frequency of the outer electric field coincides with the frequency of the orbit, but when it coincides with the frequency given by the transitions and observed in the spectrum.

When we discussed this problem at length Bohr tried to say as soon as the electric field varies with time then forces of radiation come in and therefore it may not be possible to calculate the result in a classical way. But of course at the same time he saw that it is rather artificial to invoke the forces of the radiation at this point. Therefore, we were soon inclined to say that there must be something wrong with the mechanical model of the discrete stationary state. There was one very decisive paper which has not been mentioned yet. It was a paper of Pauli on the H_2^+ ion. Pauli thought that we can possibly apply the Bohr–Sommerfeld quantization rules when we have a well-defined model with periodic orbits, like in hydrogen; but perhaps not in a model which is so complicated like, say, the helium atom, where two electrons move around the nucleus, because there we would get into all the terrible mathematical difficulties and complications of the three-body problem. On the other hand, when we have two fixed centers, two hydrogen nuclei and one electron, then the motion of the electron is still a nice periodical motion and can be calculated. For the rest the model is already rather complicated; hence, it can be used as a check as to whether the old rules really apply in such an intermediate case. Pauli did work out this model and found that he actually did not get the correct energy of the H_2^+ by his calculations. So the doubts against the use of classical mechanics for the calculation of discrete stationary states increased, and attention was moved over more and more to the transitions between the stationary states. One understood that in order to get the whole explanation of the phenomena it was not sufficient to calculate the energy. One also had to calculate transition probabilities. We knew from Einstein's paper of 1918 that the transition probabilities are defined as quantities referring to two states, initial and final state. Bohr had pointed out in his correspondence principle that these transition probabilities could be estimated by connecting them with the intensities of higher harmonics in the Fourier expansion of the electronic orbit. The idea was that every line corresponds to one Fourier component in the expansion of the electronic motion; from the square of this amplitude one can calculate the intensity. This intensity of course is not immediately connected with Einstein's transition probability, but is related to it, so it allows some estimate of Einstein's quantities. In this way the attention gradually moved over from the energy of the stationary state to the transition probability between stationary states, and it was Kramers who seriously started to study the dispersion of an atom and to relate the behavior of Bohr's model under radiation with the Einstein coefficients.

Kramers was guided by the idea of virtual harmonic oscillators in the atom corresponding to the harmonics in writing down a dispersion formula. Then Kramers and I also discussed scattering phenomena where the frequency of the scattered light is different from the frequency of the incident light. Here the scattered light quantum is different from the incoming quantum because during the scattering the atom makes a transition from one state to another. Such phenomena had just been discovered by Raman in the band spectra. When one tried to write down formulas for the dispersion in these cases one was forced not only to speak about the transition probabilities of Einstein but also to speak about transition amplitudes; one had to give phases to these amplitudes and one had to multiply two amplitudes—say the amplitude going from state m to state n with the amplitude going from n to the state k or so, and then to sum over the intermediate states n. Only when we did that did we get reasonable formulas for the dispersion.

So you see that by fixing the attention not to the energy of the stationary state but to the transition probabilities and to dispersion one eventually came into a new way of looking at things. Actually, as I just said, these sums of products which Kramers and I have written into our paper of the dispersion were almost already products of matrices. So it was only a very small step from there to say: Well, let us abandon this whole idea of the electronic orbit and let us simply replace the Fourier components of the electronic orbit by the corresponding matrix elements. At that time I must confess I did not know what a matrix was and did not know the rules of matrix multiplication. But one could learn these operations from physics, and later it turned out that it was matrix multiplication, well known to the mathematicians.

By this time you see that the idea of an electronic orbit connected with the discrete stationary state had been practically abandoned. The concept of the discrete stationary state however had survived. This concept was necessary and had its basis in the observations. But the electronic orbit could not be connected with observations. Therefore it had been abandoned and what had remained were these matrices for the coordinates.

I should perhaps mention that already before this happened in 1925, Born in his Göttingen seminary in 1924 had emphasized that it was wrong to put the blame for the difficulties of quantum theory only on the interaction between radiation and the mechanical system. He propagated the idea that mechanics had to be revised and to be replaced by some kind of quantum mechanics in order to supply the basis for an understanding of atomic phenomena. And then matrix multiplication was defined. Born and Jordan, and independently Dirac, discovered that those extra conditions which had been added to matrix multiplication in my first paper can actually be written in the elegant from $pq\text{-}qp = h/2\pi i$. Thereby they were able to establish a simple mathematical scheme for quantum mechanics.

But even then one could not say what this discrete stationary state really was, and therefore now I am coming to the second part of my talk—the concept of a "state." In 1925 one did have a method for calculating the discrete energy values of the atom. One also had at least in principle a method for calculating the transition probabilities. But what was this state of the atom? How could it be described? It could not be described by referring to an electronic orbit. So far it could be described only by stating an energy and transition probabilities; but there was no picture of the atom. Furthermore, it was clear that sometimes there are nonstationary states. The simplest example of a nonstationary state was an electron moving through a cloud chamber. So the question really was how to handle such a state which can occur in nature. Can such a phenomenon as the path of the electron through a cloud chamber be described in the abstract language of matrix mechanics?

Fortunately at that time wave mechanics had been developed by Schrödinger. And in wave mechanics things looked very different. There one could define a wave function for the discrete stationary state. For some time Schrödinger thought that the following picture of a discrete stationary state could be developed: One had a three-dimensional standing wave which can be written as the product of a function in space and a periodical $e^{i\omega t}$ of time, and the absolute square of this wave function meant the electric density. The frequency of this standing wave was to be identified with the term in the spectral law. This was the decisive new point in Schrödinger's idea. These terms did not necessarily mean energies; they just meant frequencies. And so Schrödinger arrived at a new "classical" picture of the discrete stationary state which at first he believed could actually be applied in atomic theory. But then it soon turned out that even that was not possible. There were very heated discussions in Copenhagen in the summer of 1926. Schrödinger thought that the wave picture of the atom, with continuous matter spread out around the nucleus according to its wave function, could replace the older models of quantum theory. But the discussions with Bohr led to the conclusion that this picture could not even explain Planck's law. It was extremely important for the interpretation to say that the eigenvalues of the Schrödinger equation are not only frequencies—they are actually energies.

In this way, of course, one came back to the idea of quantum jumps from one stationary state to the other, and Schrödinger was very dissatisfied with this result of our discussions. But even when we knew this and accepted the quantum jumps, we did not know what the word "state" could mean. One could of course try—and that was tried very soon—to see whether one can describe the path of the electron through a cloud chamber by means of Schrödinger's wave mechanics. It turned out that this was not possible. In its initial position the electron could be represented by a wave packet. This wave packet would move along and thereby one got something like the path of the electron through the cloud chamber. But the difficulty was that this wave packet would become bigger and bigger so that, if the electron just ran long enough, it might have a diameter of one centimetre or more. This is certainly not what we see in the experiments and so this picture again had to be abandoned. In this situation of course we had many discussions, difficult discussions, because we all felt that the mathematical scheme of quantum or wave mechanics was already final. It could not be changed and we would have to do all our calculations from this scheme. On the other hand, nobody knew how to represent in this scheme such a simple case as the path of an electron through a cloud chamber. Born had made a first step by calculating from Schrödinger's theory the probability for collision processes; he had introduced the notion that the square of the wave function was not a charge density as Schrödinger had believed, that it meant the probability to find the electron at the given place.

Then there came the transformation theory by Dirac and Jordan. In this scheme one could transform from $\psi(q)$ to for instance $\psi(p)$, and it was natural to assume that the square $|\psi(p)|^2$ would be the probability to find the electron with momentum p. So gradually one acquired the notion that the square of the wave function, which by the way was not the wave function in three-dimensional space but in configuration space, meant the probability for something. With this knowledge we returned to the electron in the cloud chamber. Could it be that we had asked the wrong question? I remembered Einstein telling me, "It is always the theory which decides what can be observed." And that meant, if it was taken seriously, that we should not ask: "How can we represent the path of the electron in the cloud chamber?" We should ask instead: "Is it not perhaps true that in nature only such situations occur which can be represented in quantum mechanics or wave mechanics?"

Turning around the question, one saw at once that this path of an electron in a cloud chamber was not an infinitely thin line with well-defined positions and velocities; actually the path in the cloud chamber was a sequence of points which were not too well-defined by the water droplets, and the velocities were not too well-defined either. So I simply asked the question: "Well, if we want to know of a wave packet both its velocity and its position what is the best accuracy we can obtain, starting from the principle that only such situations are found in nature which can be represented in the mathematical scheme of quantum mechanics?" That was a simple mathematical task, and the result was the principle of uncertainty which seemed to be compatible with the experimental situation. So finally one knew how to represent such a phenomenon as the path of the electron, but again at a very high price. Namely, this interpretation meant that the wave packet representing the electron is changed at every point of observation, that is at every water droplet in the cloud chamber. At every point we get new information about the state of the electron; therefore we have to replace the original wave packet by a new one, representing this new information.

The state of the electron thus represented does not allow us to ascribe to the electron in its orbit definite properties like coordinates, momentum, and so on. What we can do is only to speak about the probability to find, under suitable experimental conditions, the electron at a certain point or to find a certain value for its velocity. So finally we have come to a definition of state which is much more abstract than the original electronic orbit. Mathematically we describe it by a vector in Hilbert space, and this vector determines probabilities for the results of any kind of experiments which can be carried out at this state. The state may change by every new information.

This definition of state made a very big change or, as Dirac has said, a big jump, in the description of natural phenomena, and I doubt whether the unwillingness of Einstein, Planck, von Laue, and Schrödinger to accept it should be reduced simply to prejudices. The word Prejudice is too negative in this context and does not cover the situation. It is of course true that Einstein, for instance, thought it must necessarily be possible to give a kind of objective description of the state of affairs, the state of an atom, in the same sense as that had been possible in older physics. But it was indeed extremely difficult to give up this notion because all our language is bound up with this concept of objectivity. So all the words which we use in physics in describing experiments, like the words measurement or position or energy or temperature and so on, are based on classical physics and its idea of objectivity. The statement that such an objective description is not possible in the world of the atoms, that we can only define a state by a direction in Hilbert space—such a statement was indeed very revolutionary; and I think it is really not so strange that many physicists of that time simply were not willing to accept it.

I had a discussion with Einstein about this problem in 1954, a few months before his death. It was a very nice afternoon that I spent with Einstein but still when it came to the interpretation of quantum mechanics I could not convince him and he could not convince me. He always said "Well, I agree that any experiment the results of which can be calculated by means of quantum mechanics will come out as you say, but still such a scheme cannot be a final description of Nature."

Now we come to the third concept I wanted to discuss, the concept of the elementary particle. Before the year of 1928, every physicist knew what one meant by an elementary particle. The electron and the proton were the obvious examples, and at that time we would have liked simply to take them as point charges, infinitely small, just defined by their charge and their mass. We had to agree reluctantly that they must have a radius, since their electromagnetic energy should be finite. We did not like the idea that such objects should have properties like a radius but still we were happy that at least they seemed to be completely symmetrical like a sphere. But then the discovery of the electronic spin changed this picture considerably. The electron was not symmetrical. It had an axis and this result emphasized that perhaps such particles have more than one property and that they are not simple, not so elementary as we had thought before. The situation again changed completely in 1928, when Dirac developed the relativistic theory of the electron and discovered the positron. A new idea cannot be quite clear from the beginning. Dirac thought first that the negative energy holes of his theory could be identified with the protons; but later it became clear that they should have the same mass as the electron, and finally they were discovered in the experiments and were called positrons. I think that this discovery of antimatter was perhaps the biggest jump of all the big jumps in physics of our century. It was a discovery of utmost importance because it changed our whole picture of matter. I would like to explain this in more detail in the last part of my talk.

First Dirac suggested that such particles can be created by the process of pair production. A light quantum can lift a virtual electron from one of these negative energy states in the vacuum to a higher positive energy and that means that the light quantum has created a pair of electron and positron. But this meant at once that the number of particles was not a good quantum number anymore; there was no conservation law for the number of particles. For instance, according to Dirac's new idea one could say that the hydrogen atom does not necessarily consist of proton and electron. It may also temporarily consist of one proton, two electrons, and one positron. And actually when one takes the finer details of quantum electrodynamics into account these possibilities do play some role.

In the case of interaction between radiation and electron such phenomena as pair production can happen. But then it was natural to assume that similar processes may occur in a much wider range of physics. We knew since 1932 that there are no electrons in the nucleus, that the nucleus consists of protons and neutrons. But then Pauli suggested that beta decay could be described by saying that an electron and a neutrino are being created in beta decay. This possibility was formulated by Fermi in his theory of beta decay. So you see that already at that time the law of conservation of particle number was completely abandoned. One understood that there are processes in which particles are created out of energy. The possibility for such processes was of course given already by the theory of special relativity, energy being transmuted into matter. But its reality occurred for the first time in connection with Dirac's discovery of antimatter and of pair creation.

The theory of beta decay was published by Fermi in 1934. A few years later, in connection with cosmic radiation, we asked the question "what happens if two elementary particles collide with very high energy?" The natural answer was that there was no good reason why one should not have many particles created in such an act. So actually the hypothesis of multiple production of particles in high-energy collisions was a very natural assumption after Dirac's discovery. It was checked experimentally only fifteen years later when one studied very-high-energy phenomena and could observe such processes in the big machines. But when one knew that at very-high-energy collisions any number of particles can be created under the only condition that the initial symmetry should be identical with the final symmetry, then one had also to assume that any particle was really a complicated compound system, because with some degree of truth one can say that any particle consists virtually of any number of other particles. Of course we would still agree that it may be a reasonable approximation to consider a pion composed only of nucleon and antinucleon and we should not consider higher compositions. But that is only an approximation, and if we have to speak rigorously then we should say that in any one pion we have a number of configurations of several particles up to an arbitrarily high number of particles if only the total symmetry is the same as the symmetry of the pion. So it was one of the most spectacular consequences of Dirac's discovery that the old concept of the elementary particle collapsed completely. The elementary particle was not elementary any more. It is actually a compound system, rather a complicated many-body system, and it has all the complications which a molecule or any other such object really has.

There was another consequence of Dirac's theory which is important. In the old theory, let us say in nonrelativistic quantum theory, the ground state was an extremely simple state. It was just the vacuum, the empty world, nothing else, and had therefore the highest possible symmetry. In Dirac's theory the ground state was different. It was an object which was filled with particles of negative energy that could not be seen. Besides that, if the process of pair production is introduced one should expect that the ground state must contain probably an infinite number of virtual pairs of positrons and electrons or of particles and antiparticles; so you see at once that the ground state is a complicated dynamical system. It is one of the eigensolutions defined by the underlying natural law. If the ground state is to be interpreted in this way, one can further see that it need not be symmetrical under the groups of the underlying natural law. In fact the most natural explanation of electrodynamics seems to be that the underlying natural law is completely invariant under the isospin group while the ground state is not. The assumption that accordingly the ground state is degenerate under rotations in isospace enforces the existence of long range forces or of particles with rest mass zero following a theorem of Goldstone. Coulomb interaction and photons should probably be interpreted in this way.

Finally Dirac—in consequence of his theory of holes —in his Bakerian lecture in 1941 propagated the idea that, in a relativistic field theory with interaction, use should be made of a Hilbert space with indefinite metric. It is still a controversial question, whether this extension of conventional quantum theory is really necessary. But after many discussions during the last decades, one cannot doubt that theories with indefinite metric can consistently be constructed and can lead to a reasonable physical interpretation.

So the final result at this point seems to be that Dirac's theory of the electron has changed the whole picture of atomic physics. After abandoning the old concept of the elementary particles, those objects which had been called "elementary particles" have now to be considered as complicated compound systems and should be calculated some day from the underlying natural law, in the same way as the stationary states of complicated molecules should be calculated from quantum or wave mechanics. We have learned that energy becomes matter when it takes the form of elementary particles. The states called elementary particles are just as complicated as the states of atoms and molecules. Or to formulate it paradoxically: Every particle consists of all other particles. Therefore we cannot hope that elementary particle physics can ever be simpler than quantum chemistry. This is an important point because even now many physicists hope that some day we might discover a very simple way to describe elementary particle physics like the hydrogen spectrum in the old times. This, I think, is not possible.

In conclusion I would like again to say a few words about what had been called the "prejudices." You may say that our belief in elementary particles was a prejudice. But again I think that would be too negative a statement, because all the language which we have used in atomic physics in the last 200 years is based directly or indirectly on the concept of the elementary particle. We have always asked the question: "Of what does this object consist and what is the geometrical or dynamical configuration of the smaller particles in the bigger object?" Actually we have always gone back to this philosophy of Democritus; but I think we now have learned from Dirac that this was the wrong question. Still it is very difficult to avoid questions which are already part of our language. Therefore it is natural that even nowadays many experimental physicists—even some theoreticians—still look for *really* elementary particles. They hope for instance that the quarks, if they existed, could play this role.

I think that this is an error. It is an error because even if the quarks would exist we could not say that the proton consists of three quarks. We would have to say that it may temporarily consist of three quarks, it may also temporarily consist of four quarks and one antiquark, or five quarks and two antiquarks, and so on. And all these configurations would be contained in the proton; and again one quark would be composed of two quarks and one antiquark and so on. So we cannot avoid this fundamental situation; but since we still have the questions from the old concepts, it is extremely difficult to stay away from them. Very many physicists have looked for the quarks and will probably do so in the future. There has been a very strong prejudice in favour of quarks in the last ten years, and I think they would have been found if they existed. But that is a matter to be decided by the experimental physicist.

There remains the question: What then has to replace the concept of a fundamental particle? I think we have to replace this concept by the concept of a fundamental symmetry. The fundamental symmetries define the underlying law which determines the spectrum of elementary particles. Now I will not go into a detailed discussion of these symmetries. From a careful analysis of the observations I would conclude, that besides the Lorentz group also SU_2, the scaling law, and the discrete transformations P, C, T are genuine symmetries. But I would not include SU_3 or higher symmetries of this type among the fundamental symmetries; they may be produced by the dynamics of the system as approximate symmetries.

But this is again a matter which should be decided by the experiments. I only wanted to say that what we have to look for are not fundamental particles, but fundamental symmetries. And when we have actually made this decisive change in concepts which came about by Dirac's discovery of antimatter, then I do not think that we need any further breakthrough to understand the elementary—or rather nonelementary—particles. We must only learn to work with this new and unfortunately rather abstract concept of the fundamental symmetries; but this may be bad enough.

*Reprinted by permission from *The Physicist's Perception of Nature*, edited by J. Mehra (Reidel, Dordrecht, Holland, 1973).

It is nothing short of a miracle that the modern methods of instruction have not yet entirely strangled the holy curiosity of inquiry; for this delicate little plant, aside from stimulation, stands mainly in need of freedom; without this it goes to wrack and ruin without fail.

—Albert Einstein, quoted in W.H. Cropper, *The Quantum Physicists* (Oxford University, London, 1970)

Early Work in Electron Diffraction*

SIR GEORGE THOMSON
Corpus Christi College, Cambridge University, Cambridge, England

(Received June 26, 1961)

Both the work of de Broglie and a number of scattering experiments were forerunners of the two pieces of work that established the existence of electron refraction. The history of the experiments is discussed, as are some of the difficulties, both experimental and conceptual, which had to be overcome before de Broglie's hypothesis could be verified.

IT is difficult for a young physicist, and by a young physicist I am afraid I have to mean one under 50, to realize the state of our science in the early 1920's. The classical theory of light still held, supported strongly by the success of x-ray diffraction for crystal analysis. On the other hand, Planck's theory of radiation was accepted, the abnormalities for specific heats at low temperature were obviously part of the same story, and the recent work of Arthur Compton, of which we have just heard, had shown that x rays sometimes at least behaved like particles. Then above all, the detailed, even if slightly imperfect, successes of the atomic theory of Niels Bohr and Sommerfeld showed that all was not well even with that ancient foundation of science, Newtonian mechanics. It seemed contrary to logic. It was not just that the old theories of light and mechanics had failed. On the contrary, you could say that they had succeeded in regions to which they could hardly have been expected to apply, but they succeeded erratically. The statement of the elder Bragg at this period has been quoted very often, but so very well sums up the situation that I shall inflict it upon you again: light behaves like waves on Mondays, Wednesdays, and Fridays, like particles on Tuesdays, Thursdays, and Saturdays, and like nothing at all on Sundays. And over the whole subject brooded the great mystery of h.

At that time we were all thinking of possible ways of reconciling the apparently irreconcilable. One of these ways was supposing light to be perhaps particles after all, but particles which somehow masqueraded as waves; but no one could give any clear idea as to why this was done. The first suggestion I ever heard which did not stress most of all the behavior of the radiation came from the younger Bragg, Sir Lawrence Bragg, who once said to me that he thought the electron was not so simple as it looked, but he never followed up this idea. However, it made a considerable impression on me, and it pre-disposed me to appreciate de Broglie's first paper in the Philosophical Magazine of 1924, which was taken from the thesis of his University of Paris doctorate. I was sufficiently impressed by de Broglie's theory to publish a note on it in the July issue of the same journal in 1925. This note is an example of a thoroughly bad theoretical paper, and I only mention it to show that I did in fact take some notice of de Broglie's paper when it was first published. There was some talk of this theory at the meeting of the British Association held in August, 1926, and I began then to think about the waves producing diffraction effects. A German, Dr. Elsasser, had published a very short note on this point a bit before. I had not seen this nor, as far as I can recall, had it been mentioned, but it is possible, and I think not unlikely that some of the people whom I had heard talking about this, and with whom I had spoken, had in fact read this note.

C. J. Davisson was at this meeting, but I do not recall any conversation, certainly not on this subject. His discovery of electron diffraction arose from a study of the scattering of electrons and the production of secondary electrons from metals. It was part of a more or less routine study of the behavior of electrons colliding with metals.

The first paper on nickel, with Kunzman, appeared in *Science* of November, 1921. It described some surprisingly sharp peaks of reflected electrons which Davisson considered could be explained by scattering, classical of course, from

* Paper delivered as part of a program on "Topics in the History of Modern Physics" on February 3, 1961, at a joint session of the American Physical Society and the American Association of Physics Teachers, during their annual meetings in New York City.

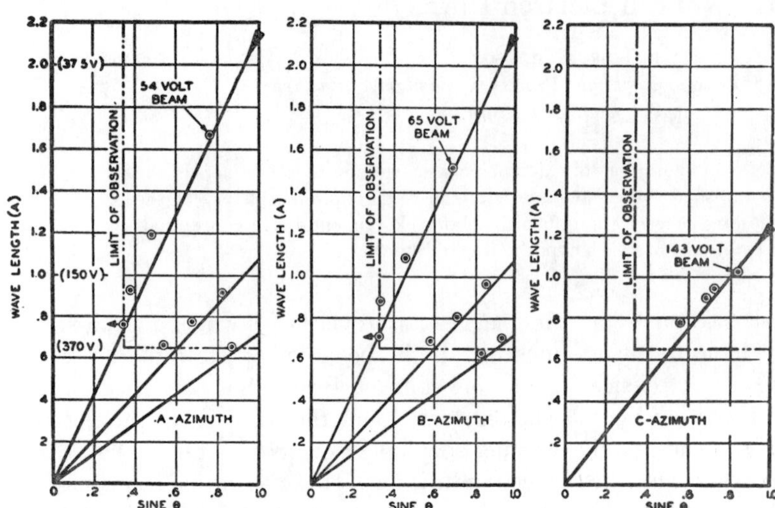

FIG. 1. Functional relations between wavelength and scattering angle.

two shells of electrons. In 1925 after studying some other metals, he came back to nickel. Until then the experiments had all been on ordinary polycrystalline metals, engineering metals, so to speak. But by an accident a specimen became oxydized and in restoring the surface by heating, it re-crystallized into a very small number of largish crystals, virtually a single crystal. There was a complete alteration in the pattern of scattering, and the former theory had to be abandoned. Davisson was inclined at first to replace it by one of transparent directions, as he called them, in the lattice, but partly, I believe, as a result of the visit to the British Association meeting at Oxford he came to believe that the effect was connected with the theories of de Broglie and Schrödinger. Schrödinger had by that time taken up the de Broglie theory and turned it into more conventional mathematics. This was the theory that any particle—electron, photon, or any other—had essentially linked up in it wave properties, and that the relation between the wavelength and the momentum was given by the famous formula "wavelength is equal to h divided by the momentum." As a result of careful research, Davisson and Germer got the first electronic beams in positions which could be explained in this way on January 6, 1927. Figure 1, taken from a somewhat later paper, shows the kind of thing which they got. The experiment consisted of bombarding a target of nickel, of a single crystal of nickel cut to an octohedral face, with electrons incident normally upon it; the reflected electrons were then analyzed. Only the ones of 90% energy or more were recorded, and a plot was made between the angle at which they came off and what is effectively the de Broglie wavelength, though expressed here as the square root of the volts. These straight lines represent, with no adjustable constant, the calculated lines on which first, second, and third order reflections should occur, and you will notice the very close agreement (Fig. 1). This agreement was obtained without adjustable constants, it was extremely good, and the only difference was that the positions of the points along the lines were not quite what would be expected. The reason for this is that with the rather low energy electrons that Davisson used (order of a few hundred volts) the so-called inner potential of the metal, which is the order of 25 v, comes in as quite an appreciable distortion. This was not fully understood at the time. It is not, I think, always realized nowadays what a supreme experimental feat these early experiments were. Slow electrons are most difficult to control, and if the result is to be of any value, the vacuum has to be what would still be considered a pretty good vacuum and what was then quite outstanding. In fact, very few people since that time have succeeded in this field of slow electron diffraction, though there are some signs that it is coming back. Davisson's results were first published in a letter to *Nature* in April, 1927.

Very shortly, I believe in a matter of a few days, after the Oxford meeting in 1926, I was visiting (as I very frequently did) at the Cavendish Laboratory, though I was then working at Aberdeen, and I saw some experiments on the scattering of electrons in helium. These showed certain effects, particularly directions of strong emission, which recalled to me the halos one gets in a Young's eriometer. This device was a class experiment which I had done for Part II of the Natural Sciences Tripos not so very many years before, and I remember my feeling of annoyance that I had not had the sense to do the experiment immediately. Now by a curious irony it turned out later that the results I had seen were quite erroneous and entirely instrumental in origin, so I think it is not fair to tell you the name of the person who, in fact, obtained them. They were withdrawn though, as a matter of fact, helium can, in suitable conditions, give effects of this kind.

When I returned to Aberdeen I thought, "Well, it has apparently been done with a gas, but let's try it on solids." So I put a young research student named Andrew Reid on to the problem, using thin filaments of celluloid. Figure 2 shows the apparatus (not quite what he used, but one that I used later on). It was an extremely easy experiment to do because we had just been working for the last few years on the scattering of positive rays, and practically all we had to do was to disconnect one terminal of the induction coil which supplied the voltage and exchange it for the other, reversing the polarity. The cathode rays go through a thin narrow tube B, replaced later on by two holes, and then they strike on the specimen (C), a thin film through which they are transmitted. They then can strike the screen (E) where the fluorescence can be observed, and the photographic plate (D) can be lowered for a permanent record. Almost at once Reid got photographs apparently showing diffuse halos. I was a little sceptical as photographs of scattered positive rays which we had been working on had apparently shown sharp edges which on examination merely turned out to be contrast effects. However, the photometer confirmed the reality of the electronic halos. They were what one might expect from molecules of definite size oriented at random, and the diameter varied with the energy of the electrons in the way to be expected on the de Broglie formula. The size was also about what one would expect for a long-chain carbon compound, but that of course was only rough. Figure 3 shows the best we ever got of the celluloid rings. In fact, it is so much better than most of the others that I think the celluloid must have recrystallized in some special form—the others were much fuzzier. We published a note in *Nature* under our joint names in June, 1927, two or three months after Davisson and Germer's first publication. The structure of celluloid was then unknown, and for a proper test it was necessary to use crystal diffraction by a known lattice. Now metals suggested themselves; their structure was well-known and simple, but this was before the techniques of sputtering and evaporation for making very thin films had been developed, or at least before I knew how to do it. The films which I used, this time working by myself, were of gold, aluminum, and platinum and were prepared by the chief mechanic of the laboratory, the late Mr. C. G. Frazer, to whom the success of the experiment is entirely owing, because this is the difficult part. He was a mechanic of the old school; he had been trained as a watchmaker, and he had that astonishing gift (which later came out) of being able to produce a complicated and perfectly working piece of apparatus after one had just talked to him and made some scribbles on dirty pieces of paper on his bench. Those of you who have had the immense good fortune

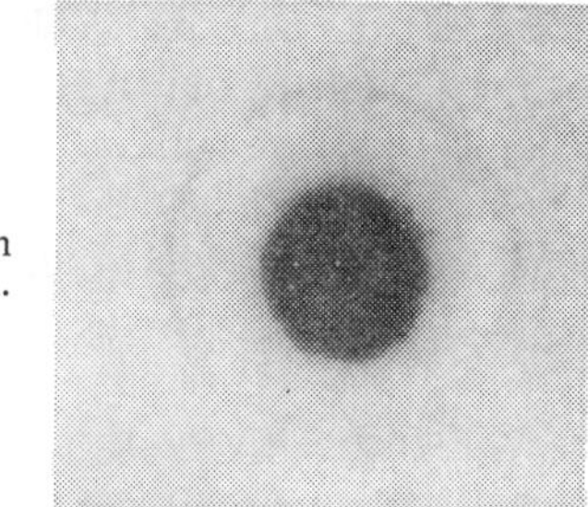

FIG. 3. Diffraction pattern from celluloid.

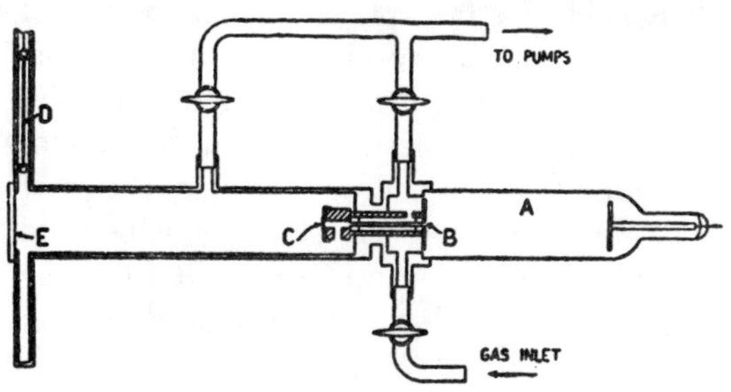

FIG. 2. Reid's apparatus for scattering from celluloid.

to work with such a man will know that, whatever may be true of wives, the price of such mechanics is above rubies.

The first films we tested were of gold, and they at once gave rings which agreed quantitatively and qualitatively with the theoretical expectation for a face-centered cubic lattice of side 4.06 A. Figure 4 shows the actual photographs. You will notice that there are two, put side by side, with different voltages as measured by the spark gap, the equivalent spark gap of the induction coil. You see the larger rings associated with the lower energy.

These three metals all agreed in detail with the predictions of de Broglie. That is to say, the rings were all of the right sizes, both in relative size as agreeing with the structure of the metal and in absolute size as calculated from de Broglie's formula. I published a note in *Nature* in December, 1927 on the platinum ones, and the Royal Society paper on the experiments with gold and aluminum appeared in February, 1928—it was two or three months later than Davisson and Germer's paper which appeared in the *Physical Review* late in 1927. The objection was raised to these experiments that they might in some way (it was not indeed explained how) be connected with the production of Bremsstrahlen, that is x rays, by the collision of the electrons with the films. I was able to disprove this in a second paper by deflecting the electrons after they had passed through the films and before they had reached the photographic plate. I simply put on a magnetic field and showed that the electrons which formed the rings (the diffraction rings as I called them) were equally bent with the quite numerous electrons that had gone through holes in the film which, although carefully made, was so heavily pitted with holes that nine-tenths of the electrons went straight through. That disposed of the view that they could be due to x rays, which would not, of course, have been deflected.

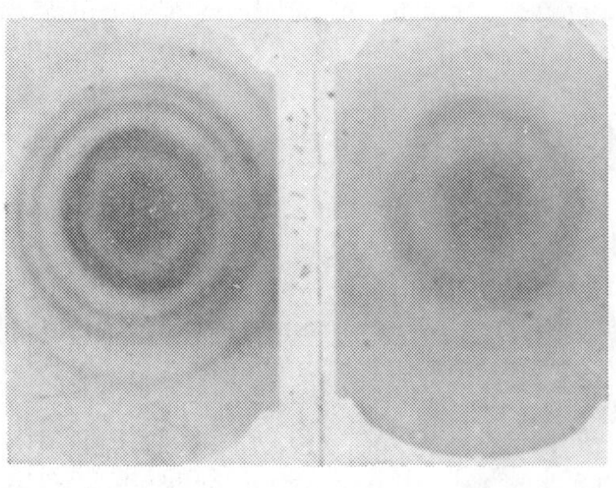

(a) (b)

FIG. 4 Diffraction patterns from gold. Spark gap: (a) 20.7 mm; (b) 10.7 mm.

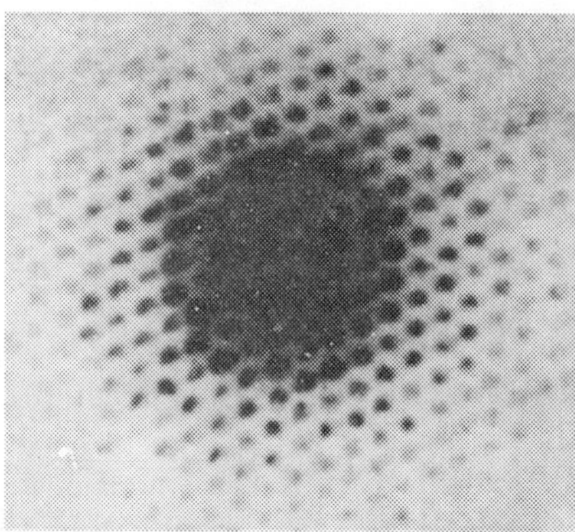

FIG. 5. Cross-grating pattern from mica.

We then started work with spluttered films, for which I got a most valuable hint of technique by word of mouth from Joliot. These electrons were in the range of 20–60 kv, a range that can easily be produced by an induction coil, an induction coil somewhat smoothed with a not very large condenser, which was all I had at the time.

This was followed up by work by Kikuchi in Japan, who used mica films made by cleavage of the mica, and showed that they, too, in transmission gave a similar effect. Figure 5 is a reproduction taken from Kikuchi's paper which was published in 1928 and shows the cross-grating patterns you get from the very thin mica film. Rather thicker films produced rings of electrons for which the third condition of interference is satisfied as well as those for the cross-grating, and also pairs of light and dark parallel lines which Kikuchi justly explained as the results of double scattering of the electrons in the films. About the same time Nichikawa and Kikuchi, using a technique somewhat similar to that of Davisson, first got the diffraction by reflection of fast electrons, that is to say electrons which did not pierce the specimen but came out from the same side of it as that from which they had entered.

Later on in 1929, Frazer produced the first

good reflection camera, which enabled a specimen to be set at suitable small angles of the order of a degree to the beam of electrons, be rotated in its own plane, also rotated so as to alter the angle of incidence, and be withdrawn and put back into place. This, with some modification, I believe has been the model for those that have followed.

Early Work on the Positron and Muon*

C. D. Anderson
California Institute of Technology, Pasadena, California
(Received August 4, 1961)

Early work on the positron and muon is described in an informal way, with emphasis on those aspects of the work which normally would not find their way into the literature.

TODAY I plan to discuss briefly two discoveries with which I have had some association, the discovery of the positron and of the muon.

First, with respect to the positron, it has often been stated in the literature that the discovery of the positron was a consequence of its theoretical prediction by Dirac, but this is not true. The discovery of the positron was wholly accidental. Despite the fact that Dirac's relativistic theory of the electron was an adequate theory of the positron, and despite the fact that the existence of this theory was well known to nearly all physicists, it played no part whatsoever in the discovery of the positron.

The aim of the experiment that led to the discovery of the positron was simply to measure directly the energy spectrum of the secondary electrons produced in the atmosphere and other materials by the incoming cosmic radiation which at that time (1930) was thought to consist primarily of a beam of photons or gamma rays of energies of the order of several hundred millions of electron volts. Although there was no experimental evidence as to the detailed interactions between such a beam of high-energy photons and matter, it was presumed from experiments at lower energies that the dominant mechanism would be the production of high-energy secondary electrons by the Compton process, the discovery of which we have just heard discussed this afternoon by Professor Compton.

The apparatus employed was planned in 1930 by Professor Robert A. Millikan and myself and consisted of a cloud chamber operated in a strong magnetic field capable of producing measurable curvatures of electrons up to energies of a few billion electron volts.

The first result of the experiment was to show that the Compton process did not play an important role in the absorption of cosmic radiation, but that instead some new processes, presumably of a nuclear type, were operative. This was brought out by the fact that about half of the high-energy cosmic-ray particles observed were positively charged and therefore could not represent Compton electrons. At the time they were presumed to be protons resulting perhaps from photo-nuclear disintegrations. It was, of course, important to provide unambiguous identification of these unexpected particles of positive charge, and this could best be done by gathering whatever information was possible on the mass of the particles, inasmuch as the photographs clearly showed that in all cases these particles carried a single unit of electric charge. Experimental conditions were such that no information as to the particle's mass could be ascertained except in those cases in which the particle's velocity was appreciably smaller than the velocity of light, and this was true for only a small fraction of the events.

A few of the low-velocity particles were clearly

* Paper delivered as part of a program on "Topics in the history of modern physics" on February 3, 1961, at a joint session of the American Physical Society and the American Association of Physics Teachers during their annual meetings in New York City.

identified as protons. As more data were accumulated, however, a situation began to develop which had its awkward aspects in that practically all of the low-velocity cases were particles whose mass seemed to be too small to permit their interpretation as protons. The alternative interpretations in these cases were that these particles were either electrons (of negative charge) moving upward or some unknown light-weight particles of positive charge moving downward. In the spirit of scientific conservatism we tended at first toward the former interpretation, i.e., that these particles were upward-moving negative electrons. This led to frequent and at times somewhat heated discussions between Professor Millikan and myself, in which he repeatedly pointed out that everyone knows that cosmic-ray particles travel downward, and not upward, except in extremely rare instances, and that therefore, these particles must be downward-moving protons. This point of view was very difficult to accept, however, since in nearly all cases the specific ionization of these particles was too low for particles of proton mass.

To resolve this apparent paradox a lead plate was inserted across the center of the chamber in order to ascertain the direction in which these low-velocity particles were traveling and to distinguish between upward-moving negatives and downward-moving positives.

It was not long after the insertion of the plate that a fine example was obtained in which a low-energy light-weight particle of positive charge was observed to traverse the plate, entering the chamber from below and moving upward through the lead plate. Ionization and curvature measurements clearly showed this particle to have a mass much smaller than that of a proton and, indeed, a mass entirely consistent with an electron mass. Curiously enough, despite the strong admonitions of Dr. Millikan that upward-moving cosmic-ray particles were rare, this indeed was an example of one of those very rare upward-moving cosmic-ray particles.

Soon additional cases of light-weight positive particles traversing the plate were observed, and in addition events in which several particles were simultaneously emitted from a common source were observed. Clearly in both types of cases the direction of motion was known, and it was therefore possible to identify the presence of several more light-weight positive particles whose mass was consistent with that of an electron but not with that of a proton.

After the existence of positrons was clearly indicated, the question naturally arose as to how they came into being. Just what was the mechanism responsible for their production? Naturally one would look to the Dirac theory to provide this explanation.

It was not immediately obvious to me, however, as to just what the detailed mechanism was. Did the positrons somehow acquire their positive charge from the nucleus? Could they be ejected from the nucleus when there were presumably no positrons present in the nucleus? The idea that they were created out of the radiation itself did not occur to me at that time, and it was not until several months later when Blackett and Occhialini suggested the pair-creation hypothesis that this seemed the obvious answer to the production of positrons in the cosmic radiation. Blackett and Occhialini suggested the pair-production hypothesis in their paper published in the spring of 1933, in which they reported their beautiful experiments on cosmic rays using the first cloud chamber which was controlled by Geiger counters.

Soon after this, experiments in which gamma rays were used showed that a pair of electrons, one positive and one negative, could be created in the coulomb field of a nucleus in such a way that the energy required to create the mass of the pair $2\ mc^2$, and their kinetic energies as well, were supplied by the incident radiation, thus giving quantitative support to the pair-creation hypothesis.

The positron thus represents the first example of a particle consisting of antimatter. It is now generally believed that all particles have their corresponding antiparticles, and, in fact, several have recently been identified, including the anti-proton and antineutron.

Progress on the theoretical side has continued, and a theory in the form of quantum-electrodynamics has been developed out of the Dirac theory by Feyman, Tomonaga, Schwinger, and others. It describes the interactions between charged particles and the electromagnetic field in a highly satisfactory manner. Also, calculations on the basis of the Dirac theory in terms of positron-electron pairs, which helped explain certain phenomena in the cosmic radiation such as the large cascade showers, were carried out by Oppenheimer, Carlson, Plesset, and others.

If one goes back a few years, say to just after the Dirac theory was announced, it is interesting then to speculate on what a sagacious person working in this field might have done. Had he been working in any well equipped laboratory, and had he taken the Dirac theory at face value he could have discovered the positron in a single afternoon. The reason for this is that the Dirac theory could have provided an excellent guide as to just how to proceed to form positron-electron pairs out of a beam of gamma-ray photons. History did not proceed in such a direct and efficient manner, probably because the Dirac theory, in spite of its successes, carried with it so many novel and seemingly unphysical ideas, such as negative mass, negative energy, infinite charge density, etc. Its highly esoteric character was apparently not in tune with most of the scientific thinking of that day. Furthermore, positive electrons apparently were not needed to explain any other observations. Clearly the proton was the fundamental unit of positive charge, and the electron the corresponding unit of negative charge. This kind of thinking prevented most experimenters from accepting the Dirac theory wholeheartedly and relating it to the real physical world until after the existence of the positron was established on an experimental basis.

The discovery of the positron is also an example of a situation which is so often present in physics, in which the same discovery is made, or could easily have been made, in experiments simultaneously underway but carried out for quite different purposes. One such example is the famous experiment of Bothe and Becker in which a light nucleus such as Be was bombarded by α particles from a radioactive source. This experiment was first performed in 1930 by Bothe and Becker and later repeated by a number of investigators. As was shown later, this single simple experiment produced neutrons, positrons, and induced radioactivity. Had the positron not been discovered in cosmic-ray research, it would undoubtedly soon have been found in continuing studies of the experiment just described.

Concerning the muon, we know that the muon was the first particle to be discovered which has a mass between that of an electron and a proton. Originally it was known as the mesotron and later as the mumeson. More recent work on its properties has shown it is more appropriate not to classify it as a meson-type particle, and the suggestion has been made therefore to call it a muon. The term meson will be reserved for particles of intermediate mass which have a strong interaction with nuclei.

The muon, as I said, was the first particle of mass between a proton and an electron to be discovered. We now know, of course, from later experiments that a dozen or so such intermediate mass particles exist, including the pi mesons and a class of so-called strange particles called *k* particles. I shall later speak briefly about these more recently discovered particles.

The discovery of the muon, unlike that of the positron, was not sudden and unexpected. Its discovery resulted from a two-year series of careful, systematic investigations all arranged to follow certain clues and to resolve some prominent paradoxes which were present in the cosmic rays.

The gist of the matter was as follows. Professor Seth H. Neddermeyer and I were continuing the study of cosmic-ray particles using the same magnet cloud chamber in which the positron was discovered. In these experiments it was found that most of the cosmic-ray particles at sea level were highly penetrating in the sense that they could transverse large thicknesses of heavy materials like lead and lose energy only by the directly produced ionization which amounted to something like 20 million ev per cm of lead. A principal aim of the experiments was to identify these penetrating cosmic-ray particles. They had unit electric charge and were therefore presumably either positive or negative electrons or protons, the only singly charged particles known at that time.

There were difficulties, however, with any interpretation in terms of known particles, as was pointed out as early as 1934 in a paper presented to the International Conference on Physics held in London of that year.

The most important objection to their interpretation as protons lay in the fact that the energy of the electron secondaries produced by the direct impact of these particles as observed in a cloud chamber contained too many secondaries of high energy to correspond with the energy spectrum to be expected if particles as massive as protons were producing these secondaries. On the other hand the spectrum was just that to be expected if the particles producing the secondaries were much lighter than protons. Furthermore, to interpret these particles as protons would mean *assuming the existence of protons of negative charge* since these sea-level particles occurred equally divided between negative and positive charges, and at that time there was no evidence for the existence of protons of negative charge.

There were difficulties also in interpreting these sea-level penetrating cosmic-ray particles as positive and negative electrons. The most important objections to their being electrons arose from three considerations. Firstly, theoretical calculations by Bethe, Heiter, and Sauter on the energy loss of electrons led to the conclusion that high-energy electrons should lose large amounts of energy through the production of radiation, which the penetrating particles in question were observed not to do. Secondly, we had found individual cases of electrons which did, in fact, show large energy losses through radiation, in some cases 100 million ev or more per cm of lead. Clearly in these cases the electrons showed a behavior quite different from that of the penetrating particles. And thirdly, the so-called highly absorbable component of the cosmic rays and the existence of electron showers could find an appealing explanation in terms of electrons if electrons did, in fact, suffer large radiation losses at high energies as demanded by the above mentioned theory.

This then was the situation in 1934 in which the sea-level penetrating particles had this paradoxical behavior. They seemed to be neither electrons nor protons. We tended, however, to lean toward their interpretation as electrons and "resolved" the paradox in our informal discussions by speaking of green electrons and red electrons—the green electrons being the penetrating type, and the red the absorbable type which lost large amounts of energy through the production of radiation.

Evidence of an entirely new type was soon obtained. In experiments carried out on the summit of Pikes Peak in 1935 a number of cases of cosmic-ray produced nuclear disintegrations were observed from which many protons were ejected, but showing also in a few cases particles which, from ionization and curvature measurements, were lighter than protons and heavier than electrons. These observations were not conclusive evidence in themselves for the existence of a new type of particle, but they did tend to lend support to this assumption in view of the other difficulties involved in interpreting the data in terms of known particles.

The next year or so brought further evidence on all the above points and only tended to strengthen the paradox further. The hypothesis that the penetrating particles were protons was further weakened by the observation of many cases of particles which did not suffer appreciable radiation collisions and still which could not be as massive as protons, as evidenced by the ionization-curvature relations of their cloud-chamber tracks. These cases could, however, be interpreted as electrons, but only if electrons ceased to radiate appreciably above a certain energy, such as say a hundred million ev.

The crux of the matter then was whether or not electrons above a certain energy did or did not experience a large energy loss through radiative impacts. In other words the paradoxical character of our data could be reduced if one assumed that the Bethe-Heiter theory, although correct for electrons of energies below a few hundred million ev, in some way became invalid for electrons of higher energy, thus permitting high-energy electrons to have a much greater penetrating power and thus perhaps to permit interpreting the highly penetrating sea-level cosmic-ray particles as positive and negative electrons.

To test this hypothesis we inserted a bar of platinum across our cloud chamber and found that the cosmic-ray particles divided themselves into two groups, one highly absorbable and one highly penetrating in the *same energy range*. Thus the last possibility to explain the data in terms of known particles was removed, and it was necessary to assume the existence of positive and negative particles of unit charge but of a mass intermediate between that of a proton and an electron. These are the particles which are now called muons.

At about this time Street and Stevenson reported an experiment in which they observed particles which were lighter than protons and more penetrating than electrons should be if the Bethe-Heiter theory were valid. These they interpreted as particles of a new type, but this interpretation followed only if the Bethe-Heiter theory were assumed not to break down at the energies concerned in their experiment.

In discussing the discovery of the muon I have not so far mentioned anything about the theoretical aspects of the situation. We saw previously how the Dirac theory predicted the existence of positrons although it played no role in their discovery. The discovery of muons similarly was based on purely experimental measurements and procedures, with no guide from any theoretical considerations.

As with the positron, this need not have been the case. For before the discovery of the muon had been finally achieved a novel idea was published in a Japanese journal by the Japanese physicist Hiedeka Yukawa. Reasoning by analogy with quantum electrodynamics, he made the suggestion that perhaps nuclear forces (which are not electromagnetic in character) could be described in terms of a particle carrier of these nuclear forces, analogous to the photon being the carrier of electromagnetic forces. Nuclear forces, however, differ from electromagnetic forces in that they possess only a short range of action. This means that if nuclear forces are described in terms of a particle carrier, this particle carrier must have a finite rest mass unlike the photon of zero rest mass which is appropriate to the long range electromagnetic forces. Yukawa estimated from the known range of nuclear forces that this carrier should have a rest mass about 200 times that of an electron.

This novel suggestion of Yukawa was unknown to the workers engaged in the experiments on the muon until after the muon's existence was established. Although Yukawa's suggestion preceded the experimental discovery of the muon, he published it in a Japanese journal which did not have a general circulation in this country. It is interesting to speculate on just how much Yukawa's suggestion, had it been known, would have influenced the progress of the experimental work on the muon. My own opinion is that this influence would have been considerable even though Dirac's theory, which was much more specific than Yukawa's, did not have any effect on the positron's discovery. My reason for believing this is that for a period of almost two years there was strong and accumulating evidence for the muon's existence, and it was only the caution of the experimental workers that prevented an earlier announcement of its existence. I believe that a theoretical idea like Yukawa's would have appealed to the people carrying out the experiments and would have provided them with a belief that maybe after all there is some need for a particle as strange as a muon, especially if it could help explain something as interesting as the enigmatic nuclear forces. Yukawa's particle at that time did seem to coincide in all respects with the known properties of the muon.

And now having said this, my next statement will be that the muon and the Yukawa particle cannot possibly be the same particle at all. The Yukawa particle was invented to explain nuclear forces. The muon, as shown by subsequent experiments, completely ignores nuclear forces and interacts with other particles appreciably only through the electromagnetic field. It was not until 10 years later that another particle of intermediate mass, the pi meson, was discovered. The pi meson does interact strongly through nuclear forces and can indeed correspond to the Yukawa particle and is so considered in present day nuclear theory.

As we all know now, the muon is observed principally as a decay product of pi mesons. Muons may, however, also be produced in other ways, by direct pair production from high-energy photons and as decay products of many of the new so-called strange particles, the *k* particles and hyperons.

The muon even today is somewhat of an oddball particle. Its existence does not in any way seem to be justified; for example, it seems to play no role whatever in explaining nuclear forces, whereas all the other more recently discovered intermediate mass particles do seem to be active participants as carriers of the nuclear forces.

It is interesting to contrast the spirit of the

times when the positron and the muon were being discovered with the spirit of the times today. Today we include some 30 particles in our list of elementary particles, and new particles seem now to be discovered almost 2 or 3 at a time. This is certainly in sharp contrast with the reluctance and the conservatism which were present when particle numbers 3–5 were being added to the list of elementary particles some years ago.

Although much new information about the properties of these 30 elementary particles is rapidly being acquired, there is much yet to be learned as far as an understanding of their role in nuclear physics is concerned, and as far as a satisfactory theoretical description of them is concerned.

Certain real progress, however, has been made in this direction through the interesting theoretical ideas of Pais, Gell-Mann, Feyman, Nishijima, Lee, Yang, and others. But a complete understanding will have to await new and better experiments and clearly new and more novel ideas.

Everyone has been hoping for and searching for a basically simple scheme of things, but this has not been achieved. Even the word "elementary" when applied to the 30 particles loses its conventional meaning. Perhaps in the future a simplification will be reached. Most of us have faith that it will. For the time being, however, I think the situation is best summed up in terms of a modern definition of the word "elementary" which I believe is due to Robert Oppenheimer. An elementary particle is something so simple that one knows nothing whatsoever about it.

Introductory note

A. O. Barut and A. A. Bartlett
Department of Physics and Astrophysics, University of Colorado, Boulder, Colorado 80309
(Received 25 October 1977; accepted 22 November 1977)

This paper is the text of an after-dinner talk given by Professor Edward U. Condon on March 8, 1969 at the close of a conference which was held at the University of Colorado to commemorate the 40th anniversary of the discovery of quantum mechanical tunneling.

The manuscript was typed by Condon himself and we have changed nothing except to correct a few obvious misprints and add the references. We believe that this paper will be of considerable value to those who are interested in the history of physics and in the interactions of science and society. It is, in part, a tribute by Condon to his colleague Ronald Gurney.

We note with pride that Ed Condon (1902–1974) and George Gamow (1904–1968) two of the independent principals in this great discovery, were members of the Faculty of the Department of Physics and Astrophysics of the University of Colorado, Boulder during the last years of their lives.

Quantum mechanical tunneling has been one of the most fruitful concepts in modern physics. It came at a time when the new quantum mechanics had been enjoying a very considerable success in explaining atomic phenomena but it was not clear at that time that quantum mechanics could be applied to nuclear phenomena. It was a very significant step forward when it was found that the basic principles of quantum mechanics could provide an understanding of the systematics of the alpha particle decays of heavy radioactive nuclei. In the years since its discovery the concept of tunneling or barrier penetration has proven to be of great importance in many branches of physics, both from a fundamental point of view and from the point of view of applications. In particular, in solid-state physics, tunneling played a major role in the works of Brian Josephson, Leo Esaki, and Ivar Giaever for which they received the 1973 Nobel Prize in Physics.

It seems appropriate to bring Professor Condon's text to the physics community as we approach the 50th anniversary of the epochal discovery of tunneling.

We express our thanks to the family of Professor Condon for giving their permission for this publication.

Tunneling—how it all started

E. U. Condon[a)]
Joint Institute for Laboratory Astrophysics and Department of Physics and Astrophysics, University of Colorado, Boulder, Colorado 80309

A personal account of the early history of tunneling and the contribution of Ronald W. Gurney.

In the talks during the day you have heard of the origin of the quantum mechanical idea of tunneling and up-to-date accounts of some of the many and various important applications which it has in atomic and nuclear and solid state physics. I will try to give you a personal anecdotal account of what I know of the early history and in particular try to give you an appreciation of the contributions of my distinguished colleague, Dr. Ronald W. Gurney.

I first became acquainted with him in Princeton in the summer of 1928 when I first went to Princeton as an assistant professor. Gurney was there from the Cavendish on some kind of post-doctoral arrangement and was just beginning to study quantum mechanics, having been mostly in experimental work on radioactivity under Lord Rutherford (not a Lord yet). I had just moved to Princeton from my first academic post as lecturer in physics at Columbia in the spring of 1928.

Two papers had just appeared that dealt with the leakage of particles through potential barriers. These were by Robert Oppenheimer and R. H. Fowler[1] and Lothar Nordheim.[2] Both had apparently independently discovered this peculiar feature of quantum mechanics of one-dimensional motion, and both had applied it to the question of pulling of electrons out of cold metals with intense electric fields, a subject that was being very much studied experimentally at that time by R. A. Millikan and C. C. Lauritsen.

Gurney had noticed these two papers in the library of the Palmer Physical Laboratory and had had the idea that maybe barrier leakage could account for the emission of alpha particles through the Coulomb barrier surrounding the nuclei of naturally radioactive elements. He had first gone to H. P. Robertson with the idea, because Robertson had already been there a year before I came and Robertson had discouraged him after a superficial examination of the matter.

Gurney was an extremely shy individual, but a persistent one, not easily put off with discouraging opinions that were not clear to him. Several weeks went by before he happened to meet me and he came one afternoon to my office and broached the same idea to me. It was at once clear that the idea had possibilities. One could estimate the barrier leakage factor by using a crude Brillouin-Wentzel-Kramers integration of the wave function across an assumed barrier which was a Coulomb potential on the outside. It did not much matter what one used for the shape of the inner part, so we just cut it off at an assumed nuclear radius that seemed reasonable.

We did not have very clear ideas on the leakage of a wave

 0002-9505/78/4604-0319$00.50

packet. We knew that this was what the problem called for, but we did not know how to do it. We contented ourselves with a steady-state solution based on the properties of the radial function for an energy that gave maximum amplitude inside and minimum amplitude outside. From this we got an early crude formula for the decay constant, and made some rough evaluations from the known Coulomb potential and the known alpha particle emission energies. Within a few days we had written a letter to the Editor of *Nature*, dated July 30, which appeared in the September 22 issue.[3] This letter set forth the idea in rather qualitative terms only, and we at once set about to work out more quantitative details. The concluding sentence of that letter was one which appealed to both of us. I read:

"Much has been written of the explosive violence with which the α-particle is hurled from its place in the nucleus. But from the process pictured above, one would rather say that the α-particle slips away almost unnoticed."

I have mentioned Gurney's shyness. He was a young bachelor then who always wore tennis sneakers and he would slink along the wall of the halls in Palmer lab instead of walking down the middle as most persons do. When we were working together I would see him flit swiftly by the open door of my office without seeming to look in. Thus I would know that he wanted to communicate with me, so I would go and hide somewhere for five or ten minutes and then when I came back, I would find a note from him on my desk, in which he had told me what he wanted to tell, or asked what he wanted to ask. Gradually of course we became acquainted and after about a month communication between us took a more normal turn.

He was a great lover of music and an accomplished piano player. Oddly enough in spite of his shyness in the laboratory, he was an enthusiast for English country dancing and soon organized a faculty dancing class in which he was the chief instructor in this art of the young faculty couples in Princeton at that time. At this business I was shy as he was in the laboratory, seldom going to such parties.

Although I still have some of the original notes involved in the preparation of our paper, they are incomplete and disorganized. I remember that we worked hard on it through the fall. That year the fall meeting of the National Academy of Sciences was held in Schenectady in October and I gave a paper on the new theory there. Irving Langmuir acted as chairman of the session, and he discussed it quite fully. However he upset me considerably by asking if we could explain the continuous β-ray spectrum of such cases. I was embarrassed because that was the first that I had heard of the continuous β spectrum which later loomed so large in neutrino theories and weak interaction and all that. But the α-particle part was well received by the physicists present. It was my first introduction to the strong group then at the General Electric laboratory in Schenectady, which included Coolidge, Hull, Dushman, and others.

You must realize that even the probability interpretation of the squared wave amplitude of the wave function was a fairly new idea at that time and since there was no knowledge whatever of the structure of nuclei then, the whole thing became a rather sensational application of quantum mechanics to the nucleus and did much to gain acceptance for the general idea that quantum mechanics could be applied to the nucleus. Remember that the neutron was not discovered until four years later so nuclei were made of protons and electrons at that time!

I also went in November to the fall meeting of the American Physical Society and presented a paper there on the same subject giving a little more detail because the calculations had gone further. That meeting was in Minneapolis. The then new physics building of the University of Minnesota was dedicated on this occasion. It was also known that Van Vleck was leaving Minnesota for a professorship at the University of Wisconsin and on this occasion there were preliminary discussions about my being his successor which did lead to my going to Minneapolis in the fall of 1929 for one year, after which I returned to Princeton.

Jack Tate's letter accepting our fuller paper is dated November 28, 1928 and the paper appeared in the February 1929 issue of the Physical Review.[4]

In the meantime, of course, Gamow's first paper in the Zeitschrift für Physik[5] had appeared almost simultaneously with our letter in Nature.[6] George gave numerous talks on the subject in various European and British colloquia, so his contribution to the subject became better known than ours. In those days American physics did not amount to much and the European physicists did not pay much attention to the Physical Review.

Gurney's fellowship expired and he left Princeton about early January of 1929. He went to the Institute of Physical and Chemical Research in Tokyo, where he stayed for a time before going to one of the maharajahs in India where he had a cousin who worked as that official's financial manager before later returning to England. I do not have his exact schedule in my records.

I should not have mentioned Robertson's refusal to see the barrier leakage idea, were it not for the fact that I have now to make a similarly derogatory comment about my failure to appreciate an important point in nuclear physics.

Soon after we had mailed off the February 1929 paper in late November, we began to think about the application of barrier penetration ideas to the possibility that nuclear transmutation could occur at lower energies than by classical mechanics because the incoming proton or α particle could leak inward through a barrier instead of needing enough energy to go over the top. In a way we recognized the importance of this idea, but we were both then under the influence of very snobbish ideas that one ought not to make a paper out of anything that was such an obvious corollary to the general idea of barrier leakage. So we did not write a paper on that. Gamow did.[7]

I remember at that time Gurney also thought of the idea of resonance penetration of the barrier, the idea not merely that the barrier becomes more and more penetrable as the bombarding particle's energy is raised, but that it is penetrated extraordinarily easily when the energy is near to one of the quasistable resonance levels at which it could be captured inside the nucleus. This time it was my turn to fail to appreciate the truth as well as the importance of this idea of Gurney's. I failed to do the rough calculations on a rough model of a light nucleus. Thus I failed to realize that down in this part of the system of nuclei the levels are sufficiently broad to be observable. I was hung up on the idea of the extreme narrowness of the levels in uranium and radium, and failed to recognize what a range of variation an exponential factor can give when the exponent is greatly changed.

Because of this failure of mine I talked him out of publishing in December 1928, the idea of resonance penetration in artificial disintegration experiments. When he went to Tokyo and was no longer subject to my bad influence, he did write a letter to Nature, dated Tokyo, February 20 (1929) which was published in the issue of April 13, 1929.[8] I had not seen it for years and just yesterday went to our library and made a copy to see whether he had stated the idea as clearly as I believed from memory that he had.

He had. After recalling the general idea of barrier leakage he writes (and this is three years before the famous Cockroft-Walton experiments on artificial disintegration and before any of the enormous body of later experimental work was done):

"The object of the present note," writes Gurney, "is to direct attention to the possibility of resonance phenomena if we take into account the solutions of the Schrödinger equation which for certain ranges of energy give psi-functions the amplitude of which inside the nucleus is large compared to the outside. For this seems to indicate that variation of the velocity of the incident particle may be accompanied by an enormous fluctuation in the probability of penetration when the energy approaches and enters the range of energy corresponding to one of the possible quasidiscrete levels. A systematic examination of thin films of various elements might disclose such a fluctuation, if the experimental difficulties can be overcome." He continues to discuss the details of the experimental suggestion.

So far as I know this paper has always been overlooked and he has not been credited in the literature with having made this suggestion. Future histories of the subject should correct this error. Although I had remembered that he wrote such a letter, I had not remembered that he had stated the idea of resonance levels so clearly nor at such an early date.

After returning to England, Gurney became interested in electrochemical topics from a quantum mechanical viewpoint and wrote two books as well as many papers on the subject. Then he went to Bristol and began a collaboration with Nevill Mott which clarified many things about ionic crystals when a photographic film is exposed to light. This work led to his publication jointly with Mott of the book entitled "Electronic Processes in Ionic Crystals."[9]

When World War II broke out with the overrunning of Poland by the German armies, Gurney was studying biophysics in Stockholm. He had married several years before. His wife, Natalie, was not at all shy. She is a lovely, active extroverted person who loved to give big parties, and herself a graduate student in political science specializing in Southeast Asia. The older persons present will remember that the winter 1939–40 was called the season of the "phony war" because there was so little action after Poland was conquered, there being a complete lull in military action until June 1940 when the Nazi armies enacted for the British the tragedy of Dunkirk, and followed by a complete conquest of France. Although the North Sea was mined in that phony war winter, Natalie took a ship from Stockholm to London to visit friends and relatives for Christmas 1939 and returned to Stockholm the same way after a holiday.

But by the summer of 1941 the Nazis had overrun Denmark and Norway, invaded Russia and started the blitzkrieg bombardment of England. People then expected that momentarily Britain would be overrun and conquered. Hitler's armies seemed invincible.

The British government ordered its nationals to return home, or at least in this case to get out of Sweden. But by this time, the only way to get out of Sweden was by ship to Leningrad and then by railroad train across Russia and Siberia, by ship to Japan, and on to America by ship from Japan to San Francisco, which is what Ronald and Natalie did. At that time I was associate director of research at Westinghouse in Pittsburgh deeply immersed in getting our microwave radar program going, making almost weekly trips to MIT to coordinate our effort with the Radiation Laboratory which was directed by Lee DuBridge for the NDRC, later the OSRD.

The Gurneys came to Pittsburgh and stayed at our house for a time. Then they went on to Washington and reported for duty to Sir Charles Darwin who was then Britain's scientific liaison officer there. He communicated with London and it was decided that he should go into military physics work at the Aberdeen Proving Ground in Maryland instead of returning to England. It was here that our colleague, Professor Richard N. Thomas, got to know Gurney well for he spent a large part of the war years on ballistics research there, having newly graduated with a BA from Harvard. In my files I have materials attesting to the importance of Gurney's contributions to the war effort at Aberdeen, but I will not go into detail on that as no tunneling was involved.

Toward the end of the war, I do not know just when, Gurney was assigned to the staff of the Metallurgical Laboratory, a part of the Manhattan District at the University of Chicago. In accordance with the well-known principles of compartmentalization of knowledge, I do not know what he did there, but he worked effectively there until the end of the war and longer.

Then in the summer of 1946 when Brookhaven National Laboratory was being organized on Long Island with Professor P. M. Morse as its first director, Morse sought to have Gurney come to his staff, and such a position was accepted. He resigned from Argonne, gave up his apartment in Chicago and was honored at farewell parties.

Now I come to the sad part of my story, one more chapter in the horrible series of national disgraces which involved many American and foreign scientists in security troubles. Morse found that he could not get Gurney cleared for Brookhaven, so that job fell through, but as proof of the total irrationality of the whole situation, he was still cleared for Argonne and could pick up again his old job which he did for a time.

Then he got a two-year appointment at the Johns Hopkins Laboratory. During this time he had his first stroke which I am sure was aggravated by the nervous worry associated with his Kafka-like situation in which security doubts continued to hang over his head without there being any way at that time they could be brought to a head.

While at Hopkins, Natalie began to do graduate work with Owen Lattimore, the distinguished American scholar of Mongolian politics who was himself under investigation about that time—see his book, "Ordeal by Slander."[10] While in Washington I became well acquainted with Lattimore as I have tried to do with most of the persons that have been attacked in the wave of hysteria which began to develop in 1946 and did not begin to abate until eight years later. I remember how Lattimore told me of a seminar course he gave one of these summers at Hopkins in Mongolian politics which met under the shade trees in the backyard of his home in Baltimore. That is such a narrow academic specialty that all the people in it are closely acquainted, so the three or four students were somewhat mystified at one student who registered for the course, who attended regularly but quite clearly did not have the native intelligence or the prerequisites for graduate work in Asian political science. The others took him out one night and got him drunk and he admitted to them that he was the FBI man assigned to keep an eye on Lattimore.

These anecdotes are not funny because they cause the victim a great deal of worry and expense and often loss of employment. Lattimore was of the Walter Hines Page school of international affairs which Hopkins abolished after Lattimore became notorious under circumstances which strongly suggested that these two things were interrelated. Later Lattimore went to England and has a distinguished professorship there, I think in the University of Manchester. America did many things like that to weaken her academic assets at that time.

After Hopkins, the Gurneys went to the University of Maryland. By this time, Ronald's health was quite poor. He had extremely high blood pressure after his first stroke and had to conserve his health more carefully. During this period, he managed to get some formal charges out of the Air Force on the basis of which he was given a security clearance hearing. His health was so poor that the hearing board allowed Natalie to do most of the testifying for him. The result was a complete clearance for access to classified military work such as he had had at Aberdeen and at Argonne during the war, and the lack of which barred him from going to Brookhaven at a time when it was being boasted that that laboratory was not engaged in secret work. The fact that it really was, being itself kept secret.

At the end of the year or two in Maryland, there happened one of the most awful incidents on the whole hysterical period. The then president of the University of Maryland, a former football coach, abruptly and with a loud public fanfare fired six members of the faculty, including Ronald. All of these faculty members had had loyalty hearings of one kind or another and all had been cleared (the cases were independent and unrelated to each other). But the Maryland president loudly told the press, patriotic fellow that he was, that "I don't want anyone on my faculty about whom there has ever been the slightest doubt!" So these cleared people found themselves out of work at a time when it was impossible to find employment anywhere after having been stigmatized in this way.

The Gurneys took an apartment on Riverside Drive in New York and tried to make a life, living near Columbia University, with him eking out a small income by doing some industrial consulting work. His health continued to decline. Old friends on the Columbia faculty were so frightened by the mood of the times that the Gurneys hardly ever saw any of them. The crowning blow came when Columbia denied Ronald the use of the physics department library for some months until I vehemently intervened. Finally during the night of April 14, 1953, he had another and final stroke and died in his wife's arms.

She called the family physician who made the necessary arrangements that night. Early that morning, however, the doctor was interviewed by two FBI men who wanted to know whether Gurney had talked before he died. He had not.

I have never been able to understand what these crazy fellows suspected Gurney of doing. I feel certain that he was not engaged in any political work or any espionage work, and the same goes for Natalie, but it must be admitted that in the mood of those times a graduate student of Owen Lattimore's was a pretty suspicious person.

I feel very bad about the fact that there was little or nothing that I could do for Natalie after Ronald's death. You see my own clearance had been suspended by the Navy security officers at Buffalo and I was marking time before being given a new hearing under the new and improved procedures that were being worked out by Eisenhower's staff. He had campaigned against Adlai Stevenson on the promise that he would not be soft on communism as he and his vice-presidential candidate repeatedly said or at least inferred that Truman had been. Even so I could have offered Natalie more personal consolation in her sorrow than I did. I was not afraid but I was too terribly distracted at the time to recognize even the simple duties of civilized friendship. Natalie tried in various fruitless ways to clear Ronald's name after his death and was told by rude and heartless officials that there was no procedure for clearing dead men since they had no "need to know."

After a while she gave up America and returned to London where she has made her home ever since these horrible happenings. She was raised in the tradition of civilized British liberalism and I do not believe she ever realized the depths of degradation and fear into which American academic people had sunk in that period.

In the meantime she has remarried. Her husband, Peter Taylor is a psychiatrist attached to one of the major London hospitals. My wife and I keep up a correspondence with them and have visited them in London. Gradually the wounds heal a little. But America still has some of her best scientists living abroad as scientists-in-exile as reminders of that shameful period. I think particularly of Dave Bohm and Bernard Peters who are professors in London and Copenhagen, respectively, who are not allowed to come to this country to see their aging parents.

Perhaps it is not proper of me to present such a serious topic as an after-dinner talk at what has otherwise been such a pleasant occasion. But with the passage of time, I see the name of Ronald Gurney being more and more forgotten. I have even met young people who always refer to Condon and Gurney, rather than the other way round, and suppose that Gurney was merely some transient student of Condon's who did not turn out well, when as a matter of fact it was he and not me who thought of barrier leakage as the model to explain α radioactivity.

Finally I tell you all of these dismal things not to wallow in despair, but because I still hope, sixteen long years after the peak of the fever was reached in 1954, that the American government may not only cease such persecutions in the future, which it pretty well seems to have done, but may even go a little farther and make amends to those which it so wrongly persecuted for so many years.

Thank you for your careful attention. Possibly governments are not human enough to be able to admit that they ever make errors. But some of those who hold prominent positions in the present government had earlier had prominent roles on the shameful events of the years 1946–54, and maybe they could, as human individuals, do a little something on behalf of the government of which they are now a part.

[a]Born Mar. 2, 1902; died Mar. 26, 1974.
[1]R. Oppenheimer and R. H. Fowler, Proc. Natl. Acad. Sci. **14**, 363 (1928).
[2]L. Nordheim, Proc. R. Soc. A **119**, 173 (1928).
[3]R. W. Gurney and E. U. Condon, Nature **122**, 439 (1928).
[4]R. W. Gurney and E. U. Condon, Phys. Rev. **33**, 127 (1929).
[5]G. Gamow, Z. Phys. **51**, 204 (1928).
[6]G. Gamow, Nature **122**, 805 (1928).
[7]G. Gamow, Z. Physik **52**, 510 (1928-29); **53** 601 (1929).
[8]R. W. Gurney, Nature **123**, 565 (1929).
[9]N. F. Mott and R. W. Gurney, *Electronic Processes in Ionic Crystals* (Oxford, New York, 1940).
[10]O. Lattimore, *Ordeal by Slander* (Little, Brown, Boston, 1950).

Time

Time is that great gift of nature which keeps everything from happening at once.

–C.J. Overbeck

A Study of the Discovery of Fission

Esther B. Sparberg
Hofstra University, Hempstead, New York

(Received 13 June 1963)

Nuclear fission was discovered twenty five years ago by Hahn and Strassmann. Was the road to the discovery as torturous as some have contended? The roots are examined; the surprising climax and its impact are described. Some scientists have commented on the discovery in retrospect. The discovery of fission is scrutinized both in the light of these comments, and of the general characteristics of scientific discovery.

THE discovery of fission was announced by Otto Hahn and Fritz Strassmann twenty five years ago in *Die Naturwissenschaften*.[1] So unexpected was the news, and so sensational, that it immediately provoked a tremendous outpouring of scientific papers. Just three and a half months later, Feather asserted in a review article, ". . . so much has been published that rigorous selection is necessary in any report of the subject."[2] Turner[3] estimated that nearly one hundred papers had appeared on the subject within the year. Reflecting the excitement of the period were the many practically simultaneous discoveries made in 1939.[4]

On this twenty fifth anniversary year,[5] it seems appropriate to review the discovery of fission. Lise Meitner[6] has asserted that "viewed in the light of our present knowledge, the road to that discovery was astonishingly long and to a certain extent the wrong one . . .." Was the path really so devious? Has Meitner, who played a leading role in the discovery of fission, been engaging in "Monday morning quarterbacking?" What were the roots of the discovery? What factors contributed to the successful solution of a difficult problem? Did the experimenters devise new techniques? Were any radical ideas formulated during the course of the investigations? What characteristics of scientific discovery described by students of the subject can be noted?

[1] O. Hahn and F. Strassmann, Naturwiss. **27**, 11 (1939).
[2] N. Feather, Nature **143**, 878 (1939).
[3] L. A. Turner, Rev. Mod. Phys. **12**, 1 (1940).
[4] H. H. Goldsmith, Sci. Monthly **68**, 292 (1949).
[5] It is also just twenty years since Otto Hahn received the Nobel Prize for the discovery of fission.
[6] L. Meitner, Advan. Sci. **19**, 363 (1963).

ROOTS OF THE DISCOVERY

The discovery of fission[7] was not an accident.[8] Nor did it result from serendipity.[9] It represented the climax of a crescendo of activity in nuclear science during the thirties. Although Hahn and Strassmann were indeed surprised by their unexpected results, the hypothesis of fission was inevitable in the light of the intensive investigations in the field. If Hahn and Strassmann had not made the discovery, it would have been made by others. In fact, I. Joliot-Curie and Savitch almost made the discovery in 1938. Several weeks after the publication of the Hahn and Strassmann paper, Philip Abelson in a letter to the *Physical Review* announced that he had found what seemed to be "unambiguous and independent proof of Hahn's hypothesis of the cleavage of the uranium nucleus."[10] A series of anomalous experiments gives rise to restlessness that is soothed only by an adequate solution. To compare the discoverers of fission to the Three Princes of Serendip, who according to Horace Walpole, "were always making discoveries, by accidents and sagacity, of things they were not in quest

[7] The word "fission" was first used in the paper by L. Meitner and O. Frisch, Nature **143**, 239 (1939).
[8] An example of a popular viewpoint is in Time **33**, 21 (1939).
[9] William L. Laurence, *Men and Atoms* (Simon and Schuster, Inc., New York, 1959), p. 12. After this manuscript was sent to the editor, a viewpoint similar to the author's was expressed in an editorial by Philip Abelson. Science **140**, 1177 (1963).
[10] Philip Abelson, Phys. Rev. **4**, 418 (1939). R. Hewlett and O. Anderson, Jr., confirm that at the time of the Hahn–Strassmann discovery, Philip H. Abelson in his doctoral research at Berkeley was also so close to the discovery of fission that he almost certainly would have made it within a few weeks. Richard G. Hewlett and Oscar E. Anderson, Jr., *The New World, A History of the United States Atomic Energy Commission* (The Pennsylvania State University Press, University Park, 1962), Vol. 1, p. 12.

of"[11] seems superficial and naive, for the scientists were in quest of a solution to a very definite problem.

Their immediate problem was an attempt to clarify some previous baffling results of Joliot–Curie and Savitch who were investigating the neutron bombardment products of uranium. This area of research had been opened up by Enrico Fermi after the discovery of the neutron (1932) by Chadwick and the discovery of artificial radioactivity by the Joliot–Curies (1934). After reading about the work of the Joliot–Curies, Enrico Fermi decided to try to produce artificial radioactivity with neutrons[12] since they had the advantage of a projectile with no charge and hence were not repelled from the positive nucleus. Alpha particles from radium or radon dislodge neutrons when they strike beryllium. Before striking the target, the neutrons may be slowed down by paraffin to produce increased radioactivity in the target element. This effect had been unexpectedly discovered by Fermi and his associates.

Fermi's plan was to systematically bombard all the known elements starting with hydrogen. When uranium was bombarded with neutrons, Fermi and his associates discovered that they had produced more than one radioactive product from the reaction. The products seemed to be four in number, as revealed by four different half-lives. Since there were only three known isotopes of uranium, one of the products was mistakenly thought to be element 93. This seemed a valid assumption, because the capture of a neutron by a nucleus usually produced an unstable isotope emitting a beta particle and thereby having an atomic number greater by one. Since their investigations also showed that these newly formed radioactive species could not be the isotopes of any element from radon (86) to uranium (92), they announced in 1934[13] that they had probably produced element 93. "We did not have enough imagination to think that a different process of disintegration might occur in uranium . . . and we tried to identify the radioactive products with elements close to uranium on the periodic table . . . we did not know enough chemistry to separate the products . . . and we believed that we had four . . . while actually their number was close to fifty."[14] Yet a German chemist, Ida Noddack, pointed out that their methods did not disprove the existence of lighter elements as bombardment products, "but this thought was considered to be wholly incompatible with the laws of atomic physics."[15] It was also suggested that the unknown product might be protactinium (91).

THE BACKGROUND OF HAHN AND MEITNER'S INTEREST IN THE URANIUM PROBLEM

The published reports of Fermi's work in *Nuovo Cimento* and *Nature* were so fascinating to Meitner that immediately she "persuaded Otto Hahn to renew our direct collaboration . . . with a view to investigating these problems. So it was that in 1934, after an interval of more than twelve years, we started working together again, with the especially valuable collaboration, after a short time, of Fritz Strassmann."[6] They immediately set themselves the task of repeating Fermi's experiments to ascertain whether or not the thirteen minute element was a protactinium isotope.[16] Since they were the codiscoverers of protactinium (element 91) in 1917, they were very familiar with its chemical properties. In fact, Hahn's interest in radiochemistry had been inspired by Ramsay thirteen years earlier; Hahn relates: "The story must go back to the year 1904, for that was the year in which I was 'transmuted' from an organic chemist into a radiochemist and thus began to learn about radioactive elements."[17] He recalls that when Sir William Ramsay asked whether he would like to work on radium, he replied that he knew nothing about the element. Ramsay retorted that "it was an advantage to be able to approach the subject with a mind free from preoccupations."[18] During his studies with Ramsay in England and later Rutherford in Montreal, he discovered two

[11] Horace Walpole in a letter to Mann, 28 January 1754, as quoted in *The Reader's Encyclopedia*, edited by William Rose Benet, (Thomas Y. Crowell Company, New York, 1948), Vol. 4, p. 1012; also described in *The Oxford Companion to English Literature*, p. 710.

[12] Laura Fermi, *Atoms in the Family* (The University of Chicago Press, Chicago, 1954), p. 83.

[13] E. Fermi, Nature **133**, 898 (1934).

[14] E. Fermi, as quoted by Laura Fermi. See Ref. 12, p. 157.

[15] O. Hahn, Sci. Am. **198**, 78 (1958).

[16] Otto Hahn, *New Atoms* (Elsevier Publishing Company, Inc., New York, 1950), p. 17.

[17] See Ref. 15, p. 76.

[18] See Ref. 16, p. 142.

new radioactive species. ". . . in the course of these studies I was acquiring what was to be of the greatest help to Strassmann and myself later in discovering the fission process: namely, a thorough familiarity with methods of separating radioactive substances."[17] Rutherford[19] himself, writing of this period, related: ". . . the results so far obtained by Hahn are of the greatest interest and importance."

In 1907, Lise Meitner began the association with Hahn that lasted for thirty years until it was severed by the actions of the Nazi regime. Their investigations involved a study of the beta rays of their almost complete set of radioactive elements. In 1917, as previously mentioned, they discovered protactinium. "It was our work on the chemical properties of this substance that gave rise later to our keen interest in investigating the irradiation of uranium with neutrons."[20] Their attempts to resolve the problem of Fermi's unknown isotope seemed to confirm Fermi's views.

In these experiments, they used an indicator technique, adding a beta-emitting protactinium isotope to the unknown neutron bombardment product. Separation of the mixture was then attempted by chemical precipitation. Results indicated that neither protactinium, thorium, nor actinium was present since the natural protactinium isotope separated out from the unknown product. It was, therefore, probably element 93, since they did not consider the possibility of splitting heavy nuclei into lighter ones. Yet Meitner noted that it was disturbing to find such a long chain of successive beta disintegrations.

They continued their researches on transmutation products and found what they thought to be transuranium elements 93–98.

CURIE AND SAVITCH DISCOVER A 3.5-H PRODUCT

Meanwhile, Joliot–Curie and Savitch also became interested in this area of research and reported that they had found a new thorium isotope. Upon checking their results, the German team wrote back to Irene Joliot-Curie that they were unable to find this isotope. Meitner[21] has ruefully noted that they searched for the thorium isotope in the filtrate resulting from the precipitation of the "transuranium" elements. Since, according to their scheme, uranium, protactinium, and thorium were definitely in the filtrate, the precipitate was assumed to be composed of the "transuranium" elements. As they did not expect the filtrate to contain either the lower atomic weight, or the "transuranium" elements, they therefore examined only the precipitate; "Here was our mistake." Even when they examined the filtrate for thorium, they did not think of looking for anything else.

In the meantime, in 1937, Joliot–Curie and Savitch made a perplexing discovery. They found a new 3.5-h product[22] which did not separate with the transurnaium elements according to the Hahn–Meitner method of platinium precipitation. This isotope, instead, precipitated with lanthanum (atomic weight 139, and atomic number 57), a medium element rather than a heavy one; moreover this new product resembled lanthanum chemically.

They were puzzled at finding an element with properties resembling lanthanum among the elements supposedly beyond uranium in the periodic table. Yet they did not realize the identity of lanthanum with this element of half-life 3.5-h because of some anomalies in the chemical evidence. Hahn confirms[15] that if they had been convinced that their 3.5-h element was indeed lanthanum, they would have been the discoverers of fission.

HAHN AND STRASSMANN REPEAT THE EXPERIMENT

Upon reading the Curie–Savitch paper, Hahn and Strassmann (Meitner had left Germany in July 1938) decided to repeat the experiments, since the results were incomprehensible to them. How could an element with chemical properties similar to lanthanum be considered a transuranium element assigned to the region beyond uranium in the periodic table? In the course of these investigations, they found what they assumed to be four radium isotopes, since they

[19] Ernest Rutherford, *Radioactive Transformations* (Charles Scribner's and Sons, New York, 1906), p. 69.
[20] See Ref. 15, p. 77.
[21] See Ref. 6, p. 364.
[22] I. Curie and P. Savitch, Compt. Rend. **206**, 1643 (1938), from the bibliography of the Hahn and Strassmann paper.

precipitated out with barium. Radium and barium are chemically similar since they are in group 2a of the periodic table, and so the radium will be carried with the barium precipitate. Accidental impurities, they were sure, could not have caused the results, for barium had been precipitated according to Strassmann's suggestion as an extremely pure chloride. The idea that these radioactive isotopes might be isotopes of barium itself did not occur to them at the time, for Hahn writes[23] that "since it seemed from the state of knowledge of nuclear physics that barium could not be produced from uranium by neutron bombardment . . . we could only conclude that the products must be isotopes of radium . . . still it was a strange affair to be producing radium from uranium under the conditions of our experiment." The transmutation of radium from uranium required the successive emission of two alpha particles and slow neutrons had never before been observed to stimulate the emission of alpha particles from uranium. The results of these experiments were published in 1938.[24]

THE PAPER ANNOUNCING THE SPLITTING OF THE URANIUM ATOM

In order to clarify their results, Hahn and Strassmann continued their investigations and they separated and positively identified what they assumed to be radium 88 and its daughter product, actinium 89. The unknown radioactive products from uranium bombardment were precipitated with a barium carrier. Since they had remained with barium, they were thought to be isotopes of radium. There appeared to be four isotopes of radium among the products, as shown by the four different half-lives. From these isotopes, "radium IV" with a half-life of 250–300 h, was separated with barium. Attempts to separate the artificial "radium IV" from barium by the fractional crystallization method of M. Curie were unsuccessful. A method they developed for precipitating the unknown with the carrier as chromates rather than as chlorides or bromides failed also.

Since it seemed possible that their inability to achieve a separation might be due to the fact that only extremely small quantities of "radium IV" were present, Hahn and Strassmann decided to check their results. They used small quantities of natural radium isotopes of an intensity just detectable with a Geiger–Müller counter; these weak isotopes were found to be separable from barium. They then used indicator experiments, mixing natural radium isotopes with their artificial "radium IV." When this mixture was added to barium, again the natural radium isotopes were separated by fractional crystallization, while the artificial one remained behind with the barium. "Radium IV" must be barium.

> "We come to the conclusion: our "radium isotopes" have the properties of barium; as chemists we should really say that considering the new substances, they may not be radium but barium, because there is no question of other elements but radium or barium."[25]

As a further confirmation, they tested the decay products from "radium IV." If the isotope was barium, then lanthanum would be its daughter product. When a natural actinium isotope was mixed with one of the primary decay products of the artificial "radium IV," with lanthanum added as a carrier, the decay product remained with lanthanum, and must therefore be a lanthanum isotope rather than actinium. This was the 3.5-h product that had been found by Curie and Savitch.

THE CONCLUSIONS OF HAHN AND STRASSMANN, AND THEIR IMPACT

Since Hahn and Strassmann, using every available technique, were unable to separate their supposed radium isotope from barium, they were forced to conclude that the "radium IV" was not radium but barium.

> "As chemists we really have to change the symbols of Ra, Ac, and Th to Ba, La and Ce instead, which is based on the scheme used in the above briefly described experiment. As nuclear chemists we are in a way close to physics, and according to all experience up to now in nuclear physics, we cannot decide to make this contradictory jump. It could be that a series of strange accidents has distorted our findings."[26]

Their conclusions were couched in these cautious words because the idea that a nucleus could be broken up had never occurred to anyone;

[23] See Ref. 15, p. 80.
[24] O. Hahn and F. Strassmann, Naturwiss. **26**, 755 (1938).
[25] See Ref. 1, p. 14.
[26] See Ref. 1, p. 15.

neutron bombardment products were always considered to be very close in atomic number to the parent atom. Hahn explains,[27] "Our overcautiousness stemmed primarily from the fact that as chemists we hesitated to announce a revolutionary discovery in physics." It was not until they again checked their results, published in their next paper, that they showed beyond a doubt that their "radium" isotopes were actually barium, and that an atom could be split.

Despite the hesitant manner in which Hahn and Strassmann first expressed their findings, the immediate impact of this paper on the scientific world was enormous. The results were widely known even before publication, for Hahn at Christmas 1938 had written with great excitement to Meitner in Sweden of the results of the experiments. What did she, as a physicist, think of the results?

> "On reading the letter, I myself was thoroughly excited and amazed and also, to tell the truth, uneasy. I knew the extraordinary chemical knowledge and ability of Hahn and Strassmann too well to doubt for one second the correctness of their unexpected results. These results, I realized, had opened up an entirely new scientific path—and I also realized how far one had gone astray in our earlier work!"[21]

Meitner immediately informed her nephew Otto Frisch of the astonishing news. Frisch confessed:

> "When Lise Meitner told me that, I would not listen at first; the barium nucleus is only about half as heavy as uranium, and I could not see how half of the uranium nucleus could have been chipped away by the impact of a single neutron. But as we talked, a new picture gradually took shape . . .."[28]

The nucleus was compared to a liquid drop with reduced surface tension due to the repulsion of the component protons. If the movement of the nucleus was violently increased by a captured neutron, the nucleus would form a "waist" and divide into two lighter nuclei. "In view of the similarity of this process to cell division, we called it (at Frisch's suggestion) 'fission' . . ..[29] Because the products of the nuclear reaction have less total mass than the reactants they realized that energy was released, which they estimated at 200 MeV. About three weeks later, on 16 January 1939, Meitner and Frisch sent a letter to *Nature* setting forth their theoretical explanation of fission.[7] Frisch quickly followed with another letter confirming that he had been able to detect the energy released during fission.[30]

Niels Bohr learned of the new developments from Frisch just prior to Bohr's departure for the United States, and he spread the news in America of the splitting of the uranium atom. The discovery and immediate theoretical explanation of Meitner and Frisch were announced by Bohr on 26 January 1939 at a conference on theoretical physics at the Carnegie Institution of Washington. So electrifying was the effect of the report that many immediately returned to their laboratories to measure the ionization energy of the fission products. The general public was informed of the scientific ferment by *Time Magazine* in an article which read, "Last week, the Hahn report reached the United States, and physicists sprang to their laboratories to see whether they could confirm it. Early this week, the laboratories of Columbia, Johns Hopkins and the Carnegie Institution announced confirmation."[31]

THE FIRST DETECTION OF THE DISINTEGRATION OF URANIUM AND SCIENTIFIC DISCOVERY

This account of the investigations that produced the discovery of fission illustrates the effect of the "relentless pressure of accumulating knowledge"[32] on the emergence of the novel concept that atoms may be split in transmutation. As long ago as 1793, Goethe[33] declared: "The most beautiful discoveries are made not so much by men, as by the period . . . they mature in the course of time . . .." The discovery of artificial radioactivity, the identification of the neutron and its immediate use as a tool of research, and the continued refinement of radiochemical techniques were the principal agents. During the four years of explorations in this particular area of research, the near-birth of the idea of fission was prevented by uncertainties

[27] See Ref. 15, p. 82.
[28] O. Frisch, *Atomic Physics Today* (Basic Books, Inc., New York, 1961), p. 20.
[29] See Ref. 28, p. 365.
[30] O. R. Frisch, Nature **143**, 276 (1939).
[31] Time Mag. **33**, 21 (1939).
[32] A. J. Ihde, Sci. Monthly **67**, 429 (1948).
[33] J. W. Goethe, cited by L. L. Whyte, Harper's Mag. **20**, 25 (1950).

in experimental techniques, and by the then-accepted concept that an atom could not be divided; but the labor pains caused by anomalous experimental results finally became severe enough to force the new concept to emerge into the scientific world. When the chemical evidence was indisputable, the birth was hesitantly announced by Hahn and Strassmann.

The influence of the generally accepted theories of the time upon scientific progress may be examined in the light of the Hahn and Strassmann findings. The suggestion that an atom could be divided had simply not occurred to anyone except Ida Noddack, whose proposal was ignored, and so no investigator looked for medium atomic weight species among neutron bombardment products. As noted previously, Lise Meitner regretfully commented that they did not examine the filtrate remaining from the precipitation of the "transuranium" elements because they did not believe it contained any elements below 90. Yet, many of their tests did point to the impossibility of separating neutron bombardment products from their lower atomic weight carriers—and each time this problem was attributed to experimental difficulties. Finally, the accumulated evidence could not be denied. Hahn has reiterated, "These tests were in opposition to all the phenomena observed up to the present in nuclear physics."[34] Even Einstein asserted, ". . . this was not something I could have predicted."[35]

Preconceived ideas may indeed act as a brake to progress, but they are also part of the supporting framework upon which new knowledge is built. The four years that elapsed before the puzzling and intriguing problems were solved were in part caused by the barricade of preconceived ideas. Contributing equally to the delay was the difficulty of the experimental techniques.

The final identification of barium as a transmutation product resulted from the painstaking application of nuclear science methods by researchers with long experience in the field. The most difficult problems that had been surmounted were radiochemical ones—the separation and identification of tiny quantities of radioactive materials. Although no innovations were evident in the Hahn and Strassmann experiments, some refinements had been made by Hahn, Meitner, and Strassmann. Hahn's consistent interest in radiochemistry is shown by his researches on the formation of mixed crystals between radium and barium just prior to these investigations.[36]

Fermi's mistaken identification of element 93 was caused by these experimental difficulties. The state of the art in 1934 was not sufficiently mature to provide the right answers. The four years between the investigations of Fermi and those of Hahn and Strassmann may be considered the period of the discovery of fission, and illustrates well the suggestion of Kuhn[37] that "the process of discovery is necessarily and inevitably one that shows structure and that therefore extends in time." The discovery had its roots in the puzzling products of neutron bombardment, and its logical culmination in the identification of barium.

Was the road to the discovery so "astonishingly long" and "to a certain extent wrong" in the light of present knowledge, as Dr. Meitner has asserted? The path was certainly not direct—but is the path to a discovery ever straightforward? A glance at the history of science would reveal that any scientific discovery was preceded by a period of groping; that blind alleys often delayed the final solution. In the light of the state of knowledge after the discovery has been made, the investigators appeared at times like sleepwalkers. Wouldn't the platitude about the superiority of hindsight to foresight be relevant here?

A question that may next be considered is the effect of nonscientific factors on the emergence of productive scientific ideas. That fission was discovered in a most rigid and authoritarian political environment raises the question of the relationship between political freedom and scientific progress. This is a difficult question to answer in terms of the discovery of fission, for the German scientists at the Kaiser Wilhelm Institute seemed somewhat insulated from the political Zeitgeist. Yet, they were not completely isolated, for Lise Meitner had been forced to leave

[34] See Ref. 16, p. 23.
[35] Albert Einstein, *Out of My Later Years* (Philosophical Library, Inc., New York, 1950), p. 188.

[36] See Ref. 16, p. 22.
[37] T. S. Kuhn, Science **136**, 764 (1962).

Germany. Despite the fact that the Nazi leaders had ordered the subjugation of all science to the needs of the state, Hahn and Strassmann were engaged in fundamental research that had no immediate applications. Fermi and his associates were also able to pursue fundamental investigations at the great physics school of the University of Rome in Fascist Italy. These activities indicate that islands of intellectual honesty may remain uninundated by their surroundings, which is a disappointing conclusion to those who would prefer to equate progress in basic science with political freedom. These "island" institutions were headed by brilliant men, who were very much interested in both their faculty and their students, and they were operated in an atmosphere of warm communication and relative leisure, where investigations were carried out without outside pressure. The general climate in the nonacademic world was one of great respect for learning; in spite of the anti-intellectual forces in the Fascist countries, the traditional regard of the Europeans for the academic profession and for scholarship could not be so easily eradicated.

The discovery of fission was propelled by the times; it was also propelled by men, and illustrates well the statement of Boring[38] ". . . great men make great discoveries. Yet it may be that the Zeitgeist determines the great discovery, and that he who makes the discovery, the Zeitgeist's agent, is great merely because the times employed him." Otto Hahn was a recognized leader in the field of radiochemistry, and had worked in this area for over thirty years. His interest in identifying uranium bombardment products was natural, and his attempts to unravel puzzling experimental results were persevering and meticulous. The triumphant discovery of barium resulted from his finesse, and genius, in that particular area of research. His contributions were officially recognized when he received the Nobel Prize in 1944 for the discovery of fission. To neglect to note here that the identification of barium was merely the apex of a pyramid of investigations would be inexcusable. Many others played essential roles in the development of the story. Among them, as mentioned previously, were Meitner, Fermi, Joliot–Curie, and Savitch.

[38] E. G. Boring, Proc. Am. Phil. Soc. **94**, No. 4, 341 (1950).

The great impact of the announcement of fission illustrates yet another characteristic of scientific discovery—the surge of experiments and related ideas produced in the scientific world by the unexpected birth of the new concept. The concept of the splitting of the uranium atom by slow neutrons led immediately to the realization and experimental demonstration that (1) a large amount of energy was released during the fission process, (2) neutrons were liberated thus suggesting the possibility of a chain reaction, (3) uranium 235 was the isotope responsible for fission, (4) delayed neutrons are emitted, (5) elements 93 and 94 would be produced if uranium 238 absorbed neutrons, (6) element 94 (plutonium) was fissionable. The successful operation of the first chain-reacting pile in December 1942, accelerated by the needs of World War II, was the victorious culmination of four years of concentrated research.

Thus, the unexpected and surprising identification of barium initiated a torrential outpouring of related investigations; the fission of uranium had been produced by the inevitable convergence of new ideas and techniques, utilized by skilled investigators operating within a general and scientific Zeitgeist that influenced both the manner and date of the successful conclusion.

ACKNOWLEDGMENTS

I wish to thank Professor Frederick L. Fitzpatrick of Teachers College, Columbia University, for his interest and help. Much of the historical research on which this paper is based is part of a doctoral dissertation written under a grant by the Science Manpower Project under the direction of Professor Fitzpatrick. I am indebted to Dr. Frances S. K. Sterrett of Hofstra University for translating the passages quoted from the Hahn and Strassmann paper.

Genesis of the Transistor

Walter H. Brattain
Whitman College, Walla Walla, Washington 99362

As an experimental physicist, I have always been interested in exploring the frontiers of science to obtain new information that will contribute to a better understanding of nature. Frequently new inventions must await expansion of our fundamental knowledge before they can become a reality. The research that led to the discovery of the transistor effect and the invention of the transistor at Bell Telephone Laboratories in 1947 is a good example.

Walter H. Brattain was born in Amoy, China, on 10 February 1902. Educated in the State of Washington and receiving a B.S. degree from Whitman College in 1924, he was awarded the M.A. degree by the University of Oregon in 1926, and the Ph.D. by the University of Minnesota in 1929. Dr. Brattain was a member of the Bell Laboratories staff from 1929 to 1967. Upon his retirement in February 1967, he became visiting Professor of Physics at Whitman College in Walla Walla, Washington. The chief field of his research has been the surface properties of solids. His major contribution to solid state physics was his discovery of the photoeffect at the free surface of a semiconductor, and the invention of the point-contact transistor jointly with John Bardeen. Dr. Brattain is a member of the National Academy of Sciences and many scientific societies. He shared the Nobel Prize in Physics for 1956 with Dr. Bardeen and William Shockley for their investigations of semiconductors and the discovery of the transistor effect.

I would like to tell you the story as I remember it, for it may answer the question so often asked by science students, "How was the transistor invented?"

Before I begin, however, you may wish to know what a transistor is. The transistor's basic function is to regulate flow of current in response to an input signal. Briefly, it performs in a solid what the electron tube does in a vacuum: it conducts, modulates, switches, and amplifies electrical signals.

The transistor, like the vacuum tube, is an active circuit element. In a transistor we have a semiconductor in which both negative and positive charge carriers (electrons and holes, respectively) exist. The ability of the semiconductor solids to conduct electricity depends on the number of these holes and electrons.

Power from a local power source can be used to make a steady current flow through such a solid. If, for example, the electrical potential at one junction of contact is such that the flow of current is limited by the number of holes available and a potential is applied at a neighboring junction to increase the number of holes, this will increase the current flow. This process enables the transistor to amplify a weak, intelligible signal with very little expenditure of power.

Semiconductors such as germanium and silicon are ideal as active circuit elements because the number of electrons or holes they contain is very small. A few added holes (positive carriers) or added electrons will make a significant difference. Actually, added holes result also in added electrons or vice versa, since a conductor cannot become charged. In a metal there are already so many electrons that it is almost impossible to change the numbers.

Although the vacuum tube continues to play an important role in electronics, the transistor has several inherent advantages. It operates without any warm-up time and requires merely a fraction of the power that a vacuum tube needs. It is extremely reliable: when properly manufactured and operated within its design limits, the transistor can operate trouble-free for decades. It can also transmit very high frequency signals at very high speeds, making it suitable for use in modern switching and computer systems, and can be made smaller than a pinhead, allowing extreme miniaturization.

Semiconductors, neither good insulators nor good conductors, first gained notice because of their photoelectric properties and their use in rectifiers where they conduct more easily in one direction than the other.[1] As early as

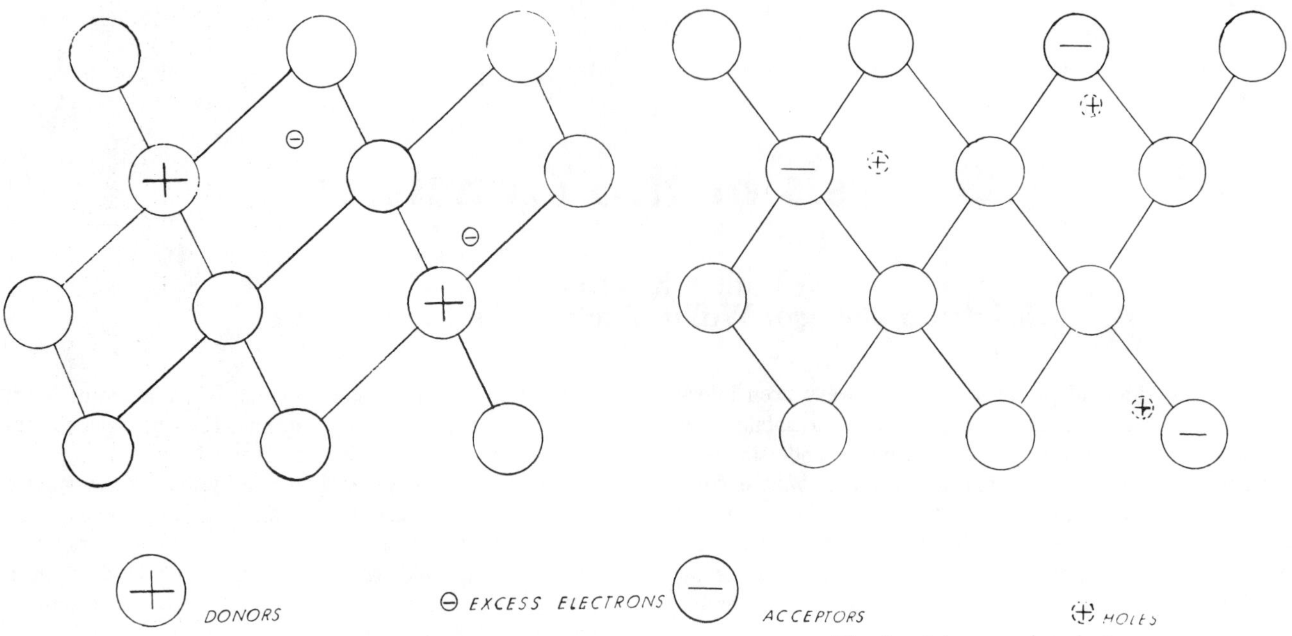

Fig. 1. *n* type semiconductor.

Fig. 2. *p* type semiconductor.

1920, serious attempts to understand the electrical properties of these solids were begun. Many scientists all over the world were interested in the problem. The advent of quantum mechanics gave us the first real understanding of how atoms stick together in solid crystals. Quantum theory also gave us understanding of how electrons in metals were free to move and conduct electricity, and A. H. Wilson showed us how this theory could explain insulators and semiconductors. Valence electrons in insulators and semiconductors are not free to move under the influence of an electric field. However, if a few valence electrons are excited thermally or by light, they can move; and the vacant holes in the bonding structure can also move and act as positively charged carriers of electric current.

In 1931, my supervisor and colleague at Bell Telephone Laboratories, J. A. Becker, and I first started work on copper oxide, one of the then better known semiconductors. (The Bell System was interested in this material because of its rectifying properties and possible use as a modulator of electrical communications signals.) There were two main problems—flow of current inside the solid and flow of current at contacts between the semiconductors and metals. At such contacts current would flow more easily in one direction than the other (retification), and also at such contacts illumination of the surface would sometimes produce a voltage (photo emf). A light shown on the main body of the semiconductor would also increase its electrical conductance (photoconductivity).

Early in 1940, Mervin J. Kelly, who was then Director of Research for Bell Telephone Laboratories, called us into his office to witness a demonstration by Russell S. Ohl, a member of the staff who was working with silicon, a then little known semiconductor. Ohl showed us a small rectangular piece of a black solid with two metal contacts. When light from a flashlight illuminated a narrow region near the middle of this piece of silicon, a photoelectromotive force (emf) of about 0.5 V was developed. This was hard to believe. In the first place, the contacts were not being illuminated and the photo emf was ten or more times larger than any we had ever seen. Moreover, the silicon was black; that is, opaque to visible light. In fact, I did not believe what I saw until Ohl gave me a piece to work with in my own laboratory.

This was the first *p-n* junction. Some time before this, G. C. Southworth had been having trouble getting something to detect the short radio waves (microwaves) with which he was working. Vacuum tubes did not work well at these wavelengths. He remembered the old cat's whisker detectors that were used before vacuum tubes were invented and decided to give them a try. He went to the second-hand radio market in the Cortland Street section of New York and there found some cat's whisker detectors, dusted them off, and tried them. They worked.

Hearing about it, Russell S. Ohl, of Bell Telephone Laboratories, became interested in the phenomenon. These detectors were semiconductors, made of either silicon or galena (lead sulfide). Ohl decided to concentrate on silicon. The silicon you could buy was very nonuniform in its detection properties. This detection involved rectification of the radio wave at the point of contact of the metal cat's whisker with the silicon. Sometimes the rectification would be in one direction and sometimes in the opposite direction, sometimes not at all. Ohl asked the metallurgists at Bell Labs to see if they could not make the silicon more uniform by purifying it. J. H. Scaff and H. C. Theuerer soon found they could purify the silicon by melting it in high vacuum. Sometimes the ingots that they made would rectify all one way and another ingot would rectify all the other way. Those that conducted best when the silicon was negative were called *n* type (Fig. 1) and the other *p* type (Fig. 2). One ingot that

Scaff and Theuerer gave Ohl was all *n* type on one end and all *p* type on the other end, and a piece cut out to include the boundary between these two regions was the one Ohl had demonstrated.*

Scaff and Theuerer went on to find that the conductivity in silicon was mainly due to very small traces of impurities. These impurities were elements in the fifth column of the periodic table, like phosphorus, which gave excess electrons when added and made silicon *n* type. Elements in the third column such as indium gave excess holes and made silicon *p* type.**

With the advent of World War II it became important to make good silicon detectors for radar. This work was carried forward by Scaff, Theuerer, Ohl, and by many others outside the Bell Laboratories. Much independent work was done in England and at the Radiation Laboratory at MIT. Lark-Horovitz and his group at Purdue worked with germanium, another semiconductor like silicon (both are in the fourth column of the periodic table). Others, including myself, worked on entirely different war problems.

Just before the War, N. F. Mott in England and W. Schottky in Germany independently developed the first good theory of rectification; B. Davydov in the USSR also contributed. When two electrically conducting solids are in contact and in equilibrium there is an electrostatic difference in potential between the two solids. This means that there must be a charge double layer at the contact. Since the free charge density, electrons or holes, in the semiconductor is very small this charge layer must penetrate into the semiconductor. This theory suggested that if the potential difference is the right sign there will be a potential hill at the surface of the semiconductor for, say, electrons and if the semiconductor is *n* type the contact will consequently rectify. If it was *p* type it would not rectify, and one would have to use a metal giving a potential difference of the opposite sign to get retification.

Another point of interest was that this rectification at the metal semiconductor contact was analogous to the recification obtained in a vacuum tube diode; many people thought of putting in a third element like a grid in the vacuum tube triode, to make an amplifier, but none of the schemes worked.***

* When *n* and *p* type silicon are in contact as in Ohl's junction the holes tend to diffuse over to the *n* side charging it positively, and electrons diffuse the other way charging the *p* side negatively. This results in an equilibrium electrostatic potential preventing any further diffusion. If the density of holes and electrons were both equal on the two sides of the junction there would be no diffusion; therefore, no charging and no potential. Shining light on the junction creates equal numbers of extra electrons and holes, thus tending to reduce the tendency to diffuse and thus reduce the electrostatic potential. This change in potential from the equilibrium value is the photo emf. W. H. Brattain, *Science,* **126**, 151 (1957).

** These impurities were below the level of spectroscopic detection at that time. The odor of the ingots as they came out of the oven was traced to phosphorus by Scaff and Theuerer.

*** R. Hilsch and R. W. Pohl succeeded in putting a grid in an alkali halide crystal but the crystal would only respond to signals longer than 1 sec.

In January 1946, because of Bell Labs' interest in the use of semiconductors in circuit devices and the possibility of a solid state amplifier, scientific research to enable us to understand semiconductors was resumed in earnest. The research team was headed by William Shockley and S. O. Morgan. Shockley, a theoretical physicist, worked as part of the research team he headed. John Bardeen, also a theoretical physicist, joined the group (Fig. 3).

At the first meeting of the group it was realized that in spite of all the work done before and during the war, we were still far from a real understanding. One reason was that copper oxide and other semiconductors on which the early work had been done were very complicated solids. Silicon and germanium were the simplest, and the decision was to try to understand these first. Our work was directed toward a fundamental understanding of the problem though we were well aware of the technical importance of a semiconductor amplifier if one could be made.

Besides those already mentioned there was G. L. Pearson, who was primarily interested in the bulk properties of semiconductors, R. B. Gibney, a physical chemist, and H. R. Moore, a circuit expert who aided greatly in making measurements and devising novel circuits for our experiments. I was primarily interested in what went on at the surface where contact was made or at the boundary between *n* and *p* types. Of this group only Pearson and I had worked extensively with semiconductors before the war.

Scaff and Theuerer were in a position to furnish us with polycrystalline ingots of either *n* or *p* type silicon or germanium of any specified resistance. This, of course, was a great help. Based on the Mott-Schottky theory of rectification, Shockley had come to the conclusion during the War that it should be possible to control the density of electrons near the surface of the semiconductor by means of an electric field applied between the surface and a

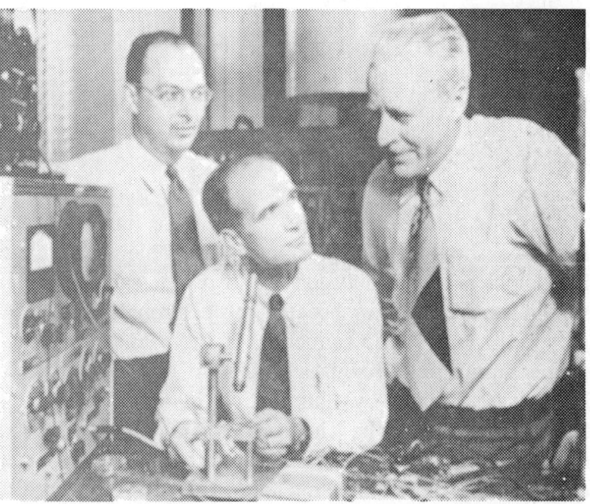

Fig. 3. Nobel Prize winners John Bardeen (left), William Shockley, and Walter H. Brattain at Bell Telephone Laboratories in 1948 with apparatus used in the first investigations that led to the invention of the transistor. The trio received the 1956 Nobel Physics Award for their invention of the transistor.

metal electrode insulated from the surface. If this were true one could vary the conductivity in the surface of a thin wafer semiconductor by means of the field and thus make an amplifier. Many experiments were devised by Shockley to test this hypothesis, but the effect was several orders of magnitude less than predicted.

Moreover, several predictions based on the theory did not agree with experimental results. In order for two electrical conductors to be in equilibrium there must, in general, be an electrostatic potential difference between them (contact potential). The prediction was that there should be such a potential between *n* and *p* type silicon and between *n* and *p* type germanium. Experiments showed that this potential was very small, almost zero. Different metals should have different contact potentials with respect to the semiconductor. If the sign of this potential was not right there should be no rectification at the point of contact; yet, experimentally, all metal points worked more or less equally well. Additionally, S. Benzer at Purdue found that contact between two pieces of germanium, one *n* type and one *p* type, did not act as one would expect. I believe the group as a whole slowly realized that these results were all of a piece, and it was Bardeen who successfully explained them all by applying to this problem the concept of surface states; that is, that the electrons could be trapped at the semiconductor surface, and that the semiconductor was in equilibrium with its surface before any electrical contact was made to it. This, of course, implied a space charge layer in the surface of the semiconductor equal and opposite to the charge trapped on the surface. Consequently, the electrostatic potential change between the interior of the semiconductor and the surface which was necessary for rectification was a property of the semiconductor and its surface—independent of the metal contact. This theory immediately suggested new experiments.

When any novel ideas or new experimental results were obtained by any member or part of the group it was our practice to call the group together for a presentation and discussion. These sessions were inspirational, and it was during such sessions that many of the better suggestions for further work were made. Two suggestions for experiments came about as a result of the meeting and discussion that occurred on the occasion of Bardeen's presentation of his theory. Shockley suggested that, if trapping centers for electrons on the surface were limited in number, one should be able to measure some small change in contact potential between *n* and *p* type samples of, say, silicon as the samples became more strongly *n* type and *p* type through the introduction of more and more of the proper impurities. I suggested that, if extra electrons and holes were excited by illumination in the surface region (where there must be an electric field due to the space charge), the electric field would tend to separate the excited electrons and holes, thus changing the surface charge and contact potential. Both these experiments were successfully performed.

Another suggestion was to try to reduce the temperature low enough so electrons trapped at the surface could be frozen and a field effect observed. Experiments by Pearson and Bardeen showed that this was the case.

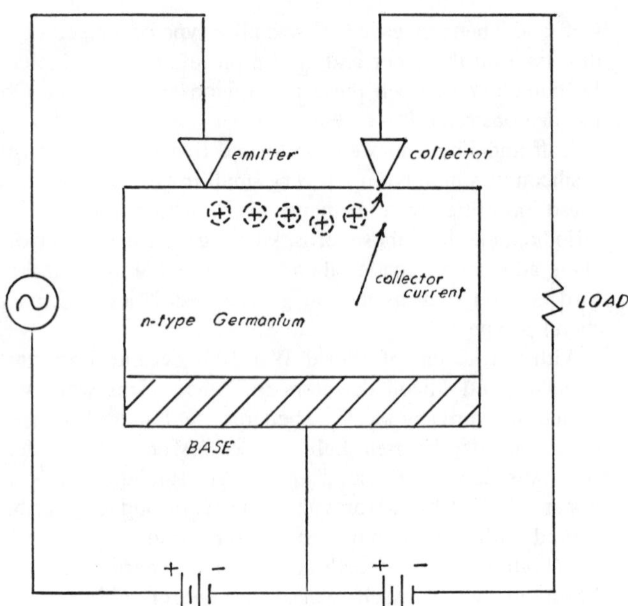

Fig. 4. Point contact transistor at work.

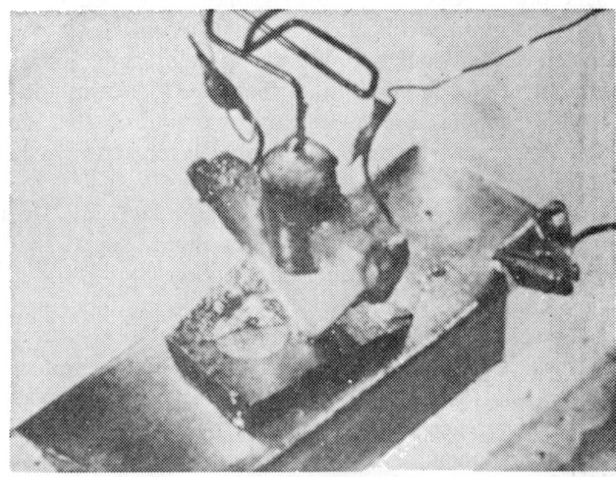
Fig. 5. The original transistor.

Another experiment that I tried was to measure the change of potential at the germanium or silicon surface as a function of temperature. Condensation of moisture from the air on the cold semiconductor surface interfered with this experiment. As a result it was decided to try immersing the system in an insulating liquid. The apparatus had been arranged to measure contact potential and photo emf's, and when the liquids were tried large changes in photo emf's were observed. Some of the liquids tried (such as water) were not strictly insulating but were electrolytes. When I was showing these phenomena to Gibney he suggested varying the potential between the semiconductor surface and the reference electrode. When using an electrolyte we could make the photo emf very large by this means. By changing the sign of the potential we could make the photo emf go through zero and change sign. It was recognized that this was, in essence, Shockley's field effect. By using the electrolyte we could vary the space charge layer and potential inside the semiconductors at the surface.

These results were presented to the group as a whole and one morning one or two days later Bardeen came into my office with a suggested geometrical arrangement to use this effect to make an amplifier. I said let's go out in the laboratory and do it. We covered a metal point with a thin layer of wax, pushed it down on a piece of *p* type silicon that had been treated to give an *n* type surface. We then surrounded the point with a drop of water and made contact to it. The point was insulated from the water by the wax layer. We found as expected that potentials applied between the water and the silicon would change the current flowing from the silicon to the point. Power amplification was obtained that day!

Bardeen suggested trying this on *n* type germanium, and it worked even better. However, the water drop would evaporate almost as soon as things were working well, so at Gibney's suggestion we changed to glycol borate, which hardly evaporates at all. Another problem was that amplification could be obtained only at or below about 8 Hz. We reasoned that this was due to the slow action of the electrolyte. Optimum results were obtained with a dc negative bias on the electrolyte when using *n* type germanium. Under these conditions we noticed an anodic oxide film being formed under the electrolyte. We decided to evaporate a spot of gold on such a film and, using the film to insulate the gold from the germanium, use the gold as a field electrode to eliminate the electrolyte. The film was formed, the glycol borate washed off, and the gold spot with a hole in the middle for the point was evaporated. When this was tried, an electrical discharge between the point and the gold spoiled the spot in the middle; but by placing the point around the edge of the gold spot, a new effect was observed. In washing off the glycol borate, we had inadvertently washed off the oxide film which was soluble in water. The gold had been evaporated on a freshly anodized germanium surface. When a small positive potential was applied to the gold, holes flowed into the germanium surface greatly increasing the flow of current from the germanium to the point negatively biased at a large potential! Four days later, on 23 December 1947, two gold contacts less than two thousandths of an inch apart were made to the same piece of germanium; and the first transistor was made (Figs. 4

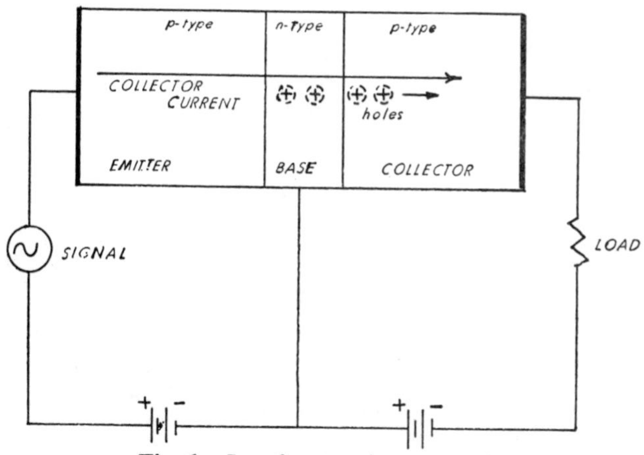

Fig. 6. Junction transistor at work.

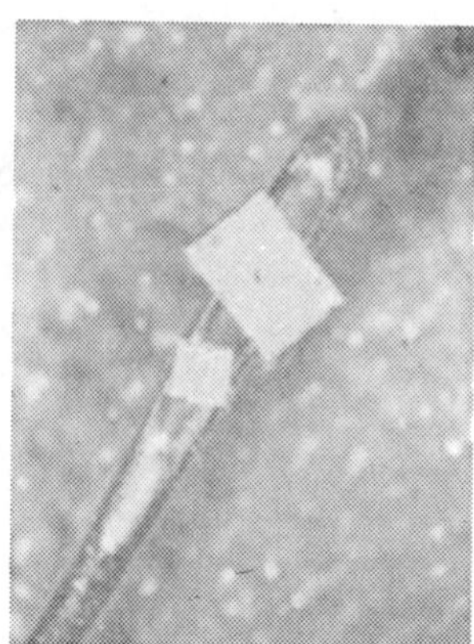
Fig. 7. Microminiaturized modern transistors enlarged 200 times.

and 5). This was an eventful day, and here is how I recorded it in my notes:

"Using the germanium surface (see top of page 197) and the gold contacts the following circuit was set up. This circuit was actually spoken over and by switching the device in and out a distinct gain in speech level could be heard and seen on the scope presentation with no noticeable change in quality. Various people were present and witnessed this test and listened—of whom some were the following: R. B. Gibney, H. R. Moore, J. Bardeen, G. L. Pearson, W. Shockley, H. Fletcher and R. Bown. H. R. Moore assisted in setting up the circuit and the demonstration occurred on the afternoon of 23 December 1947."

The transistor was named in my office by J. R. Pierce. Bardeen and I wanted a name that would fit in with varistor and thermistor, but we were more or less devoid of ideas when I presented the problem to Pierce. Pierce knew that the point contact transistor was the dual of a vacuum tube, circuit wise. After some thought Pierce mentioned the important parameter of a vacuum tube—transconductance—then a moment later its electrical dual—transresistance. Then he said "transistor" and I said, "Pierce, that is it."

The original transistor was called a point-contact transistor because it was essentially a wafer of germanium with two pointed wire contacts made close together on one side. Shockley went on to work out the theory of the *p-n* junction and to predict the *p-n-p* or *n-p-n* transistor (Fig. 6).

When in 1950 G. K. Teal and J. B. Little successfully made single crystals of germanium, it soon became possible for Morgan Sparks to make a junction transistor at Bell Labs. A few years after the alloy transistor was developed, and later the diffused base transistor made it possible for a single wafer of semiconductor material to contain thousands of transistors.

The same knowledge and theories that led to the transistor's development have led to the development of whole new families of solid state devices. This in turn has brought about a remarkable technology called microelectronics. This new technology has shrunk transistors and other circuit elements to dimensions almost invisible to the unaided eye. For example, a complete microcircuit consisting of ten to twenty transistors and forty to sixty resistors can be interconnected and built onto a piece of silicon only about a tenth or twentieth of a square inch (Fig. 8).

Today, then, the transistor is no longer a single device. It is a family of devices made in dozens of different ways and used in almost every kind of electrical circuit. When the work that led to the transistor effect was being done, however, my chief interest was in understanding solid state surfaces. The transistor actually came into being because of our work in solid state physics, specifically in the field of semiconductor physics.

The transistor came about because Bell Laboratories undertook to develop fundamental knowledge to a stage where human minds could understand phenomena that had been observed for a long time.

I believe the discovery of the transistor effect and the invention of the transistor illustrate the power of scientific research in the areas of science relevant to a purpose. My personal satisfaction came mainly from the basic discoveries we made, but I must also admit no little pleasure in the fact that my work also has had widespread practical application for the benefit of man.

Reference

1. References to many of the events mentioned in this article can be found in G. L. Pearson and W. H. Brattain, Proc. Inst. Radio Engrs. 43, 1794 (1955), where the early history is told in greater detail.

Physics in American Colleges Before 1750

John J. McCarthy
St. John's University, Brooklyn, New York

A RECENT article[1] states that science instruction in this country began around 1750. It is difficult to state definitely when science actually entered the curriculum, but this date is probably not far wrong.

While the present article will be confined to the teaching of physics or natural philosophy, it should be realized that it, together with astronomy, comprised all the instruction in science commonly given in the colleges during this period. The secondary schools did not have any science courses until the founding of the first academy in 1749.

This period might be divided into two parts: The first began with the opening of Harvard in 1638 and extended to the beginning of the downfall of Aristotelian physics, which was certainly in progress by 1687; the second, a transitional period, during which physics was evolving into a science, ended around 1740.

New Englands First Fruits[2] mentions physics, astronomy, and "The Nature of Plants" as part of the curriculum of Harvard in 1642. It even lists, in Latin, fifteen theses "Physicas" which may have been debated at the first Harvard commencement in 1642. Translating a few of these, we learn that "Form is an accident," "Form is the principle of individuation," "Whatever is moved is moved by another (thing)," and "Putrefaction in a damp place arises from external heat," were then considered within the domain of physics. That the subjects of early Harvard theses were stable medieval disputations has been pointed out by Morison[3] and by Walsh.[4] Harvard in its early years was anxious to impress its English friends by duplicating the curriculum of the English universities, and this was the kind of natural "science" they were then teaching.

While "The Nature of Plants" seems to have disappeared from the curriculum, physics and astronomy remained. However, natural philosophy, as physics was called during the eighteenth and most of the nineteenth centuries,

[1] D. Roller, Am. Phys. Teacher **6**, 244 (1938).
[2] London, 1643.
[3] S. E. Morison, *Harvard College in the Seventeenth Century* (Harvard Univ. Press, 1936), p. 227.
[4] J. J. Walsh, *Education of the Founding Fathers of the Republic* (Fordham Univ. Press, 1935).

included all natural phenomena—both animate and inanimate, terrestrial and celestial.

The character of the physics textbooks of this period was quite unscientific.[5] Peripatetic or scholastic physics, as Aristotelian physics is called, is generally characterized as an attempt to determine the "why" of nature. Textbook writers in those days selected passages from Aristotle to which they added comments from various authorities. Aristotle's views in the realm of physics were little more than opinions based on common observations. Starting with these, both he and his commentators hoped to arrive at some universal truth by means of deduction.

It is impossible in this brief discussion to outline the contents of these Aristotelian textbooks fully. The general idea was to explain the *raison d'etre* of the entire universe from the stars to the center of the earth, with the physiology of animals and plants, and a little psychology thrown in for good measure. Considerable space was used in trying to define "time," "space," "causality," "motion," and the like.

Latin was the language in which the Aristotelian texts were written and the college laws compelled its use in the classroom. No mathematics, problems, experiments or experimental results, and few illustrations, graced these books. The teacher had no need to demonstrate the principles contained in them—in fact, no apparatus was available for such demonstrations—and the student was excused from the solution of physics problems as well as from laboratory work. The methods employed in class were to read the text, and to outline, discuss and dispute the principles contained in it. At graduation some of these principles might be used as theses to be defended or refuted according to the rules of syllogistic argument.

The first physics at Yale was also Aristotelian. "Pierson's Manuscript of Physicks," which seems to have been copied and studied by the students in lieu of a regular textbook, was based on an Aristotelian text and the notebook of Dr. Abraham Pierson.[6] Schwab[7] speaks of the use of the less metaphysical works, Jean Leclerc's *Physica* and Jacques Rohault's *System of Natural Philosophy*, in the 1720's.

Like Harvard and Yale, William and Mary also taught Aristotelian physics. The college laws of 1736 state:[8]

> Forasmuch as we see now dayly a further progress in Philosophy, than could be made by Aristotle's Logick and Physicks, which reigned so long alone in the schools, and shut out all other; therefore we leave it to the President and Masters, by the advice of the Chancellor, to teach what Systems of Logick, Physicks, Ethicks, and Mathematicks, they think fit in their schools.

Princeton seems to have taught scientific physics from its opening in 1747. This is concluded from the fact that a Princeton student of 1750 wanted a copy of Benjamin Martin's *Philosophia Britannica*,[9] a truly scientific work. Also a college publication of 1752 mentions the need of ". . . a proper Apparatus for Philosophical Experiments; . . ."[10] The University of Pennsylvania and all other colleges established thereafter taught scientific physics from the beginning.

When did the period of Aristotelian science end? Morison[11] places the change in astronomy at Harvard as beginning before 1659, with the introduction of the works of Copernicus, Galileo and Kepler, and the consequent banishment of the Ptolemaic system. With regard to physics he has concluded[12] that Charles Morton's semi-scientific *Compendium Physicae* was in use there before 1687. As was pointed out previously, Yale departed from Aristotelian physics in the 1720's, William and Mary in 1736.

The foregoing dates do not mean that Aristotelian science was entirely forgotten. The attempt was made to teach both Aristotle's physics and the new physics formulated by Newton and the rest—somewhat like the attempt in our day to teach both the classical

[5] See S. E. Morison (reference 3, pp. 225–26) for a discussion of the textbooks that were probably used at Harvard in this period.

[6] L. F. Snow, *The College Curriculum in the United States* (New York: By the Author, 1907), pp. 32–33.

[7] J. C. Schwab, "The Yale College Curriculum 1701–1901," Educational Review **22**, 5 (June, 1901).

[8] Statutes of the College of William and Mary in Virginia 1736, *Bulletin of the College of William and Mary*, Vol. VII, No. 3 (Jan., 1914), p. 14.

[9] John MacLean, *History of the College of New Jersey from its Origin in 1746 to the Commencement of 1854* (J. B. Lippincott and Co., 1877), Vol. 1, pp. 141–42.

[10] *A General Account of the Rise and State of the College Lately Established in the Province of New Jersey* (Princeton, 1752), p. 7.

[11] Reference 3, p. 216.

[12] Reference 3, pp. 238–49. The *Compendium Physicae* was not printed but was copied by the students.

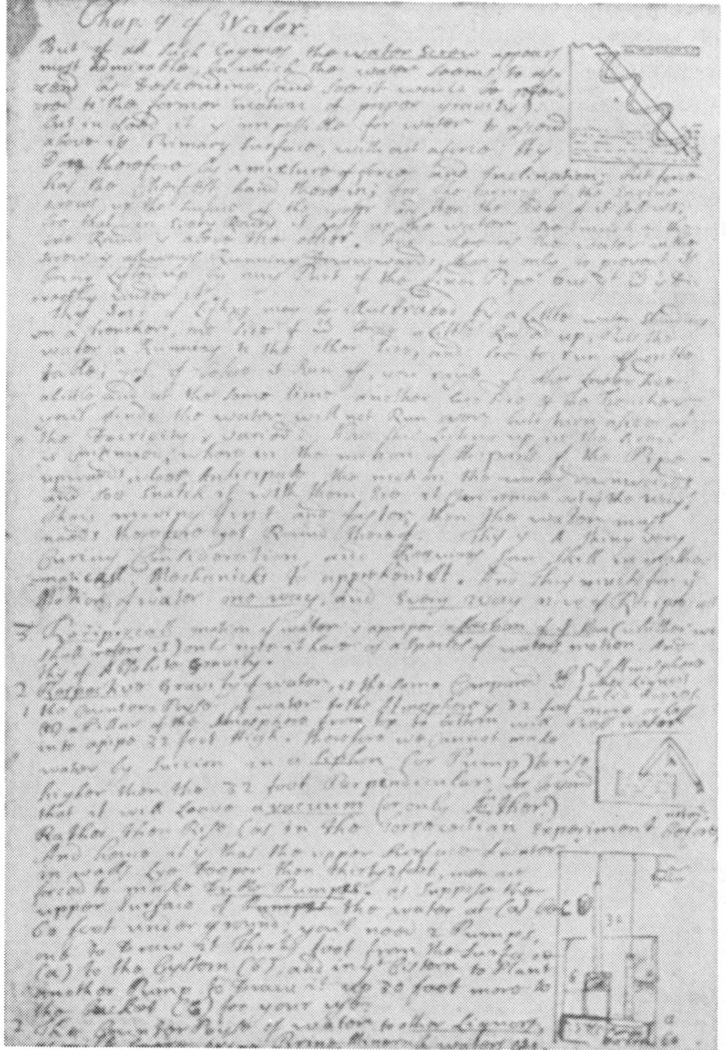
Chap: 4 of Water.

An extract from Morton's *Compendium Physicae*, copied by Nathaniel Eells of the Harvard class of 1699. [Yale University Library.]

and modern theories of radiation. For this reason it seems appropriate to call this a transitional period, since in it physics was evolving into a science.

Some portions of the textbooks[13] used during this transitional period were scientific in that they were based on experiments, observations and mathematics, whereas other portions were Aristotelian. The following quotation[14], from Rohault's preface to his book, previously mentioned, might help to show the general feeling toward Aristotle in this period:

> Further you will not find a great many Things in this whole Treatise contrary to Aristotle; but you will find more than I could wish that are contrary to most of the Commentators upon him: And besides you will meet a great many Things, which neither Aristotle nor his Followers have treated of at all, which I have however judged more useful than many others which Philosophers have wholly imployed themselves in. And in all this I did not think it very ill in me to depart from some particular Notions, when I found that these Notions were disagreeable to Truth.

Rohault follows the practise of the Aristotelian commentators in giving space to the opinions of various authorities. Thus he has chapters headed: "Of the Elements according to the Opinions of the Ancients," "Of the Elements of the Chymists" and "Of the Elements of Natural Things." Considerable attention is devoted to the definitions of form, essence, alteration, corruption, etc., as in the Aristotelian books. Like them also is the teleological note which is present in many parts of the book.

On the other hand, the section on geometrical optics differs but little from that found in recent textbooks. It has many diagrams, illustrating the principles of mirrors and lenses, which plainly suggest to the reader that he may check up on these principles himself. The astronomical section is also scientific. Some of the principles are demonstrated mathematically, usually in the footnotes.

Another thing, which plainly indicated that physics was becoming a science in the early half of the eighteenth century, was the increasing mention of "philosophical apparatus." An account[15] of the burning of William and Mary

[13] Reference 3, pp. 232–34. On these pages are listed the transitional books probably used at Harvard.

[14] Translated by John Clarke (London, 1723). Some material was added by him.

[15] *The History of the College of William and Mary from its Foundation 1693 to 1870* (John Murphy and Co., 1870), p. 35.

College, in 1705, mentions the destruction of the philosophical apparatus; Harvard[16] received some apparatus from a benefactor in 1727; Yale[17] was promised some in addition to what it had already in 1719. While philosophical apparatus included such things as globes, telescopes, and surveying instruments, the apparatus previously mentioned in connection with Harvard and Yale was intended for experiments in physics.

The use of apparatus produced one important innovation in the teaching of physics, namely, the demonstration. It arose from the belief that when a student saw a principle demonstrated by means of apparatus, he would better understand it. Eventually this led to the establishment of laboratory work in the nineteenth century.

With the advent of textbooks written in English, the necessity of reading the text in class vanished. Students were now expected to memorize the assignment and be prepared for a barrage of questions from the teacher. As physics became more and more scientific, the importance of the disputation, with its emphasis on the opinions of authorities and on deduction, declined.

Another sign of the times was the extension of science teaching to the secondary schools through the academy movement. The first academy is generally conceded to have been the one founded in Philadelphia in 1749 with Benjamin Franklin as one of its promoters. It evolved into the University of Pennsylvania. Franklin's advocacy of scientific education is so well known that it is unnecessary to dwell on it here.

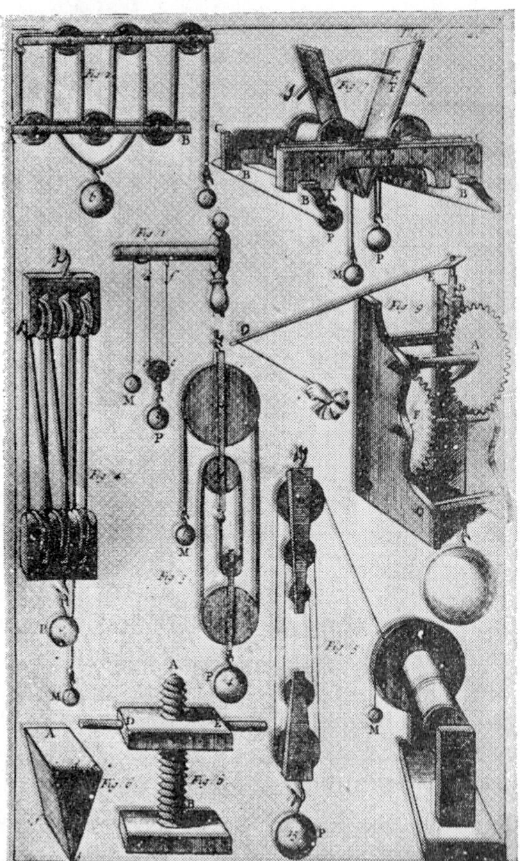

A plate from van s'Gravesande's *Mathematical Elements of Natural Philosophy* (1726). [Columbia University Library.]

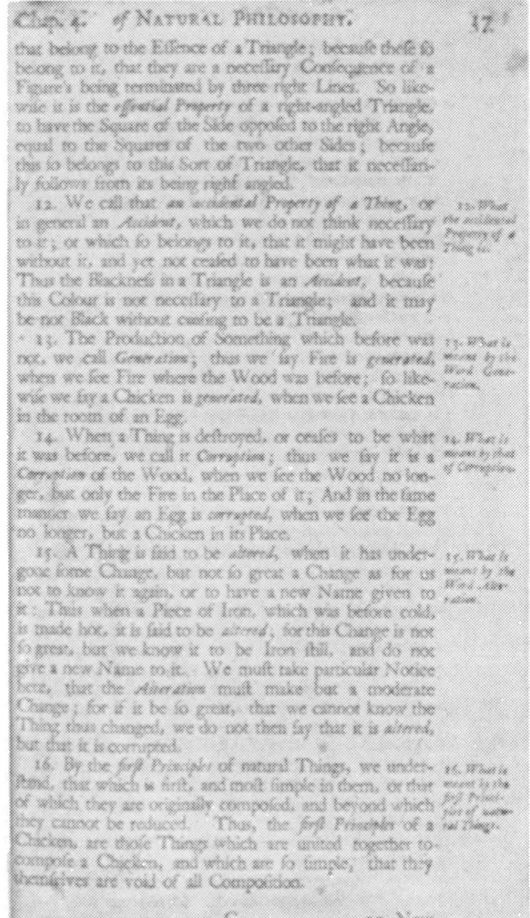

Chap. 4. of NATURAL PHILOSOPHY. 17

that belong to the Eſſence of a Triangle; becauſe theſe ſo belong to it, that they are a neceſſary Conſequence of a Figure's being terminated by three right Lines. So likewiſe it is the *eſſential Property* of a right-angled Triangle, to have the Square of the Side oppoſed to the right Angle, equal to the Squares of the two other Sides; becauſe this ſo belongs to this Sort of Triangle, that it neceſſarily follows from its being right angled.

12. We call that *an accidental Property of a Thing*, or in general an *Accident*, which we do not think neceſſary to it; or which ſo belongs to it, that it might have been without it, and yet not ceaſed to have been what it was: Thus the Blackneſs in a Triangle is an *Accident*, becauſe this Colour is not neceſſary to a Triangle; and it may be not Black without ceaſing to be a Triangle. *12. What the accidental Property of a Thing is.*

13. The Production of Something which before was not, we call *Generation*; thus we ſay Fire is *generated*, when we ſee Fire where the Wood was before; ſo likewiſe we ſay a Chicken is *generated*, when we ſee a Chicken in the room of an Egg. *13. What is meant by the Word Generation.*

14. When a Thing is deſtroyed, or ceaſes to be what it was before, we call it *Corruption*; thus we ſay it is a *Corruption* of the Wood, when we ſee the Wood no longer, but only the Fire in the Place of it; And in the ſame manner we ſay an Egg is *corrupted*, when we ſee the Egg no longer, but a Chicken in its Place. *14. What is meant by that of Corruption.*

15. A Thing is ſaid to be *altered*, when it has undergone ſome Change, but not ſo great a Change as for us not to know it again, or to have a new Name given to it: Thus when a Piece of Iron, which was before cold, is made hot, it is ſaid to be *altered*, for this Change is not ſo great, but we know it to be Iron ſtill, and do not give a new Name to it. We muſt take particular Notice here, that the *Alteration* muſt make but a moderate Change; for if it be ſo great, that we cannot know the Thing thus changed, we do not then ſay that it is *altered*, but that it is corrupted. *15. What is meant by the Word Alteration.*

16. By the *firſt Principles* of natural Things, we underſtand, that which is firſt, and moſt ſimple in them, or that of which they are originally compoſed, and beyond which they cannot be reduced. Thus, the *firſt Principles* of a Chicken, are thoſe Things which are united together to compoſe a Chicken, and which are ſo ſimple, that they themſelves are void of all Compoſition. *16. What is meant by the firſt Principles of natural Things.*

C 17. Now

A page from Rohault's *System of Natural Philosophy* (1723). [Columbia University Library.]

[16] Benjamin Peirce, *A History of Harvard University* (Brown, Shattuck and Co., 1833), p. 151.

[17] F. B. Dexter, *Documentary History of Yale University* (Yale Univ. Press, 1916), p. 193.

The transitional period ended at Harvard and Yale before 1740. Harvard[18] was probably using van s'Gravesande's *Mathematical Elements of Natural Philosophy*, a scientific work, before 1737. Yale students appear to have used the 1731 edition of this work.[19] William and Mary[20] was teaching scientific physics by 1758, but whether it went through a transitional period between 1736 and 1758, the writer is not able to say definitely.

An examination of several editions[21] of the van s'Gravesande book plainly indicates that it relies upon experiments to prove the truth of its principles, rather than the opinions of authorities. Despite its title, it consists almost entirely of descriptions of experiments and uses mathematics sparingly. This textbook was one of the first to confine itself to the field of inanimate nature. In order to explain the experiments properly, it made use of many more illustrations than did the books of the transitional period. Like the other early scientific textbooks in physics, it gave much attention to astronomy, optics and mechanics. Sound, electricity and heat received a few pages each, while magnetism is barely mentioned.

The reason for the stress on astronomy, optics and mechanics lay in the fact that they were well developed and could readily be explained by observation and experiment. Sound does not constitute a very big part of a general physics book even today. While there was some factual information available on heat, electricity and magnetism, theoretical knowledge in those fields was in a very low state. This may have caused the author to neglect them.

This brings to an end the account of physics teaching in American colleges prior to 1750. It was in this period that physics changed from a metaphysical subject, not at all unlike that taught in the medieval universities of the thirteenth century, to a full-fledged science.

[18] Reference 16, p. 237.
[19] Reference 7, p. 5.
[20] G. W. Ewing, "Early Teaching of Science at the College of William and Mary," Bulletin of the College of William and Mary, **32**, 4, Apr. (1938), p. 5.
[21] The editions published in London in 1726, 1737 and 1747; translated from the Latin by J. T. Desaguliers.

Joseph Henry - America's premier physics teacher

Barbara Myers Swartz

On April 28, 1826 the Trustees of the Albany Academy appointed Joseph Henry as Professor of Mathematics and Natural Philosophy for the coming year. He was to start on September first, at an annual salary of $1,000.00, "payable quarterly." For the 28-year old Henry, the move launched a teaching career, first in Albany and then at Princeton, that spanned more than two decades, and an interest in education that lasted a lifetime.

To the scientific community in the 19th century, both in the United States and abroad, Henry was one of the most important American scientists. Yet far too many today would echo the words of the Princeton trustee in 1832 who asked, "Who is Henry?" Physics students and their teachers recall the unit of inductance named for the man. Many know of Henry's experiments which seemed to parallel or follow Faraday's in the first half of the 19th century. However, few are aware of Henry's role as a physics teacher or his contributions to physics teaching, both at the equivalent of junior-senior high school and in college. Many of his experiments and methods are still used today, and many of the "new discoveries" made earlier in this century and associated with teaching were really "old Henry" views and practices. His teaching deserves a new look.

Learning by doing

Henry was a firm believer in "learning-by-doing" long before Dewey's popularization of the term at the beginning of this century. Henry's purpose was two-fold: 1) to spark the student's interest; and 2) to provide practical experience in applying principles to everyday life. In teaching science, where limited classroom apparatus made individual learning-by-doing impractical in his day, he had students assist him in demonstrations – to operate complicated batteries, or to pull mightily on a rope attached to the keeper in an effort to separate it from the magnet. In demonstrating magnetic induction, Henry let the students "see" the electricity transfer from one wire to another, in some cases with an arc from the secondary, in others by using the secondary coil to activate a clapper which then struck a bell. Henry found that the "exhibition was rendered more interesting by causing the induction to take place through a number of persons standing in a row between the two conductors," evidently forming part of the secondary circuit. They could actually feel the electricity pass through their bodies "with the effect of a moderate charge from a Leyden jar." He found that when seven feet separated the primary from the secondary, there was a moderate shock felt in the tongue. At three to four feet the shocks were severe, and at one foot the shock was "too severe to be taken."*

*Henry was able to demonstrate induction in many spectacular ways, partly because of the techniques he had developed for producing large coils and magnets. In our present terms, we would describe Henry's discoveries in terms of mutual and self-inductance. Some of his arrangements of iron and coils we would recognize as transformers, with the primary pulsed by sudden connection to a galvanic battery. He also induced current in distant wires or coils by discharging Leyden jars through long primary wires or coils, thus anticipating the discoveries of Hertz.

Barbara Swartz *received her Ph.D. in history from SUNY at Stony Brook. Her creative work includes a biography, research and papers on the American revolution and the westward migration, and raising six children. She has taught history in both schools and colleges, and also holds offices in several government environmental agencies. (Box 1074, Setauket, New York 11733)*

Fig. 1. The young Joseph Henry, probably painted by Julius Ames about 1829. This photographic reproduction and all of the other photographs appearing with this article are from the Smithsonian Institution Photography Collection, Record Unit 95.

Henry also believed that learning-by-doing was a better preparation for applying principles to practical problems. His earlier experience as a surveyor had shown that applying mathematics with sextant and compass was more involved than simply calculating formulas in a notebook. A summer geological expedition with the famed geologist, Amos Eaton, had reinforced this conviction as he absorbed Eaton's own learning-by-doing technique among the students. By the time Henry began teaching at the Albany Academy, he was convinced of the need to teach "mensuration of superfices* particularly as applied to survey and . . . the practical use of the quadrant, the compass and other mathematical instruments by the actual employment of them in the field."

Henry worried just as much about the opposite extreme – all practicality and no scientific theory upon which the practical applications depended. He regretted the "enormous expenditure of time, of ingenuity and of fortune" which were "yearly lavished on the most futile attempts to innovate and improve." He observed, for example, that "there are at the present time many ingenious and illiterate mechanics engaged in attempts to invent self-moving machines," a fact in his view "that does not tell much for the diffusion of knowledge among the mechanics of this country." He recognized that "without a knowledge of science the artisan is little more than a labouring machine," incapable of developing improved designs on his own. "Let him be required to act in some untried exigency where he has not precedent," Henry warned, "and see how soon his slender resources will fail." Science, on the other hand, could teach not only what has been done and what we can expect further, "but it also points out to us what projects are at variance with the laws of nature and what we can never hope to attain." For Henry, science offered the key to the future, a lesson which he sought to impart both as teacher and later as shaper of the newly organized Smithsonian Institution.

*A branch of geometry dealing with the measurement of surface areas.

Use of visual aids

Unlike many of his colleagues, Henry used visual aids to assist him in teaching, at a time when such aids were almost nonexistent. Although the Albany Academy had purchased aids early in its history, most schools had no reference books, no critical editions of literary works, no maps, charts or globes, not even pictures on the walls. Henry stressed the importance of such apparatus to improve teaching. He urged the trustees of the Albany Academy, and later those at Princeton and at the Smithsonian Institution, to provide good demonstration equipment for the teacher.

At first not used to working with visual aids, Henry would append notes in the margin of his lectures: "exhibit a drawing of the apparatus." He devised a demonstration tool consisting of balls, to represent atoms, connected by springs, to represent attraction and repulsion, in order to show schematically how "bodies of atoms are kept in equilibrium." (In deference to the fact that the atomic theory was not yet universally accepted, however, he taught his students that "if matter be not thus constituted, the facts still correspond.")

Henry made extensive use of the blackboard, which had been virtually unknown in classroom teaching before 1820. During a visit to West Point in 1826, he observed that "one article very necessary in teaching *chemestrys* is found in this room VIZ. a black board on which the student is taught the atomic theory and all algebraic formula in chemistry. Indeed it appears to be one of the principles of teaching in this institution that everything as far as practical should be demonstrated on the blackboard." He marveled that "the student is even required to draw all articles of chemical apparatus and explain them in this way."

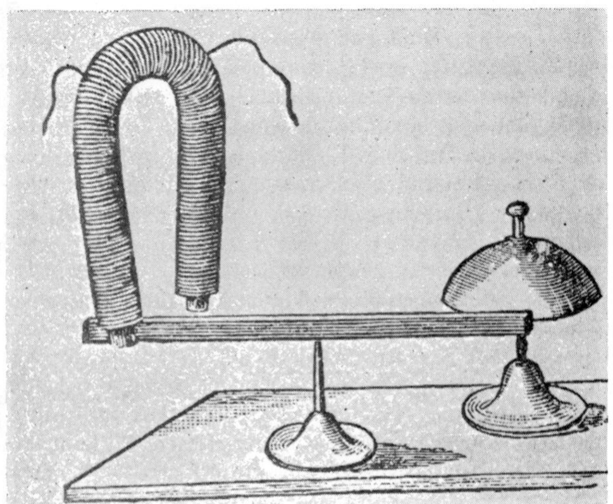

Fig. 2. When Henry activated the magnet, the metal rod swung on the pivot and tapped the bell.

Fig. 3. Classroom at the Albany Academy where Henry taught for six years. Note the blackboards, an innovation in Henry's day.

He did more than marvel at the blackboard method, however: he enthusiastically put it to use, whether to write mathematical rules, draw illustrations, or summarize his lectures. Meticulously copied lecture notes and elaborate drawings of Henry's demonstration apparatus appear with monotonous regularity in the numerous notebooks of his students. Preserved in the Princeton University Archives, the notebooks enable the researcher to observe in what years Henry deviated from his standard lecture series or added new topics as new discoveries were made by the scientific community.

His standard physics course

He deviated rarely. His standard physics course (called Natural Philosophy then), whether in 1832 or 1836 or 1843 or 1847, covered a set list of topics in 65 lectures: *Mechanics* (both "rational and physical") including friction, machines, statics, and dynamics; *Hydrostatics and Hydrodynamics,* including eight propositions relating to hydrostatics and the equilibrium of fluids, followed by motion of liquids, spouting fluids, oscillation of liquids, and water power; *Pneumatics,* covering aerostatics (which treats of the equilibrium of gases) and aerodynamics (including the elasticity of air, ways to induce motion in air, wind and wind power), forcing pumps, engines, the water clock, diving bell, and the aerial ocean around the earth; *Heat* "including the steam engine;" *Sound* (and the "doctrine of vibrations"); *Electricity and Magnetism,* comprising 12 lectures and including such topics as galvanism, electromagnetism, atmospheric electricity, and lightning rods; *Sight and Radiant Heat*; and *Meteorology.*

Henry generally began his lectures in a rather pedestrian way, with a definition of terms: "*Science*, Gentlemen, is the knowledge of the *laws* of phenomena, whether they relate to *mind* or *matter.*" "Philosophy is the knowledge of the laws which regulate the phenomena of nature whether in the intellectual or moral world." "The law of Nature is the relation which pervades a class of facts. Without law, the universe would give no evidence of design." "Chemistry is distinguished from Natural Philosophy inasmuch as it investigates the particular nature of bodies." He then discussed the "four methods by which man arrived at the knowledge of the Laws of Nature," (observation, experiment, induction, and deduction), followed by the art of thinking, the distinction between knowledge and wisdom, and the importance of punctual attendance.

But while the notebooks made dry reading, there is also evidence that he livened up each class with interesting background information to help put the topic into some perspective. After a recapitulation of the preceding lectures he introduced each new topic with historical discussions and reference to important discoveries in physics and chemistry: Newcomen and Watts with the steam engine; the pyramids when talking about applied mechanics; Galileo's discovery of the pressure of the atmosphere when discussing pneumatics. Occasionally on the blackboard there would be a comic illustration of the topic for the day. Steam, for example, was introduced with a figure in tails riding a cylinder of gas, with a cloud coming out of the rear of the "machine," quite graphically demonstrating propulsion.

Demonstrations – the core of his teaching

To Henry, experiments were vital in teaching. In 1826 Henry wrote that "in illustrating . . . every other part of mechanical and natural science well selected experiments are of the utmost value."

Indeed, demonstrations were the core of his teaching. Aware that "this science does not afford to the lecturer the advantage of fixing the attention of his audience by a series of brilliant experiments as in chemistry," Henry strove to find ways of capturing the student's imagination through his demonstrations. Perpetually hampered by lack of enough money to finance purchases of equipment, Henry, like all good teachers, made use of common materials ready at hand: sponges, cotton, sand, silk thread, ink, balls, wires, ropes, springs, pulleys, wheels, sealing wax, fur, kites, drums, quilt paper boxes, bricks, cranks, blow pipes, sheets of glass and copper. His measuring equipment consisted of verniers, calipers, comparators, gauge plates, compasses, bulbs, a micrometer headscrew, a spherometer, and a dividing machine. Where equipment did not exist, as in the new science of electromagnetism, he constructed his own ingenious – and haywired – apparatus, such as electromagnets and complex batteries to use with them.

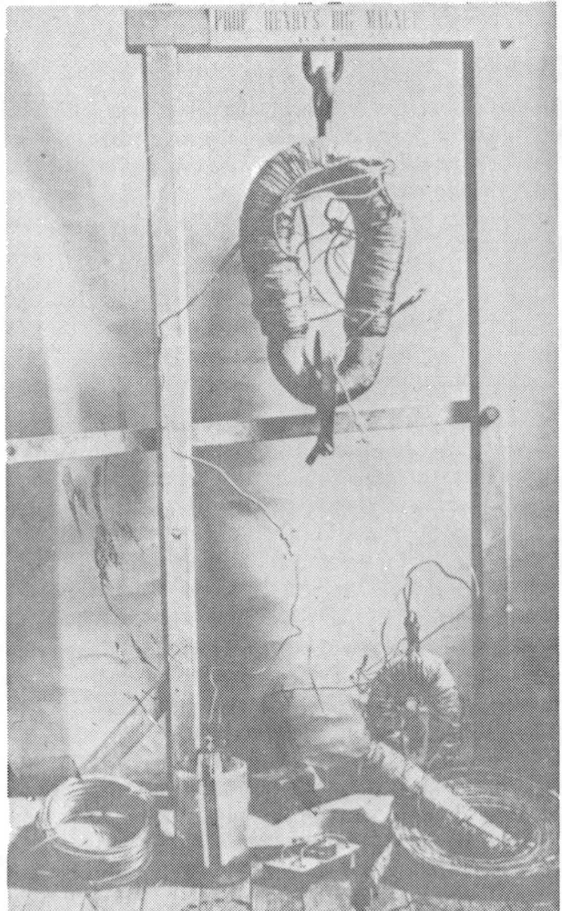

Fig. 4. Collection of some of the apparatus which Henry used at various times in his teaching and research. One of his batteries is the second object from the left. The two objects in the center on the table are what we would call double-pole, double throw relay switches which enabled him to reverse the polarity of the magnet as the two rods on each side dipped into cups of mercury. On the left and the right side of the picture are two different types of insulated coils, one of ordinary thin wire and the other a flat ribbon of wire, each wound with an insulating material. A second magnet stands under the frame to the right.

The problems of apparatus . . .

Henry's magnets were a triumph of ingenuity and persistence over obstinate workmen, unavailable materials, and inadequate knowledge and experience. While teaching in Albany, Henry constructed several small magnets for industrialists as well as other college professors. This was a time-consuming process, for he wrote, "I can get nothing made in Albany in the philosophical line except I stand continually over the workmen during the operation or unless, which is most often the case, I do the work entirely myself."

Materials were in short supply, producing varying results. When possible he substituted other materials. There was frequently a shortage of sheet zinc for his batteries and when he substituted cast zinc, he found it "to produce a different result from the other," for the electromagnet "would operate only with comparatively little power." Subsequently, when there was "no sheet zinc in Albany," he "sent for a supply" from New York City, but in the winter when that supply was gone, he wrote that "I shall not be able to procure any until the river opens." Sometimes there was no appropriate substitute and then "want of material prevented the results being pushed further."

Sometimes Henry used a square bar of Swedish iron about three inches on a side to form his magnet. Sometimes the magnet was "formed of a bar of soft iron 30 inches long and bent into the form of a horse shoe." On occasion he had to purchase a bar of iron and take it to the local forge "as we have no convenience for bending Large Irons." And sometimes he used a "bar of American iron which . . . was unusually hard" and "by no means selected on this account but was taken because it happened to be the only piece the proper size to be procured in Albany at this time."

Once the iron was in hand, the magnets had to be wound, a process that "requires considerable attention in winding and in insulating the wires in order that the greatest effect may be produced with a small galvanic power." To insulate the copper wire, "the ordinary size used for house bells," Henry used a method similar to that for covering "bonnet wire . . . by stretching a strand of wire of 30 feet between a small steel swivel fastened in a vise and the spindle of a common spinning wheel. While one person turned the wheel rapidly [rotating the wire] another guided the silk along the wire and in this manner a strand of 30 feet could be covered with cotton or silk in about 8 or 10 minutes." Using 26 of these covered strands, he wound the iron base for his magnets, leaving each of the ends protruding from the windings. These ends could be connected later in a variety of ways as Henry saw fit.

Henry found the process of winding the magnets "a very tedious one" which "occupied myself and two other persons every evening for two weeks." He learned through experience that he had to be careful about the ends of the wires so there would be no short circuit, for "by not attending to these particulars in one instance a magnet which was partially wound for me by a mechanic entirely failed."

A "small galvanic battery soldered to the ends of the wires" passed a stream of electricity through the wires around the horseshoe, producing temporary magnetism in the soft iron. To build his galvanic battery, Henry used zinc plates, 9 in. x 12 in., surrounded by a copper envelope which was open at the top and bottom, giving him an effective surface of 1½ sq ft. Eleven plates constituted a sub-battery, while eight sub-batteries made up a single whole battery. In his "leisure" he made the "neces-

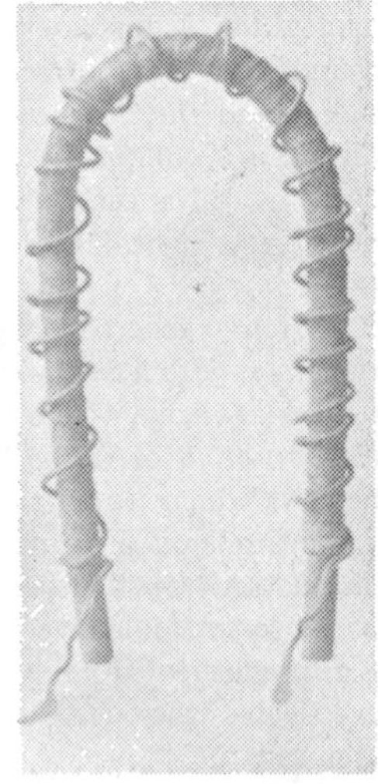

Fig. 5. An "old" type of magnet similar to the one used by the English scientist, William Sturgeon, who varnished the iron to form an insulated horse shoe, and then coiled a single strand of wire around the iron. Henry developed a technique for covering the wire with insulating material and could then create a stronger magnet with the many turns of wire.

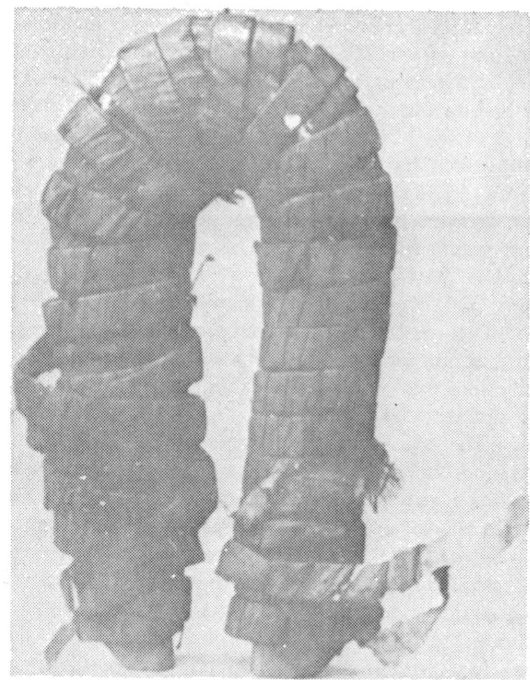

Fig. 6. One of Henry's magnets wound with a wider ribbon of wire covered with insulating material.

sary experiments relative to the proper size of the battery."

To activate the battery, he used a crank or windlass to lower or raise cups of acid into which any single pair or combination of plates could be immersed. The acid was "of such a strength and quality as to act powerfully and suddenly on the zinc." He didn't know the "precise strength of the acid used," for the "acid and water was not measured," but "Mr. Sturgeon recommends nitric acid with six or eight times its weight of water only."*

Henry found a "saturation of magnetism . . . in the greatest degree at the first moment of immersion of the battery," for the "action of the acid did not continue more than half a minute." Henry learned too, that if the magnetic action were to be continued for some time, it would be "necessary to employ a larger battery and to immerse this

Fig. 7. A Cruikshank battery, one of several different types used by Henry. Zinc and copper "sandwiches" fit into grooves inside the trough, and when Henry wished to activate the battery, he poured dilute nitric acid into the trough.

*William Sturgeon was an English scientist whose early magnet was exhibited at the Columbia College of Physicians and Surgeons in New York about 1824 or 1825. Figure 5 is a photograph of his type of magnet.

but partially at first and gradually let it into the acid so that while the power decreases in the part [first] immersed a fresh portion will continually come in contact with the acid." He also made it a practice not to use the battery for several hours before an experiment "in order to produce the greatest effect." Even so, "less than a pint of dilute acid acting on two hands' breadth of zinc would lift 750 pounds."

. . . And the possibilities of apparatus

Using different magnets and different batteries, Henry was able to sustain great weights. The "power of the magnet may be shown to a class . . . by piling on the scale beneath the magnet about three-quarters (say 1500 pounds) of the maximum weight which it will support." Then, "after showing that the magnet fairly sustains this, by slowly withdrawing the acid from the battery we suffer the whole to fall five or six inches." He found that "this never fails to produce a great sensation among the audience as before the fall they can scarcely believe the magnet supports the weight." He urged, however, that "pieces of plank or timber should be placed on the floor so that the fall may not be more than five or six inches." Sometimes he varied the procedure and used students pulling on a rope which looped through several pulleys to demonstrate the ability of the magnet to sustain heavier weights.

To demonstrate reverse polarity, Henry needed two batteries: the one permanently attached to the magnet plus another "attached by means of thimbles of mercury in such a manner that the galvanic current from it may circulate in an opposite direction." After loading the armature with several hundred pounds of weight, he "let an assistant quickly raise the jar containing the acid so as to immerse suddenly the first battery at the same instant you withdraw the poles or wires of the second battery from the amalgamated thimble." He noted that "when this is properly managed the weight will continue to adhere although there is a moment of time when the horseshoe is devoid of magnetism." His students continued to marvel years later that "it could be magnetized, demagnetized and remagnetized so rapidly that a weight of hundreds of pounds could not detach itself from the grasp of the magnet in the interval of reversing the currents."

In order to "render the fact of the actual change of polarity evident to a large class," Henry used magnetic needles, about ten or twelve inches long. He "found it most convenient to make the needles of pieces of watch spring tied together (but magnetized separately)," tucking a small brass cap between them for a bearing. He supported them "on a stand with a fine sewing needle as the pivot," and attached a card marking each pole with North or South. He placed the magnetic needles on each side of the electromagnet and "these at the instant of the change of polarity turn half way round and present their opposite poles to the magnet."

Electromagnetic induction was a favorite subject of Henry's demonstrations. He could induce current from one wire into a parallel wire, whether straight or helical, although the "effect was most strikingly exhibited when the conductor was a flat ribbon, covered with silk, rolled into the form of a helix." With such an arrangement, "brilliant deflagrations and other electrical effects of high intensity were produced." Short wires (1 ft) produced no shock or

Fig. 8. Another of Henry's magnets, with a series of weights suspended from the "keeper" or armature, as Henry called it.

Fig. 9. Two double-pole,double throw relay switches of different design used by Henry at various times, to reverse polarity in a magnet, for example.

Fig. 10.

sparks, but a long wire (30-40 ft) produced both, when the battery ends were withdrawn from the mercury cups. In an open secondary the "snap of the spark could be heard in an adjoining room with the door closed," if the wire "ribbon" were an optimum thickness (1½ in.) and length (96 ft) and coiled. (Note the coiled wire "ribbon" on the right in Fig. 5.)

On occasion he would send a pulse of current through a wire outside the building which induced a pulse in a parallel wire inside the building, or in a similar arrangement, from one room to another. On other occasions he strung two parallel wires outside the building, one in front and the other in back, separated by several hundred feet and terminating in metal plates in the ground. Using a battery of several Leyden jars, he could pulse electrical current through one of the wires. The induced current in the other wire was sufficient to magnetize nearby needles, even though a large structure and considerable distance intervened.

But while electromagnetism fascinated Henry, he used other demonstrations to illustrate other principles of physics and chemistry. To demonstrate porosity and absorption, he filled a glass full of water. Then he carefully lowered a sponge or cotton into the glass, without spilling a drop, "since the body has absorbed all the water

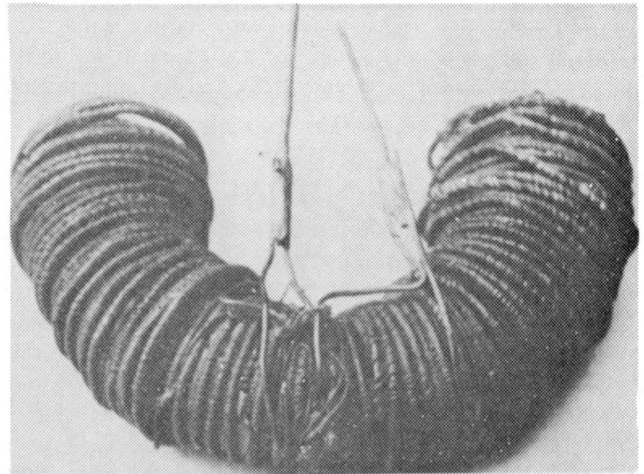

Figs. 10 and 11. A number of permanent magnets, fastened together. Note that a piece was broken out from each end of one of the magnets. An armature which could fit into these slots was wound with a coil of wire (Fig. 11). When the armature and coil were quickly withdrawn, the induced voltage produced a spark across the projecting ends of the coil. The equipment is in the Princeton Collection.

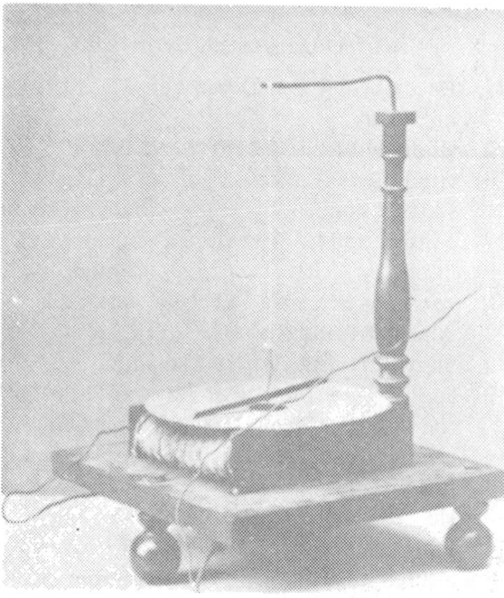

Fig. 12. An early galvanometer to measure electric current. In his early experiments, Henry did not measure the current used in his magnets, but rather spoke about the "strength" of the magnets in terms of the weight each could support.

that was displaced by its being placed in the vessel." To demonstrate centrifugal force, he spun a dish of mercury suspended on a twisted string. As the string untwisted, the velocity increased and the mercury rose from the bottom of the dish to form a ring of metal around the circumference.

To illustrate capillary action he put blue ink in a bowl into which he inserted a glass tube filled with sand which drew the ink up into the tube. Henry used an Atwood's machine to demonstrate the law of falling bodies. To measure moisture in the air, Henry constructed a primitive hygrometer by taking a rope "perfectly clean from grease and other substances which might prevent the absorption of the moisture." Fastening one end, he passed the rope "which should be of considerable length," over a number of pulleys "so as to lessen the friction." On the free end he fastened a pointer which could move up and down against an index as the rope expanded or contracted "in proportion as the atmosphere is moist or dry."

The state of education in Henry's day

In assessing Henry's role as teacher and educator, we must be aware of the state of education and science in his day and avoid judging his contributions by the standards and vantage point of today's complex scientific world. Although great strides had been made in providing public education, it was hardly "universal" by even the middle of the 19th century. Reading, writing, and ciphering were the usual limits of school education for the majority of those who attended formal classes, and only a few northern cities had public high schools.

Teacher training was abysmal. The first private teacher-training school opened as an experiment in Vermont in 1823, and the first public teacher-training school appeared in 1839 in Massachusetts. Until 1829 there were no manuals on "how to teach" and the new teacher of necessity learned by doing. About 1820, when Henry applied to teach in a district school in Albany to help finance his own education, qualifications for teaching were meager: a good moral character; a "thorough knowledge of the common branches" (spelling, reading, and writing); and a willingness to accept the low salaries offered ($8.00 per month, sometimes "payable in money or wheat, according to the terms of the contract"). Low salaries generally also meant that teachers were very young, often no more than 16, and sometimes not as old nor as large as their students.

The learning environment was far from ideal. Schoolhouses were overcrowded, without adequate light and ventilation. Classes were large, with students of varying age and size crowded together in a single room. The teacher expended an "immense amount of physical energy," Henry wrote, "in the vain attempt to keep order while his attention should be wholly occupied with the class under his immediate instruction."

Half of Henry's time at the Albany Academy was devoted to "teaching the higher classes in Mathematics," and the "other half in the drudgery of instructing a class of 60 boys in the elements of Arithmetic." While he preferred to use rewards to stimulate learning, he realized that at times they were ineffectual and recourse must be had to the rod, but then only "sparingly, and perhaps only in extreme cases, and for the purpose of eradicating a vicious habit." Although a mellower Henry paid lip-service to this view in later life, while he was at the Academy, he declared that "as usual I am busily engaged in applying BIRCH to one end and ARITHMETIC to the other."

Teachers used the Lancastrian system of instruction, with older or "faster" students called monitors teaching the rudiments to younger ones. Students laboriously copied theorems, examples, problems to be solved, and other notes from whatever textbooks were available, a method which Henry employed at both the Academy and at Princeton. The pupils memorized their lessons, then "cited" and "re-cited" for the teacher or the monitor, generally not more than half an hour per subject per day.

Less than 1% of the country's youth were able to go to college. However, between a half and three-quarters of the college-bound were from low income or downright poor families, often teaching in a district school during the winter to pay for their education. During the 18th century, the small amount of science offered in college was overshadowed by the classical curriculum with its emphasis on Greek and Latin, on literature and on moral philosophy, rather than more immediately practical subjects. Between 1815 and 1850, the very years when Henry was teaching, there was a spectacular growth both in the number of college science courses offered and in their caliber. By 1850, professors of science and mathematics represented about half of the faculty, while a third of the courses offered or required were in science. In fact, in contrast to the 18th century, when amateur scientists held the monopoly on accumulating and dispersing scientific knowledge, colleges had become the "major depository and dispensary of scientific expertise in education," according to Stanley Guralnick, an historian of science.

For too many, knowledge was considered static, to be acquired by rote drilling and memorization. There was little concern with stimulating intellectual curiosity or in making use of the student's interests to encourage learning.

At the college level there was some attempt to provide new courses to keep pace with new developments in science and to provide instruction in applied fields, as Henry did. "Preparation for life," however, was not a burning concern for it was taken for granted that a liberal education in college afforded the best "preparation for life."

Henry's philosophy of education

Such was the world in which Henry began his teaching career. Henry brought to his new profession a philosophy of education based on his own experiences and observations as a student, a tutor, and a district school teacher. This philosophy would undergo only slight modification during the remainder of his life. It was a remarkable philosophy for its time.

Borrowing from the discipline he knew and loved, Henry was convinced that the "laws which govern the growth and operation of the human mind are as definite, and as general in their application, as those which apply to the material universe." He believed that the "different faculties and powers of the mind are not simultaneously but successively developed," with "some periods of life . . . better suited to particular acquirements than others." In educating the individual, he observed, "we ought to follow the order of nature, and to adapt the instruments to the age and mental stature of the pupil." In an age which saw all children as little adults, there is a remarkably modern and Piagetian ring to this view.

For Henry the "order of nature is that of art before science." Thus, "memory, imitation, imagination, and the faculty of forming mental habits" should be the objects of early education, for it was "only in early life, while the mind is in a pliable condition," according to Henry, "that these mental faculties can most readily and most perfectly be acquired." He advocated development of language skills and rote drilling in arithmetic as the "principal and prominent object of the primary or common school," and declared that "nothing ought to be suffered to usurp their place," for, he warned, "as life advances the facility of verbal acquisitions declines."

The "higher principles of science, on which these arts depend," however, could "only be thoroughly understood by a mind more fully matured," for the "judgment and the reasoning powers are of slower growth." These should be the "last part of mental education," for they are "the latest in arriving at maturity."

He deplored the growing tendency to neglect rote drilling, as in arithmetic or grammar, and "to substitute other objects of more apparent, but of less intrinsic value," such as the once-over-lightly approach with rapid advancement to the next higher level of study. He required his own arithmetic students to read all of Euclid's books at least four times, to be sure the pupils had caught all of importance with the subsequent readings, to reinforce the learning experience through repetition, and to appreciate the beauty of Euclid's mathematical language and proofs. Henry predicted that the less rigorous methods of education would lead not only to "irreparable injury to the individual, but also to the public," since "all the practical operations of life in which these processes are concerned . . . are badly performed."

Of more serious consequence, however, was the "endeavor . . . to invert the order of nature, and attempt to impart those things which cannot be taught at an early age, and to neglect those which at this period of life the mind is well adapted to receive." He saw the "new and less laborious system of early precocious development" as retrograded, not advanced. He charged that "by this mode we may indeed produce remarkably intelligent children who will become remarkably feeble men." To an age which agonizes over the declining SAT scores and aches to explain why Johnny can't read, there is a remarkably prophetic note to his words.

In 1840 Abbott Lawrence gave Harvard $50,000.00 to start the Lawrence Scientific School and for the first time colleges acknowledged a responsibility for technical education. While many in the general public had argued for more relevant education, which emphasized vocational training and practical application rather than book learning and theories, Henry from the start advocated a practical education for his students, designed to give both skills and theoretical background. Perhaps in reflection of his own difficulties in obtaining an education and finding a niche for himself, he advocated providing every pupil with the "opportunity of passing step by step through the whole series" of graded courses. While he acknowledged that few might finish, yet "at whatever period the pupil may abandon his studies, he should be found fitted for some definite pursuit or position in life." In an age when "preparation for life" had no particular meaning in the educational system, Henry's views were decidedly in advance of his time.

Brief facts on Joseph Henry's life

1797,	December 9 (or December 17, 1799) birth in Albany
1822	graduated from Albany Academy
1826	appointed Professor of Mathematics and Natural Philosophy at the Albany Academy
1827	constructed his first electromagnet
1829	honorary LL.D. from Union College
1830	constructed first electric telegraph – at Albany Academy
1830	married cousin Harriet Alexander
1832	appointed Professor of Natural Philosophy at the College of New Jersey (Princeton)
1846	elected first Secretary of Smithsonian Institution
1849	elected president of AAAS
1851	honorary LL.D. from Harvard University
1868	elected president of U.S. National Academy of Sciences
1871	elected president of Philosophical Society of Washington, DC
1871	appointed Chairman of the U.S. Light House Board
1878,	May 14 death at Washington, DC

The state of science in Henry's day

Until well into the 19th century, a major part of American science consisted of collecting specimens, arranging them in cabinets, and displaying them to students and the public in popular "scientific" lectures. "Scientific" articles also filled the newspapers, generally about curiosities or absurdities, and couched in technical jargon calculated to impress the uninitiated. "Thus a person with a smattering of a knowledge of shells," Henry charged, "makes a speech on the subject to the geological section, is much applauded by those who know but little about the affair and in this way cheats himself and the American public into the belief that he really knows something." Other members of the scientific community extended the fields of geography, navigation, and surveying through the practical application of mathematics, astronomy, and geophysics, while only a very few were involved in basic research. But so long as every man was his own scientist, able to collect specimens and give them names and thus participate in "science," the egalitarian atmosphere of the new democracy and the emphasis on immediate practicality worked against the development of a rigorous scientific method and the acceptance of American scientists on a par with those in Europe.

Henry was disturbed by the generally low esteem in which American scientists were held by Europeans and by the caliber of American science in general. The scientific community was small, with little formal opportunity for communication. It required three weeks for the mail packet to arrive from Europe while mail service in the United States was erratic.

While he was at the Albany Academy, Henry had "access to a valuable collection of scientific works and most of the European periodical publications," the major means available for keeping abreast of new developments in his field. His was a unique situation, however, for few American institutions felt the need for or subscribed to these journals. In contrast to Europe which had an extensive network of scientific publications and high quality research, America boasted few scientific journals and none of the caliber of the European.

Henry chided Benjamin Silliman, Sr., the editor of the major *American Journal of Science and the Arts*, for publishing "little that is new or interesting" and for filling the *Journal* "with a mass of trash relative to electricity and electro magnetism which would disgrace the annals of electricity." Henry objected to "the injury done to the character of American science by such publications," not only by denying American scientists the opportunity to keep up with new developments, but also making American science a subject for potential ridicule abroad.

Convinced that a "promiscuous assembly of those who call themselves men of science in this country would only end in our disgrace," he urged that the "real working men in the way of science in this country should make common cause and endeavor by every proper means unitedly to raise their own scientific character. To make science more respected at home to increase the facilities of scientific investigations and the inducements to scientific labours." His role as a teacher and his allied contributions must be seen in this larger context.

Henry early recognized the need for contacts between practitioners of the same discipline. Throughout his life, both as a teacher and long after he had formally left the profession, he urged and assisted in the formation of scientific societies. "Frequent interchange of ideas and appreciative encouragement are almost essential to the successful prosecution of labors requiring profound thought and continued mental exertion," he noted. Observing that "those engaged in similar pursuits should have the opportunities for frequent meetings at stated periods," he found this "more particularly the case with the cultivators of the abstract science, who find but comparatively few fully capable of appreciating the value of their labors, even in a community how much soever enlightened it may be on general subjects." Apart from the stimulus such contacts afforded, it provided the chance for keeping up on developments in the field as well as focusing, through questions, the advance of knowledge.

Henry believed that active research was an important function of a good teacher and he vigorously denied that good researchers were not good teachers or vice-versa. Indeed, citing Priestley, Dalton, Davy, Oersted, Faraday and "a host of others," he declared that "it is precisely among the most celebrated explorers of science in the present century that the most successful and noted teachers have been found."

It was "equally clear that the practice of teaching is . . . not incompatible with the leisure and concentration of mind requisite for original research." More than that, the "latter must, in fact, act beneficially alike on the instructor and the instructed." He declared that "a small amount of lecturing . . . just enough by sympathetic communication with admiring pupils" can "fan, as it were, his enthusiasm," while the students "catch . . . a portion of the enthusiasm of the master, and are stimulated to exertions of which they would otherwise be incapable." In addition, the instructor would gain "in clearness of conception in the appreciation of the new truths he is unfolding by imparting a knowledge of their character to others," and in answering questions posed by the students would gain new insights into the phenomena of nature. Such a teacher could stimulate "other members of the faculty . . . to higher efforts" as well.

Henry's scientific courses themselves helped raise the caliber of science in America. By designing and teaching scientific courses which came to grips with causes and observed effects, which made room for new scientific discoveries, and which taught a rigorous scientific method, Henry with his colleagues raised a new generation of scientists with a solid scientific background, trained in laboratory skills and able to enter as full partners into the European world of science. Alfred Mayer, his "informal" student and probably his most famous protégé, is a case in point.

That most of his students chose not to enter that new exciting scientific world, however, should not negate nor obscure Henry's role. Long after he had left the teaching profession, he continued to work for solid scientific advances and rigorous education through his stubborn insistence on using Smithson's bequest to the United States for the advancement of scientific knowledge and for its dispersion both to practitioners and to the general populace. His *Annual Reports* and other Smithsonian publications provided him with a medium for reaching most scientists and for implanting the changes he thought necessary for American science. He was the leaven, not the finished loaf, and his contributions to science education enabled the profound changes to occur which we have seen during the past century.

He offered his students a window on the new world. He displayed a "faculty . . . of presenting the deepest truths with a clearness and simplicity that brought them within the grasp of ordinary minds." Such was his teaching that "his lectures always received the most profound

Fig. 13. Stained glass window in the First Presbyterian Church in Albany, New York where Henry was baptized.

attention," and "no graduate of Nassau Hall [at Princeton] in that period went forth from its walls without a profound sense of the great benefit derived from the instructions of the professor, and a warm attachment to the man." His "style was pure and simple, very terse and forcible," one student recalled, and "his manner of lecturing easy, graceful, and impressive." In fact, "so admirably were the principles of physical science expressed, so clearly were the facts presented, and so successfully were the experiments performed," in the opinion of one of his students, that "even the dullest members of the class had knowledge forced into them almost without effort on their part, and the brightest were aroused to the utmost enthusiasm."

It was in keeping with his spirit that he should devote his entire life to education of some sort. In 1854 he wrote, "he has not lived in vain, who leaves behind him . . . a child better educated morally, intellectually, and physically than himself."

Bibliography:

Most of the quotations in this paper were taken from Joseph Henry's letters, diaries, papers, etc., now housed in the Archives of the Smithsonian Institution. Nathan Reingold and his assistants are in the process of editing the vast collection, providing easy access through these splendid volumes to the Henry materials. The third volume of the projected 15-volume *Papers of Joseph Henry* recently was sent to the publishers.

Other quotations and ideas came from:
Student notebooks in the Joseph Henry Collection in the Princeton University Archives;

the *Smithsonian Miscellaneous Collections;*

the *Annual Reports to the Board of Regents of the Smithsonian Institution;*

Ellwood P. Cubberley, *Readings in Public Education in the United States;*

Clifton Johnson, *Old Time Schools and Schoolbooks;*

R. Freeman Butts and Lawrence Cremin, *History of Education in American Culture;*

Theodore R. Sizer, *The Age of the Academies;*

Nathan Reingold, editor, *Science in 19th Century America;*

Donald Zochert, "Science and the Common Man in Ante-Bellum America" and

Stanley Guralnick, "Sources of Misconceptions on the Role of Science in 19th Century American Colleges," both of which originally appeared in *ISIS*, but have been reprinted in Nathan Reingold, editor, *Science in America Since 1820.*

There is a single biography of Henry: *Joseph Henry, His Life and Works* by Thomas Coulson.

A History of the Physics Laboratory in the American Public High School (to 1910)*

SIDNEY ROSEN
Brandeis University, Waltham, Massachusetts and Harvard University, Cambridge, Massachusetts

(Received July 22, 1953)

This paper attempts to trace the development of the American high-school physics laboratory from its beginnings in the early 1800's to its culminating domination of the science-teaching scene in 1910. An effort has been made to follow lines of educational and scientific thinking during the successive stages of development, so that a total picture of the cultural and social forces that shaped this form of science education emerges.

1. INTRODUCTION

MOST of us who, since 1910, have been exposed to high-school physics courses during our public school education are quite familiar with their pattern. Individual laboratory work forms an integral part of most of these courses. There are some physicists and physics teachers who feel that in the high schools today the physics laboratory is not being used to great advantage, and who have suggested alternate programs.

It may be of interest to those who are concerned with this problem to examine the historical development of the high-school physics laboratory with a particular view to the cultural forces that shaped the concept of laboratory work for the individual student. The year 1910 has been chosen as a chronological terminus for this history. Though the last half-century has seen some changes in practice in the laboratory (particularly the addition of visual aid equipment), the theory behind the format of high-school laboratory work seems to have undergone little further development.

2. EARLY USE OF APPARATUS

Science teaching, and particularly the teaching of natural philosophy, in the secondary schools was, for the most part, a matter of textbook pedagogy until the post Civil War period. The first headmaster of the Cambridge, Massachusetts, English High School, Elbridge Smith, could recall that when the school opened in 1847:

"In science, the instruction was wholly by catechism. There were illustrative diagrams in the textbook which might, or might not, be explained and might, or might not, be transferred to the blackboard . . . Not a single particle of apparatus or a book of reference except the Bible and possibly a dictionary."[1]

The larger colleges and wealthier academies had some demonstration apparatus; but in the early days such equipment was the exception rather than the rule in the secondary schools. In a listing of academies in New England in 1839, only two out of more than fifty are recorded as having a "good chemical and philosophical apparatus." As early as 1834, however, some college professors were urging the use of demonstration apparatus for the better teaching of natural philosophy in secondary schools.[2]

In 1847, the Boston School Committee adopted a list of apparatus to be purchased from the Wightman Apparatus Company, and to be furnished to each school. The equipment cost $260.00 (plus $5.00 for packing) and included among its sixty-five items an inertia apparatus, a complete set of mechanical pulleys, simple and compounds levers, an inclined plane, a capstan, a hydrostatic bellows, a patent lever air pump, brass Magdeburg hemispheres, ten assorted electrical toys, and a dissected eyeball, showing its arrangement.[3] Figure 1 shows some typical apparatus of the period.

A familiar name in the educational literature before the Civil War was that of Josiah Holbrook, a student of Silliman at Yale who began to

* This paper was adapted, with modifications, from part of a doctoral dissertation being prepared by the author at Harvard University.

[1] E. Smith, "Early history of the high school," *Annual Report of the School Committee*, Cambridge, Massachusetts (1892), p. 54.

[2] B. Hale, "On the best mode of teaching natural philosophy," *Lectures of Am. Inst. of Instr.* (Boston, Massachusetts, 1834), p. 289.

[3] J. P. Wightman, *Catalogue of Phil., Astron., Chem., & Electr. Apparatus* (Boston, Massachusetts, 1854).

manufacture philosophical apparatus at Hartford, Connecticut, about 1829.[4] With the fervent zeal of a missionary, Holbrook sold, along with his apparatus, the *cause célèbre* of teaching science by the demonstration of phenomena. With every set of apparatus, he included not only an instruction booklet, but a complete set of question and answer manuals for teacher and student that simplified instruction to the *n*th degree. Holbrook's lectures to teacher and parent groups stimulated the widespread use of his comparatively inexpensive orreries and tellurians[5] (see Fig. 2); in 1831, he founded the Boston Mechanic's Lyceum, where science was popularized for adults. After Holbrook's death in 1854, the business was expanded, with offices in New York and Chicago, and the leading sellers of school

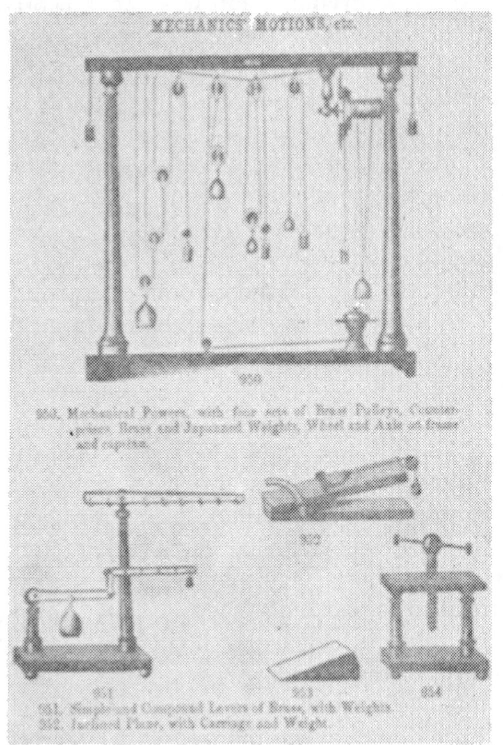

FIG. 1. A page from the *Illustrated Catalogue of Mathematical, Optical, and Philosophical Instruments and School Papers* (James W. Queen and Company, Philadelphia, 1859). This catalog also illustrated centrifugal force apparatus, air pumps, and many electrical toys.

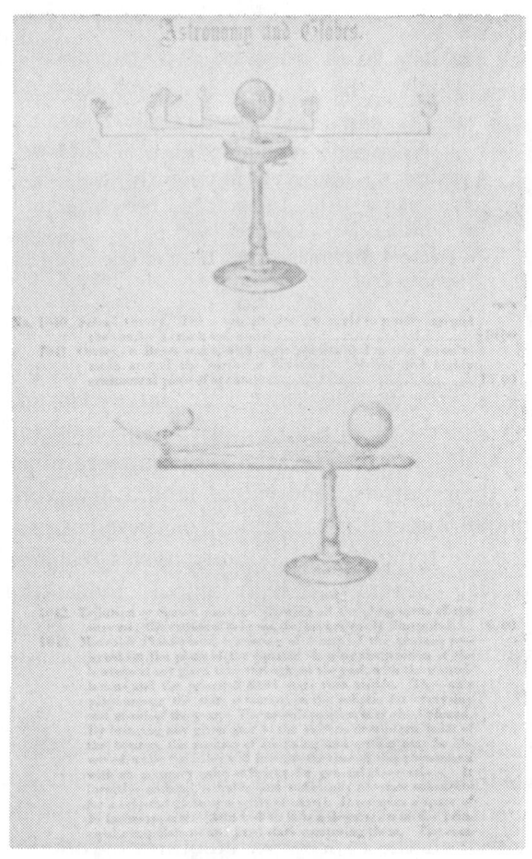

FIG. 2. A page from the Queen Company catalog showing the famous Holbrook orrery and tellurian.

equipment featured Holbrook's apparatus in their catalogs.

For those high schools (mostly in large urban areas) whose financial status permitted the purchase of apparatus, "the teaching of science passed into the illustrative stage. Principles of chemistry and physics were illustrated by air-pump, electric machine, battery and their accessories . . . superstitions were dispelled . . . ."[6] It was a rare school, however, where the pupils "were permitted and even encouraged to take the apparatus in hand and learn by manipulation as well as observation."[1]

3. THE CHANGING ATTITUDE TOWARD SCIENCE TEACHING

Massachusetts, in 1857, sparked a growing enthusiasm for science in the high school by passing a law requiring the teaching of natural

[4] J. Holbrook, Am. J. Educ. VIII **20**, 299 (1860).

[5] *Orrery*—a three-dimensional model of the solar system, cranked by hand; *tellurian*—a similar model of the sun-earth-moon system.

[6] See reference 1, p. 61.

philosophy, chemistry, and botany in high schools located in towns of 4000 population or over. During the post-bellum years, this effort was rapidly expanded by: (1) the influence of Herbert Spencer's educational philosophy; (2) the Morrill Act of 1862; (3) the influence of the German universities on American scholars: (4) the apparent success of the German *Realschule*, the French *école polytechnique*, and, by the late 1870's, the Russian and Swedish industrial schools.

The publication, in England, in 1861, of Herbert Spencer's essays on education had a marked effect on American educational thinking. At a time when "an intensely materialistic spirit reigned—the urge to exploit new sources of wealth, to make fortunes, to grasp power,"[7] Spencer's plea for the replacement of the classical curriculum by science found fertile ground. Compatible with the go-getting, business-minded pattern of American life were such statements as the following:

"Efficiency in the production, preparation, and distribution of commodities depends on those methods fitted to the respective natures of those commodities; it depends on an adequate knowledge of their physical, chemical, or vital properties, as the case may be; that is, it depends on Science."[8]

More irresistible was Spencer's crashing summation:

"For direct self-preservation, or the maintenance of life and health, the most all important knowledge is—Science. For that indirect self-preservation which we call earning a livelihood, the knowledge of greatest value is—Science. For the due discharge of parental functions, the proper guidance is to be found only in—Science . . . for the most perfect production and highest enjoyment of art in all its forms, the needful preparation is still—Science. And for the purposes of discipline, intellectual, moral, religious—the most efficient study is, once more—Science."[9]

Proponents of Spencerian "scientism" were quick to urge the adoption of school curriculums strongly scientific in nature:

"As Sparta trained her youth from earliest years for war, our public schools must inspire zealous and skillful devotion to our material industury for the more sacred uses of peace. Then we shall give to the world untold combinations of physical and chemical forces which, like the American telegraph, shall promote a civilization worthy of the highest Christian nation."[10]

Two decades later, however, more cautious educators were advising that "care should be taken to prevent the submerging of humanism by the advancing tide of science,"[11] and that "Herbert Spencer's reasoning as to what knowledge is of most worth seems to the ordinary reader to be very good logic. The strongest argument against it . . . is the fact that when tried, it is difficult to get as good results from it as from the system which he strongly condemns."[12]

In 1862 Congress, with President Lincoln's approval, passed an act granting to each state 30 000 acres of federal land for each legislator the state had in Washington. This land was to be used toward the endowment of a college of mechanical and industrial arts. Sponsored by Justin P. Morrill of Vermont, the law fostered the creation of many Midwest and Western universities and helped inject new blood into some of the older ones already in existence.

The vocational bent of the land-grant colleges and universities served to influence the curriculums of many high schools. Primarily, a door to higher education was opened for the graduate of the English course in the high school; also, emphasis began to be placed on science courses in the secondary schools. In 1875, the High School Committee of Milwaukee, Wisconsin, recommended that Greek be dropped from the curriculum "since few graduates go to the higher institutions of learning, and Greek is not required at the State University for admission."[13]

During the 19th century in America, higher education was influenced more and more by the

[7] A. M. Schlesinger, *Political & Social Growth of the American People, 1865–1940* (Macmillan Company, New York, 1941), p. 41.

[8] H. Spencer, *Education, Intellectual, Moral, and Physical* (Appleton-Century-Crofts, Inc., New York, 1900), p. 12.

[9] See reference 8, p. 84.

[10] H. O. Ladd, "Elements of natural science in our public schools," *Am. Inst. of Instr.* (Boston, 1876), p. 27.

[11] J. G. Schurman, School Rev. I, 2, 67 (1893).

[12] R. S. Keyser, School Rev. I, **3**, 137 (1893).

[13] *School Reports* (Milwaukee, 1875), p. 32.

growing number of college graduates who went to Europe to do graduate work at the German universities. These men returned to the United States consumed by a passion for a new kind of truth. Goethe in the previous century had coined an aphorism: "*Die Weisheit ist die Wahrheit.*" The German university scholars changed this attitude to: " . . . *nur in dem Streben nach der Wahrheit.*" Truth was not final; "final truth" was a superstition from which man could only be freed by the earnest scholar penetrating deeply into narrow, specialized fields, using "the special library, the laboratory, the seminar, the monograph, and the learned periodical."[14]

By the end of the 19th century, the "specialist" dominated all fields of higher knowledge, and, to use the metaphor of a famous American social historian, "wedded a skeleton bride, whose osseous kiss and rattling embrace rewarded him with an ecstasy beyond Helen's."[15] Especially in scientific investigation did the German university ideals flourish; for the laboratory and its very implications of impersonal observation of factual phenomena postulated the most complete adherence to the new attitude toward truth.

It was inevitable that the secondary school be infected with this contagious ideology; in the 1870's, the laboratory method of teaching the physical sciences began to appear in the high schools. By the turn of the century, however, some educators were aware of the difference in specialization at the higher and secondary levels. One physics teacher, writing about 1910, saw that through the infection of teachers by the reverence for research, "the idea of educating young people through science has been for the time being lost sight of . . . science teachers have been so busy trying to teach science, they have forgotten to teach boys and girls."[16]

At the same time, the *Realschule* in Germany and the *école polytechnique* in France, secondary schools whose curriculums emphasized scientific courses of a propaedeutic nature, also served to accentuate the demand for a science-based curriculum in the American high school. Physics was a required subject in these schools; in the German school, it was taught with great thoroughness and consistency; in the French, with a lighter, more philosophical and historical attitude. Along with praise for its teachers and methods, however, the *Realschule* was seen to stress knowledge too much—the student had "little self-reliance and acquaintance with the real world."[17] Nevertheless, these school systems excited commentaries and analyses in many Bureau of Education circulars issued in the latter decades of the 19th century.

The feeling for vocational practicality as the core of the high-school course was certainly heightened by the Russian Technical School exhibit of pupils' woodwork and iron handicraft at the Philadelphia Exposition in 1876.[18] The concept of a school where children could learn to work with their hands as a form of creative expression was part of Froebel's[19] educational plan. Finland, Russia, and then Sweden constructed school systems based on this kind of vocational training, involving instruction in woodwork, bookbinding, copper work, and other arts. President John D. Runkle of the Massachusetts Institute of Technology was so impressed by the Russian exhibit, he decided to make a course in the use of tools a part of the Institute's curriculum. His enthusiasm spread to the secondary schools, and soon "manual training" became part of the American school pattern. In the high school, this led to the inclusion of courses in mechanical drawing and applied physics; also, to the creation of manual arts or "Trade" schools.[20] Manual training, however, became a required course only in the elementary school and had little effect on science teaching in the high school except for the accenting of the application of physical principles to machinery

[14] M. Curti, *The Growth of American Thought* (Harper and Brothers, New York, 1943), p. 582.

[15] A. M. Schlesinger, *The Rise of the City* (Macmillan Company, New York, 1933), p. 220.

[16] C. R. Mann, "The present condition of physics teaching in the United States," *Broad Lines in Science Teaching* (Macmillan Company, New York, 1910), Chap. XIX, p. 227.

[17] J. K. Lord, Education VII, 4, 245 (1886).

[18] This was the Exposition that hailed the centennial of American independence. Though the 1876 Exposition is considered to have been mediocre compared to later expositions and fairs, it had two direct effects on future American history: (1) the amount, variety, and excellence of the foreign exhibits jolted most of the 9 million Americans who came to see out of their almost complete provincialism; (2) in an obscure corner of the Massachusetts exhibits, Alexander Bell displayed his telephone to the world.

[19] Froebel was the German educator responsible for the kindergarten. His dates are 1782–1852.

[20] The first manual training school in the United States was opened in St. Louis, Missouri, in 1880.

in some schools. An unusual suggestion for manual training in the high school did appear in 1910, as a result of a movement in the German schools called *Arbeitsunterricht* or, broadly, "constructive activity." This was simply the combining of manual training and laboratory physics, where the pupil learned to use tools in order to construct his own physical apparatus. The student would then supposedly have a feeling of "oneness" with the apparatus, and learn his physical principles more thoroughly and willingly.[21]

Physics laboratories where the individual student or groups of students could perform experiments were *rarae aves* in the secondary schools until the last decade of the 19th century. The chemistry laboratory was already in use in some of the schools by 1870. In that year, it was written of the new chemical laboratory in the Boston Girls' High and Normal School:

"All foreigners and visitors to the city were brought to see it and they opened their eyes in amazement at the strange sight."[22]

It is difficult to determine how many secondary schools had physics courses where pupils actually performed experiments individually, since the phrase "laboratory work" could also mean demonstration by the teacher.

In 1878, a Commission of Education report showed that two factors were causing resistance to the introduction of physics laboratory courses in the high schools: (1) the cost was, for many schools, prohibitive; (2) teachers trained in laboratory methods were scarce. Cost, however, was not a major factor in the large urban centers of the United States.[23]

In the rural areas, instruction in physics was probably wholly textbook teaching, with some experiments performed occasionally by the teacher; this was the situation in a great majority of the cases reported by the Bureau of Education in 1880. The rural teacher, if he had the resourcefulness and opportunity, could use the phenomena of nature, observed first-hand and roughly measured, to teach physical principles; or, the school could purchase a set of *Dr. Johnson's Substitute for Philosophical Apparatus*:

"Consisting of indestructible charts, comprising 500 diagrams, which represent over $6000 worth of apparatus, illustrating the principles of natural philosophy and astronomy . . . designed as a cheap and durable substitute for the expensive philosophical apparatus in the Common Schools and Academies. Set for $15—10 charts."[24] Chart No. 10 is reproduced in Fig. 3.

The problem of getting competent teachers was not so easily solvable. In 1871, the average high-school faculty consisted of one college graduate who taught Latin and Greek, and a few others who taught all other branches. This condition was somewhat ameliorated during the period 1880–1900 by the increase of science courses, including laboratory work, in the nation's normal schools. Such pedagogical devices did not seem to produce apt teachers. In 1887, *The Academy* editorialized that "high school principals and academic teachers are more reluctant to say anything to the public about the science teaching that is now done in their schools than in any other department of academic work . . . pupils are crammed with facts to pass the Regents examinations . . . there is a lack of teachers in secondary schools who are competent to instruct in science . . . many teachers teach science only as an assignment . . . and are incompetent to teach science because of poor training."[25] A few years later, a University of Wyoming professor wrote:

"It is impossible, under present conditions successfully to teach the sciences, especially those requiring laboratory equipment, in our public schools. We should not undertake what we cannot do well . . . the majority of our public school teachers are not prepared to give suitable instruction in all the sciences . . . it requires a specialist to see all the ultimate known facts of a single line."[26]

The problem of getting the scientific specialist to expend his efforts in the field of secondary education was appraised as early as 1872 with

[21] F. N. Freeman, School Rev. XVII, 9, 609 (1909).
[22] R. P. Williams, Science N.S. XIII, **342**, 100–104 (1901).
[23] In 1876–1877, the city of Boston spent $8465.00 on philosophical, chemical, and mathematical apparatus. In 1886, the Milwaukee High School spent $682.28 on physical apparatus alone!

[24] From an advertisement in The Am. Educ. Weekly (1873).
[25] Editorial, The Academy, II, **2**, 82 (1887).
[26] A. Nelson, School Rev. I, **8**, 471 (1893).

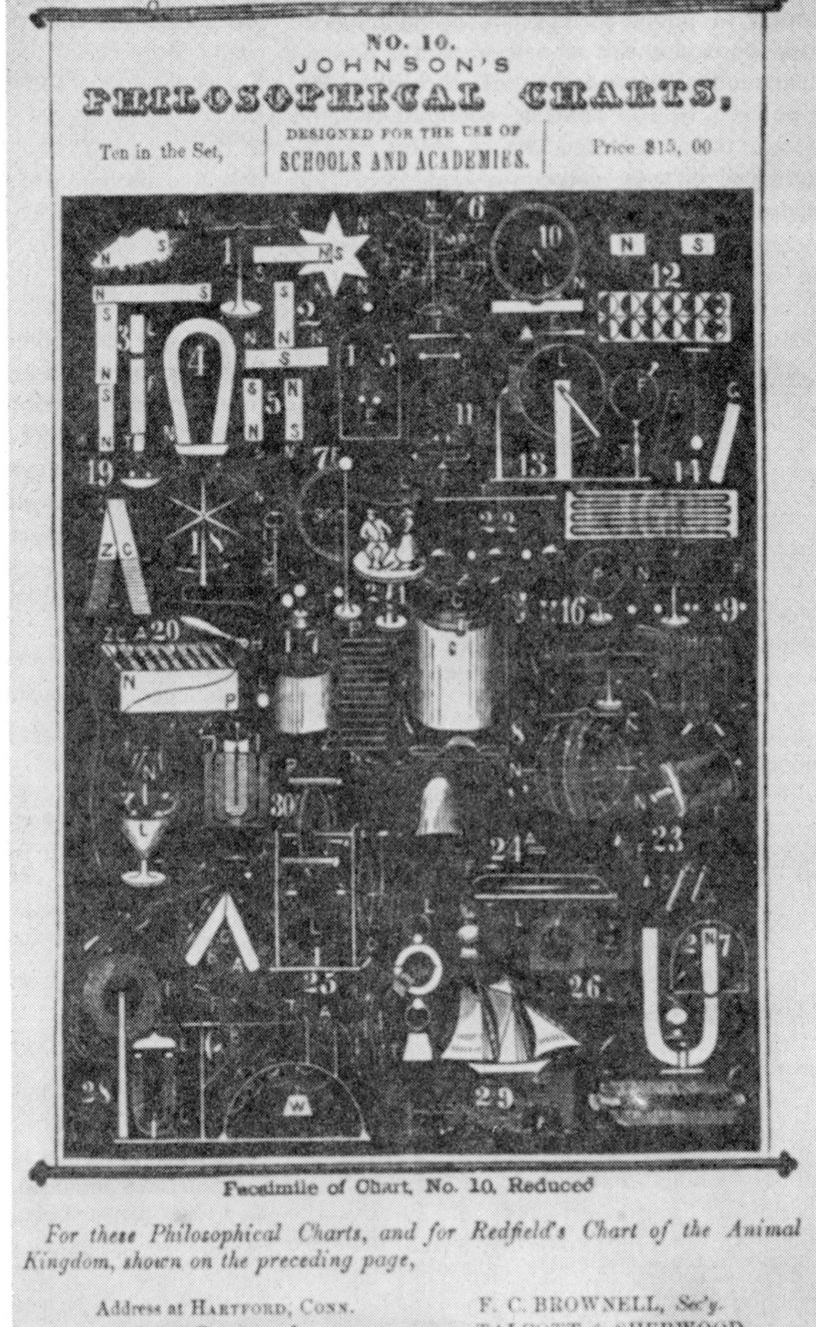

FIG. 3. One of the Dr. Johnson charts. During this period, many wall charts were printed white on black to give the effect of chalk on a blackboard.

great foresight by Daniel C. Gilman of Yale (who was to become president of Johns Hopkins in 1876):

"The great cost of high schools of science is by no means appreciated. The necessity for having men of distinction in special investigations, and for having a great many special teachers, and for having ample means of experiment and illustration,—all this is very imperfectly understood. The readiness with which men of truly scientific attainments are caught up to aid in the construction of public works, the development of mines, the exploration of new territory, the administration of great industrial establishments, and numerous other services, renders it difficult to retain them as instructors of youth on the meagre allowances commonly bestowed for educational services."[27]

In 1880, only 4 out of 176 school systems (in cities with over 7500 population) offered a course in elementary physics with laboratory work. Of some 337 secondary schools reporting to the Bureau of Education (including private schools and academies) only 7 offered such a course; the remainder taught physics either by textbook and demonstration, by textbook alone, or offered no physics at all. By 1906, the picture had changed. J. W. MacDonald, Agent of the Massachusetts Board of Education, was able to report that over 100 high schools in the state had well-equipped chemistry and physics laboratories. Perhaps, the major cause of this sudden swing toward laboratory teaching was the publication of the *Harvard List* in 1886.

4. THE FORTY EXPERIMENTS

The Harvard University Catalogue for 1870 advised students to pursue a course in elementary mechanics before entering. President Eliot, himself a chemist, had already installed at Harvard the elective system of studies, an imitation of the German university system. In 1872, he designated a new option for admission to the college: mathematics and physics. This marked the first acceptance of physics as an accredited secondary school subject by an American college.[28] In 1876, all candidates for admission to the freshman class had to present one of the following combinations of science: (a) botany; (b) physics and chemistry; (c) physics and descriptive anatomy. In 1886, physics itself was divided into a two-part admission subject: either (1) astronomy and physics, covering certain designated chapters of specified textbooks, or (2) "a course of experiments in the subjects of mechanics, sound, light, heat, and electricity, not less than forty in number, actually performed at school by the pupil . . . the Faculty requests all teachers who can command the necessary apparatus to present their pupils in the second of these alternatives." The candidate was also required to come to the University to be examined in experimental prowess at the University physics laboratory. During the same year, a pamphlet listing the forty prescribed experiments along with recommendations concerning the methodology of measurement and the time to be spent in the school laboratory was printed and distributed to the secondary schools.[29] The list of experiments is reprinted here. They were chosen from the pages of the four most popular physics textbooks in 1886: (1) Trowbridge, *New Physics;* (2) Pickering, *Physical Manipulation, Vols. I and II;* (3) Worthington, *Physical Laboratory Practice* (published in London); and (4) Gage, *Elements of Physics.*

The Harvard List of Forty Experiments

Volume

1. Direct measurement of volume.—(Density).
2. Displacement.

Density

3. Principle of Archimedes.
4. Density of solid by flotation.
5. Densimetry.
6. Density of air (rough).

Pressure

7. Estimation of pressure at different depths.
8. Comparisons of densities.—Method of balancing columns.
9. Mariotte's tube.—(Estimation of atmospheric pressure by law of Boyle and Mariotte).
10. Construction and use of barometer.

Mechanics

11. Elasticity.—Stretching of Wire.—(Relation of stress to strain.—Flexure).
12. Breaking strength of fine wires.
13. Coefficient of friction.—(Work).
14. Bent lever.—(Comparison of masses).—Moments of force.
15. Center of gravity.

[27] D. C. Gilman, "On the growth of colleges and their present tendency to the study of science," *Am. Inst. of Instr. N. S.* (Boston, 1872), p. 112.

[28] Downing, *Teaching Science in the Schools* (University of Chicago Press, Chicago, 1925), p. 6.

[29] Harvard College, *Provisional List of Experiments in Elementary Physics for Admission to College in 1887* (Cambridge, 1886).

Dynamics

16. Law of falling bodies.
17. Velocity of falling bodies.
18. Laws of pendulum.—Influence of length of arc.—Variation of length of pendulum.

Sound

19. Pitch.
20. Comparison of two tuning forks.—Graphical method.
21. Resonance tube.—Wavelengths.
22. Velocity of sound.

Heat

23. Temperature.—Fixed points of thermometer.—Melting and boiling points.
24. Expansion by heat.—Air thermometer.
25. Mixing hot and cold water.
26. Specific heat.—Method of mixture.
27. Latent heat of water.
28. Latent heat of steam.
29. Conduction of heat.
30. Loss of heat by convection and radiation.—Law of cooling.

Light

31. Photometry.
32. Law of reflection.—Angle of prism.
33. Refraction.—Principle of spectroscope.—Angle of deviation.
34. Principle focus of lenses.—(Practical examination of telescope and microscope).

Electricity and Magnetism

35. Tracing lines of magnetic force.—Electromagnet.
36. Dipping needle.
37. Electrical attractions and repulsions.—(Arrangement of series in electropositive order).
38. Distribution of electrical charge.—Proof plane.—(Absence of charge in the interior of a conductor).
39. Comparison of emf by Ohm's law.
40. Comparison of resistances by Ohm's law.

This bold dictum to secondary schools had its most immediate effect on those whose graduates were prepared for admission to Harvard (mostly Latin grammar schools and academies). In 1886, the Boston Public Latin School added to its curriculum an extra year of physics, so that the forty Harvard List experiments could be performed. Similar changes were made at the Cambridge Latin School, Milton Academy, and Phillips Academy, Andover. Few, if any, colleges were inspired to imitate Harvard immediately. In 1891, only 19 colleges in the United States required science for admission to the A.B. degree course. Trinity College, in Hartford, Connecticut, did require for admission to the Science course, in 1899, either of the following: (a) biology; (b) botany and physical geography; (c) chemistry; (d) physics—a laboratory notebook covering at least 120 hours of laboratory required; but, by 1903, the laboratory part of the physics requirement was abandoned.

President Eliot chose a young assistant professor of physics, Edwin H. Hall, to manage the administration of the new physics requirement. Hall spent much of his time composing and revising the descriptive circulars that changed as experience with the laboratory requirement increased. He administered the entrance examinations that tested the candidate's ability to handle experimental apparatus (it cost the college as much as $500—a substantial amount in those days—for this examination in 1891),[30] and eventually wrote and published a textbook manual based on the required list of experiments. It is through Hall's writings that the story of the Harvard List and its effects are best told.

During a lecture tour in France in 1915, Hall recalled that: "The influence of this "descriptive list" of experiments . . . spread throughout the country, often suggesting the formation of other lists less difficult than the original. Whatever the faults of this work, it must be recognized that it established the use of the laboratory as a most important element in the teaching of physics in the secondary schools. In this way, the teaching of physics in the secondary schools was specialized in America instead of allowing it to be, as had been for the last thirty years, the secondary work of a teacher . . . who, perhaps, added it to Latin or Greek."[31]

Between 1887 and 1900, educators jumped on the laboratory-method bandwagon. The arguments of those who felt that laboratory work was impractical, over-expensive, impossible, and too sophisticated for the high school were easily overridden by eager proponents of the new method of teaching physics. The president of Brown University, E. Benjamin Andrews, hailed the splendid reform of using "the laboratory, the demonstration, the notebook, the field trip,

[30] Editorial, School & College I, 1, 43 (1892).

[31] E. H. Hall, "American Schools," a lecture delivered in March, 1915 at the Universities of Grenoble and Montpelier, France. The above is the author's translation from the original copy preserved in the Harvard College Library Archives.

and prepared specimens to replace the dry-as-dust old textbook."[32] In the same year, 1892, the Association of Colleges in New England proposed that elementary physics be introduced into the late years of the high-school program as a substantial subject, to be taught by the experimental or laboratory method, and to include exact weighing and measuring by the pupils themselves.[33] The New Bedford, Massachusetts, School Board decided to teach laboratory physics in the high school to small classes of twenty or less and to limit this kind of instruction to boys "since the female mind, in general, was not so constituted as to apprehend the philosophy of physics."[34] Dr. Abercrombie, the head of Worcester Academy, declared stoutly that the experimental method was so valuable that "we require a year's work in laboratory physics of every pupil who would finish our course, whether going to college or not."[35] The swing to laboratory teaching in the secondary schools was imbued with the characteristic optimism of the American people; it was an educational panacea for the aches and pains of a coming scientific age:

"The laboratory has won its place in the school. Its introduction has proved successful. It is destined to revolutionize education. Pupils will go out from our laboratories able to see and to do."[36]

Though, as Hall pointed out, the laboratory course in physics was far superior to the older textbook method of teaching, one defect soon became noticeable. The laboratory method became a fad. After 1890, few new high schools were constructed without physics and chemistry laboratories:

"Empiricism is the watchword of today. 'Read Nature in the language of the experiment,' cries the reformer. The cry has been heard and heeded; and the high school or academy which is not well equipped with laboratories and apparatus is not looked upon as 'progressive,' as 'up to the times'."[37]

There were some who felt that there was an inherent danger in teaching science only by experimentation. Professor Payne of the University of Michigan warned that " . . . In this new theory of teaching, there are at least three specious fallacies: one, that all knowledge is presentative; second, that the chief and highest aim of instruction is discipline; third, that the student is to be a specialist, a scientist . . . the employment of specialists to teach the sciences in secondary schools has given currency to the fallacy that the only proper mode of teaching is by inductive, experimental research."[38]

Professor Hall, by the middle 1890's, had become aware of the failings of the experimental method of teaching physics in the high schools. He blamed President Eliot and Professor Cooke (a Harvard chemist) for "the mistake of projecting the physics course on precisely the same lines as the chemistry course."[39] Instead of tempering experimental work with informational lectures and recitations, the emphasis, Hall found, "was all on experiments and experimenting, as it should be, perhaps, in chemistry, but not in physics . . . the physicist must have time to chew the cud . . . to ignore this . . . is to make a serious mistake."[39]

The Harvard List, in 1897, was modified to emphasize, for secondary schools, a mixed course of laboratory, textbook, and lecture. The number of experiments was reduced to thirty-five to be chosen from sixty. It was also Hall's suggestion that a standardized course in physics be accepted by all college administrations for admission. Such a course would include: (1) not less than one year of physics; (2) a large amount of individual laboratory work, mainly quantitative, done by the pupil; (3) the laboratory work to consume approximately one-half the course time; (4) the rest of the time to be devoted to lecture, textbook instruction, and qualitative experiments by the instructor.[39]

This program was a restatement of the requirements suggested by the Physics, Chemistry and Astronomy Conference (of which Hall was a member) of the famous Committee of Ten.

[32] Andrews, School & College I, **1**, 1 (1892).
[33] School & College I, **1**, 119 (1892).
[34] The Academy II, **3**, 137 (1887).
[35] Abercrombie, School Rev. I, **8**, 451 (1893).
[36] LaR. F. Griffin, School & College I, **8**, 477 (1892).
[37] L. L. Conant, School Rev. I, **3**, 211 (1893).
[38] Payne, The Academy II, **1**, 12 (1887).
[39] E. H. Hall, "A story of experience with physics as a requirement for admission to college, with certain propositions," *Bulletin No. 2, Proc. of the 3rd Annual Conference, New York State Science Teachers' Association*) New York University, 1899), p. 585.

The committee, ninety members in all, headed by five college presidents, one college professor, two private school headmasters, one public school principal, and the United States Commissioner of Education, was organized in 1891 by the National Education Association. Under the leadership of President Eliot of Harvard, the Committee of Ten made a study of both the teaching of secondary school subjects and the need for standardization of college entrance requirements. The report of the Conference mentioned above, published in 1892, urged that physics and chemistry be required for admission to all colleges and universities, that these subjects be taught by a combination of laboratory work, textbook, and didactic instruction, and that one-half of the course time should be devoted to laboratory work.

The two-decade interval, 1880–1900, saw the almost complete rejection of textbook work in high-school physics teaching as a method "dead at the very root." Instead, the modern laboratory became the classroom, because it did away with the reporter and brought the student face to face with nature. Hall claimed that the popular rise of laboratory teaching in the secondary school was, in spite of European influences, an American phenomenon. In Germany and France, he found laboratory work optional for the pupil. In most cases, this meant duplicating the experiments demonstrated by the teacher, and few pupils bothered to take the course.[40]

Though 1910 was the culminating year of three decades of extraordinary advances in science teaching in the high schools, Dr. C. R. Mann, a physicist, in a summation of results, concluded that "laboratories have not solved the problems of science teaching . . . we do not know how to use laboratories most effectively."[41] He blamed this partial failure in American teaching on: (1) poor teachers and poor textbooks; (2) passing examinations as the prime reason for study; (3) concepts of physics as presented not connected enough with daily life—too little application of these concepts in the solution of the student's daily problems.

This apparent divorce between laboratory work and reality had already begun to worry science teachers. At the July, 1905 convention of the N.E.A. Department of Science Instructors, it was noted that physics was becoming less popular among high-school students because:

"In the laboratory the student is introduced at once to the difficult subject of measurement, required to make immediate use of such unfamiliar instruments as the diagonal scales, the vernier caliper, and the balance sensitive to a centigram; to report his results in terms of the metric system, to discuss errors, sources of error, percentage of error, averages, and probabilities; to deduce laws, many of which he knew before, from data that cannot be made to prove anything, and to apply these laws to a set of problems that have no apparent relation to his immediate scientific environment, or to the questions that he is so anxious to have answered."[42]

Professor John Woodhull of Columbia University attacked the average college physics course as a barren kind of pedagogy which the high school would eventually transcend:

"These college students have a starvation course in measurement called physics. Their tutors, having just passed through the same course with excessive specialization, are suspicious of that expansive thing called culture. They affect to despise, not only the public, but all departments of learning other than their own. They surpass the theologians in narrowing down their lines of orthodoxy. Some teachers of science are like polarizers. The truth which gleams in all directions is narrowed down to one plane when it is transmitted by them . . . the influence of the college . . . is toward driving culture . . . out of the schools . . . . "[43]

Woodhull wanted the laboratory in the high school to give the pupil "that self-activity . . . merely for the purpose of coming in contact with things, making their knowledge real, acquiring a certain balance of judgment which comes from actual contact with things."[43] A further breaking away of the high-school physics course from a study of abstract principles was afforded

[40] Hall and Smith, *The Teaching of Chemistry and Physics in the Secondary School* (Longmans Green and Company, New York, 1902), pp. 328–330.

[41] See reference 16, p. 228.

[42] Packard, "Physics for boys and girls: an introductory course," (Proceedings, N.E.A., 1903), p. 881.

[43] Woodhull, School Rev. XV, 2, 127 (1907).

by the industrial advances of the machine age by this time. It was common sense for an educator in 1905 to suggest:

" . . . the pupil of high-school age is more interested in the course of the many mysteries about him, and the secret of machinery, than he is in learning the mathematical relation of effect to cause. I am free to admit that I believe it advisable to leave out much which we now teach, and to devote the time to the study of the industrial phase of physics. This can be done in the class or lecture room, in the laboratory, and by trip to factories and plants."[44]

5. CONCLUSION

By 1910, the physics laboratory had become an accepted part of the high-school physics course. In spite of all attempts to standardize the laboratory course and to infuse it with new and more interesting experiments,[45] high-school teachers were criticized for two major failings: (1) the courses, averagely, were too quantitative, too dependent on measurement for its own sake; (2) the courses were too specialized, too abstract, having no relation for the student with life and the natural phenomena about him.[46] As high schools began to be interested in providing courses in *applied* physics for their pupils, with more and more emphasis on the qualitative aspects of natural phenomena, the gap between the high school and the college widened. During the first ten years of the 20th century, the number of high schools in the United States increased phenomenally.[47] Colleges faced the problem of having to buttress the increasing costs of their own expansion by gleaning more and more candidates from the graduates of the public high schools. This meant an intensification of the old curriculum problem: should the colleges accept the physics courses as organized from the public high-school point of view, or should the high schools knuckle under to the college requirements "narrowly interpreted by examiners and bigotedly enforced by readers of examination papers?"[48]

To summarize briefly, it appears that the rise of the physics laboratory in the public high school during the period *ca* 1821–1910 was influenced by four general cultural conditions: (1) the growing feeling that the scientific curriculum was more important for the demands of the age than the older classical curriculum; (2) the increasing emphasis on laboratory work and empirical research in the universities and colleges; (3) the influence of college admission requirements on the high school curriculum; and, (4) the peculiar optimistic American habit of popularizing to emotional extremes certain "progressive" ideas.

The author wishes to thank his three advisors at Harvard University, Professor Robert Ulich, Associate Professor Fletcher G. Watson, and Associate Professor I. Bernard Cohen, for their inspiration and suggestions during the writing of this paper.

[44] C. H. Perrine, School Rev. XIII, **1**, 73 (1905).
[45] Mann, School Rev. XIV, **3**, 212 (1906).
[46] H. L. Terry, School Rev. XVIII, **4**, 241 (1910).
[47] Approximately 10 500 new high schools were opened in that decade!
[48] See reference 43, p. 127.

The Teaching Laboratory

Edwin Hall
and the emergence of the laboratory in teaching physics

Albert E. Moyer

My father, in the 1930s, and I, in the early 1960s, had similar experiences in high school physics courses. We both learned Hooke's law and Boyle's law, for example, by personally experimenting with springs, weights, and mercury-filled tubes. Around 1910 my grandfather also would have studied these physical laws by actually conducting his own laboratory experiments. But this instructional continuity terminates with the generation of my great-grandfather. If he had attended an American secondary school in the 1870s, he would have approached Hooke's law and Boyle's law through "book learnin'," depending on a wordy and fact-laden text such as Neil Arnott's *Elements of Physics; or, Natural Philosophy, General and Medical: Written for Universal Use, in Plain or Non-Technical Language.*

The circumstances surrounding this major shift in American physics instruction in the period from about 1880 to 1905 are historically complex, arising from a climate of educational reform and leading to, among other things, a new-found sense of professional identity for American physics teachers. One man, nevertheless, is clearly discernible as being most responsible for the shift: Edwin Hall, Harvard professor and second recipient of the AAPT's Oersted Medal, awarded in 1937 for "pioneer work in the introduction of laboratory instruction in physics." Before detailing Hall's contribution, however, I will first survey the efforts of his predecessors who initiated the laboratory approach in the teaching of secondary school physics.

Early trends toward laboratory instruction

In 1878 the United States Commissioner of Education sent questionnaires to a large sample of the nation's high schools, academies, normal schools, colleges, and universities inquiring about their programs in chemistry and physics. Frank Clarke, professor of chemistry and physics at the University of Cincinnati, was charged with analyzing the resulting descriptive data. From the comprehensive tables in Clarke's report—which appeared in 1880 as a nationally distributed Bureau of Education "Circular"—we see that out of 607 public and private secondary schools only 11 offered a physics course with some laboratory work, and of these 11 schools, only four had a course that extended over the full academic year. Although nearly all of the 607 schools offered some training in elementary physics, the vast majority relied on textbooks, lectures, recitations, and other forms of written or spoken instruction to present the subject.

While we might expect Clarke's personal attitude regarding laboratory instruction in physics to have been in consonance with the overwhelming national indifference suggested by his data, we find instead that he was enthusiastic about laboratories. Possibly influenced by his background in chemistry, a field in which the value of the

Albert E. Moyer *is currently winding up his doctoral studies in the history of physics at the University of Wisconsin, where in 1974 he obtained an M.A. He also holds degrees in physics — from Oberlin College (B.A., 1967) and the University of Colorado (M.S., 1969) — and he previously taught physics at Lees Junior College in Kentucky. (University of Wisconsin, Madison, Wisconsin 53706)*

Fig. 1. EDWIN HERBERT HALL, born in Gorham, Maine in 1855, received his B.A. (1875) and M.A. (1878) from Bowdoin College and his Ph.D. (1880) from Johns Hopkins. He began teaching at Harvard in 1881 and completed his Cambridge career as Rumford Professor of Physics, Emeritus, dying in 1938. (Photo, ca. 1890, courtesy Harvard University Archives.)

instructional laboratory was fairly well established by 1880, Clarke insisted in his report that laboratory work should be an "essential and prominent feature" of every course in the physical sciences; he asserted that two of the goals of science education are "to train the faculty of observation and to teach . . . the experimental method of grappling with unsolved problems." For him the textbook approach, with its emphasis on facts and information, was "backward."

Clarke also found, through this government sponsored survey, that secondary-school physics courses were frequently duplicated in colleges. Seeing this as "clearly wasteful," he called for cooperation between the lower schools and the colleges, especially with regard to the establishment of uniform course standards. Thus, by 1880, the basic questions that would occupy physics educators for the next few decades were being formulated—the questions of uniformity in secondary-school courses and in college admission criteria, and the related issue of laboratory instruction.

Motivated directly by Clarke's report, the Commissioner of Education in 1883 distributed another questionnaire as a further step toward upgrading physics programs in secondary schools and toward dealing with undesirable diversity and duplication in high schools, normal schools, colleges, and universities. Charles Wead, University of Michigan professor and director of this new study, saw it as an effort to determine the aims and ideals

TABLE I. Charles Wead's 1884 "List of Fundamental Experiments in Physics."

Compare and measure lengths, volumes, and masses.
Inertia.
Composition of forces.
Parallel forces.
Centre of gravity.
Lever, inclined plane, &c.
Pendulum.
Centrifugal action.
Archimedes' principle.
Density and specific gravity.
Capillarity.
Simple barometer.
Boyle's law.
Air pump experiments.
Pumps and siphon.

Expansion of liquids and gases.
Bending of compound bar.
Verify fixed points of a thermometer.
Conduction of heat.
Temperature of mixtures of water.
Specific heat of a solid.
Latent heat of ice, steam, vapors.
Heat from friction.
Properties of permanent and temporary magnets.
Magnetic curves.
Simple galvanic cell.
Useful forms of galvanic cells.
Effects of current on magnetic needle.
Electro-magnets.
Influence of resistance of conductor.
Chemical effects of current.
Heating effects of current.
Induction.
Telegraph and telephone.
Frictional electricity; two states.
Electrical machine; Leyden Jar.

Vibration and production of waves.
Resonance.
Interference of sound (fork and jar).
Monochord.

Photometer.
Reflection; plane and curved mirrors.
Refraction of light.
Dispersion and spectrum.
Total reflection.
Lenses; construction of image.
Combination of colors.

of physics teachers around the country, thus going beyond Clarke's more descriptive endeavor. On the basis of his forthcoming report, Wead hoped that a committee could "draw up a practical scheme sufficiently definite, detailed, elastic, and progressive to secure its wide adoption in the schools of the country." In general, the 70 respondents to this 1883 poll favored such interschool coordination and standardization of course content.

In this new survey educators were asked to answer such basic questions as the following:

> What should be the prevailing character of the high school work? Inductive or deductive? For information or for discipline? With or without text-

Experiment 2. Support two iron bars, *a* and *b* (Fig. 90), bent into the form of a curve, about 8^{cm} apart, and so situated that a ball *n*, rolling down them, will be discharged from them in a horizontal direction. So connect the wires of an electric battery *c* with these bars, that while the iron ball *n* rests upon them the circuit is closed, and the iron ball *m* is supported by the attraction of the electro-magnet *e*. Now allow *n* to roll down the curved path. When it leaves the bars, the circuit is broken, *e* instantly loses its power to hold *m*, and *m* drops. But both balls reach the floor at the same instant. If the horizontal velocity of *n* is varied, by allowing it to start at different points on the bars, so as to cause it to describe different paths, the two balls will, in every case, acquire exactly equal vertical velocities.

Fig. 90.

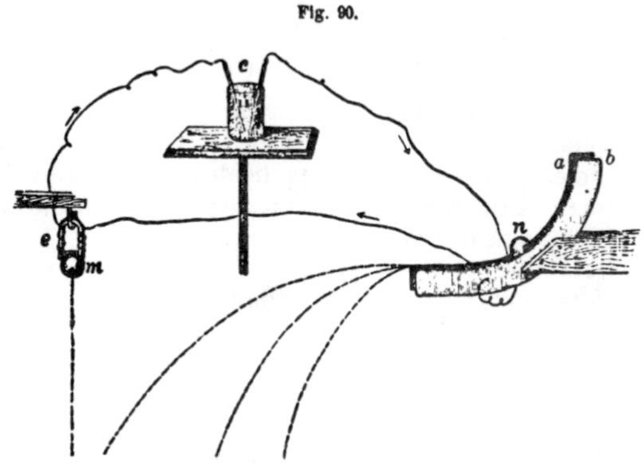

Fig. 2. Alfred Gage's pioneering textbook of 1882 contained experiments such as this "excellent verification of the second law of motion."

book? With laboratory work?

Wead discovered that most of the respondents advocated instruction which would follow an inductive rather than a deductive approach and which would foster mental discipline rather than transmit factual knowledge. In other words, they felt that a high school physics course should train students to "first observe the phenomena sharply and then seek for a cause or for the law according to which the forces act." Regarding the active participation of pupils in laboratory work, Wead similarly found that 26 out of 32 secondary school teachers believed that an elementary physics course should be "beyond all question experimental." In addition, the experiments should be "largely qualitative; but such determinations as specific gravity, the pressure of air, the focal length of lenses, and etc., are not likely to be omitted, and experience will doubtless lead . . . to making the work more and more quantitative."

Although laboratory work was widely favored, Wead realized that "unfortunately, few teachers can speak of the results of this kind of teaching, it has been tried so little." Consequently, to help overcome this unfamiliarity and inexperience, he included in his nationally distributed report a list of 47 "fundamental experiments that should never be omitted in a high school course" (Table I). Moreover, as an example of an innovative "inductive" textbook, in which experiments performed by pupils lead to the apprehension of general principles, he frequently referred to Alfred Gage's recently published *Elements of Physics* (Boston, 1882)—a book that Wead himself had helped prepare, having critically previewed the entire manuscript (Fig. 2).

Elements of Physics had a wide circulation and, as the first American text to emphasize student experiments, it probably had a wide influence. Alfred Gage, a teacher at the English High School in Boston, justified his reliance on formal experiments by pointing to the success of chemistry instructors over the prior 20 years. Although chemistry "was formerly a dull and almost profitless study," the introduction of student laboratories had rapidly transformed it into "one of the most interesting and useful" components of the high school curriculum. This accomplishment led Gage to ask what many physicists were probably asking in the 1880s: "Is there any reason why laboratory practice should not do a similar work for physics?"

Harvard and Edwin Hall

An influential group of Gage's academic neighbors in Boston shared his optimism regarding laboratories. In 1886 while revising admission standards, a committee of Harvard College administrators called for the creation of a laboratory requirement in physics for secondary school students preparing to enter Harvard. Although Edwin Hall, then a young member of the physical department, was assigned the task of designing the new program, the program had been instigated by two prominent Harvard chemists—Charles Eliot, the school's innovative president, and Professor Josiah Cooke, Eliot's prior mentor and one of the first Americans to have a laboratory for teaching college chemistry. Just as in Gage's textbook and Clarke's report of 1880, chemists helped trigger the soon-to-be-influential Harvard laboratory guidelines for high school and academy physics.

Following the administrative decision to establish a laboratory requirement (as an alternative to the traditional textbook requirement), Edwin Hall and his associates proceeded to draft for the 1886 "Harvard Catalogue" a formal statement of the new requirement, the final version of which called for: "A course of experiments in the subjects of mechanics, sound, light, heat and electricity, not less than forty in number, actually performed at school by the pupil." The experiments were to be chosen from various published manuals, and the pupil was to assemble a laboratory notebook which he would later present to his Harvard examiners. Regarding the implementation of these new regulations, Hall later recalled: "The next year ninety-one candidates offered Elementary Experimental Physics. I personally conducted each boy's laboratory examination one very hot day in Room 41 of the laboratory with only the help of the assistant janitor . . . . "

It soon became evident that this original requirement permitted too much latitude in choice of experiments. Consequently, President Eliot asked Hall to prepare detailed descriptions of 40 definitive experiments for use by secondary schools. Hall, then in his early thirties and holding the rank of Instructor, found the task difficult and felt that his superiors were becoming impatient for the final product. "My position at Harvard," he recollected, "was none too secure at that time and I saw that a crisis in my own affairs had arrived." After one last all-night work session, he finally submitted 13 of the proposed 40 experiments to President Eliot. Eliot was pleased, as Hall later indicated: "I had gained his confidence. From that

TABLE II. Edwin Hall's "Harvard Descriptive List," as lengthened and revised in 1897.

TABLE II. Edwin Hall's "Harvard Descriptive List," as lengthened and revised in 1897. Hall divided the 61 exercises into two major groups: the first 25 exercises were easier, used simpler apparatus, and in effect, constituted an abridged course suitable for younger students or for a school with limited facilities. In most schools, however, pupils progressed through both groups of experiments and consequently encountered some of the same topics at two levels of difficulty (compare, for example, exercises 6 and 7 with 32).

MECHANICS AND HYDROSTATICS.

1. Weight of unit volume of a substance.
2. Lifting effect of water upon a body entirely immersed in it.
3. Specific gravity of a solid body that will sink in water.
4. Specific gravity of a block of wood by use of a sinker.
5. Weight of water displaced by a floating body.
6. Specific gravity by flotation method.
7. Specific gravity of a liquid: two methods.
8. The straight lever: first class.
9. Centre of gravity and weight of a lever.
10. Levers of the second and third classes.
11. Force exerted at the fulcrum of a lever.
12. Errors of a spring balance.
13. Parallelogram of forces.
14. Friction between solid bodies (on a level).
15. Coefficient of friction (by sliding on incline).

LIGHT.

16. Use of Rumford photometer.
17. Images in a plane mirror.
18. Images formed by a convex cylindrical mirror.
19. Images formed by a concave cylindrical mirror.
20. Index of refraction of glass.
21. Index of refraction of water.
22. Focal length of a converging lens.
23. Conjugate foci of a lens.
24. Shape and size of a real image formed by a lens.
25. Virtual image formed by a lens.

MECHANICS.

26. Breaking strength of a wire.
27. Comparison of wires in breaking tests.
28. Elasticity: stretching.
29. Elasticity: bending; effects of varying loads.
30. Elasticity: bending; effects of varying dimensions.
31. Elasticity: twisting.
32. Specific gravity of a liquid by balancing columns.
33. Compressibility of air: Boyle's law.
34. Density of air.
35. Four forces at right angles in one plane.
36. Comparison of masses by acceleration test.
37. Action and reaction: elastic collision.
38. Elastic collision continued: inelastic collision.

HEAT.

39. Testing a mercury thermometer.
40. Linear expansion of a solid.
41. Increase of pressure of a gas heated at constant volume.
42. Increase of volume of a gas heated at constant pressure.
43. Specific heat of a solid.
44. Latent heat of melting.
45. Determination of the dew-point.
46. Latent heat of vaporization.

SOUND.

47. Velocity of sound in open air.
48. Wave-length of sound.
49. Number of vibrations of a tuning fork.

ELECTRICITY AND MAGNETISM.

50. Lines of force near a bar magnet.
51. Study of a single-fluid galvanic cell.
52. Study of a two-fluid galvanic cell.
53. Lines of force about a galvanoscope.
54. Resistance of wires by substitution: various lengths.
55. Resistance of wires by substitution: cross-section and *multiple arc.**
56. Resistance by Wheatstone's bridge: specific resistance of copper.
57. Temperature coefficient of resistance in copper.
58. Battery resistance.
59. Putting together the parts of a telegraph key and sounder.
60. Putting together the parts of a small motor.
61. Putting together the parts of a small dynamo.

*By "multiple arc," Hall meant two or more coils or curved sections ("arcs") of wire connected in parallel ("in multiple").

time on he gave me his powerful support and he took great pleasure in seeing the influence of Harvard spread throughout the country in what he believed to be a beneficent educational movement."

The experiments in the "Harvard Descriptive List of Elementary Physical Experiments" (Table II and Fig. 3) were largely quantitative, since Hall felt that "the definite tasks of a physicist almost always involve measurements." Although acknowledging that "the method of leading to principles by means of experiment . . . is no new invention," he believed that "the Harvard course is perhaps distinguished by the strictness with which this method is followed."

Even though Hall had been reluctant to enter the realm of secondary school pedagogy, he soon became the leading propagandist for the new Harvard course of experiments. In 1887, in a letter to the editor of *Science*, he made the first national announcement of the Harvard alternative to textbook physics. He argued that laboratory instruction was essential as it would provide training in observation, would supply detailed information, and would arouse the pupil's interest. Perhaps his strongest advocacy of the Harvard program came in 1891, however, when his *Text-Book of Physics* was published. This book, which included experiments that were identical to those in the "Descriptive List," went through three editions and many printings (Fig. 4).

Through Hall's effort and Harvard's prestige, the "Descriptive List of Elementary Physical Experiments" increased in popularity and soon a number of other books

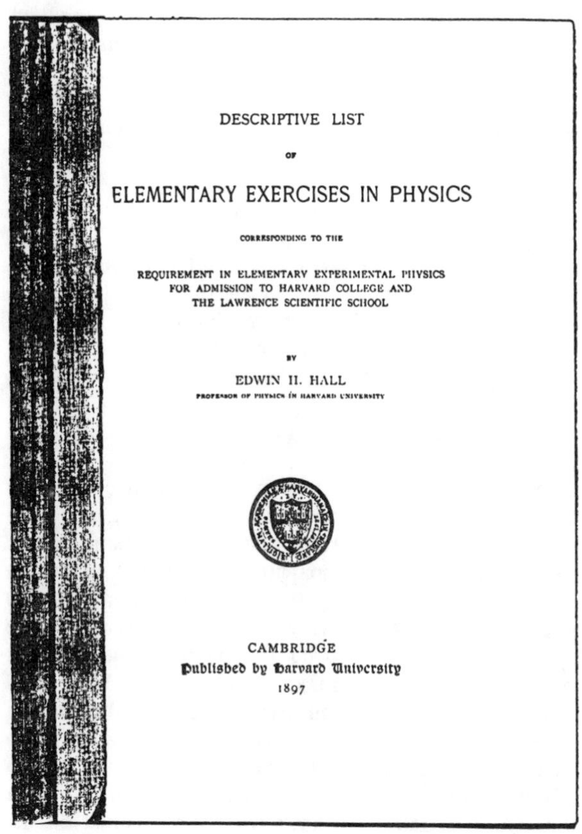
DESCRIPTIVE LIST

OF

ELEMENTARY EXERCISES IN PHYSICS

CORRESPONDING TO THE

REQUIREMENT IN ELEMENTARY EXPERIMENTAL PHYSICS
FOR ADMISSION TO HARVARD COLLEGE AND
THE LAWRENCE SCIENTIFIC SCHOOL

BY

EDWIN H. HALL

PROFESSOR OF PHYSICS IN HARVARD UNIVERSITY

CAMBRIDGE
Published by Harvard University
1897

Fig. 3. Copies of the once popular "Descriptive List" are today rather difficult to find. According to the *National Union Catalog*, this 1897 revised edition of the pamphlet is the oldest surviving edition, with copies at only two libraries — Harvard and Oberlin.

based on it appeared on the market. Writing in 1893, Assistant Professor Hall (he had been promoted) pointed out that "a large majority of those now admitted to the Freshman class have taken this laboratory course in the schools, although a text-book alternative is still accepted." In later trying to account for this rapid adoption of the "Descriptive List," Charles Mann, professor of physics at the University of Chicago, concluded in 1912 that the "List" had filled a crucial instructional void; during the late 1880s "teachers of physics were scarce, there were few laboratories and little experience to guide those who wished to introduce this work." In like manner, Hall was convinced as early as 1893 that Harvard's examinations in experimental physics "have served the very important purpose of *creating* a standard of work for the schools."

The Committee of Ten

During the final decade or so of the 19th century, the broader issues involving coordination between secondary schools and colleges were still felt to be pressing. Accordingly, in 1887 the National Educational Association took up the problem of "uniformity" in high school curricula and in college admission standards and, in 1892, convened a committee of ten reputable educators to examine this problem fully. With Harvard's dynamic and outspoken Charles Eliot as chairman, this "Committee of Ten" was insured an attentive national audience. The group's report, which was widely circulated in 1893, consisted of two sections: the general findings or recommendations of the basic Committee of Ten, and the statements of nine subcommittees or "Conferences" that had been established to represent individual disciplinary areas. Throughout both sections of this influential document, the nascent idea of teaching physics through laboratory work was given further encouragement.

CHAPTER XXVIII.

SOUND.

354. Definition.—The word *sound* refers sometimes to the sensations we receive through the auditory nerves, sometimes to the external cause of these sensations. We shall use it mainly in the latter sense.

355. Transmission of Sound.—Light comes to us perfectly well through a vacuum. Can sound do the same?

EXPERIMENT 1.

Place a small alarm-bell, in operation, on a wad of cotton beneath the bell-jar of an air-pump, and exhaust the air as rapidly and as perfectly as you can.

Does the rarefaction of the air have any effect upon the loudness of the sound as perceived by the ear?

Let the air re-enter rapidly and note the effect.

If the alarm-bell used in the preceding experiment had rested directly upon the firm pump-plate, the sound would have made its way to the outer air by way of this solid medium. The following experiment illustrates still further the capacity of solids for carrying sound.

EXPERIMENT 2.

Hold one end of a slender rod of wood 3 m. or more in length firmly between the teeth while some one sounds a tuning-fork and then holds the stem against the free end of the rod.

It is a common and interesting experiment to hold the head under water while a comrade at a distance raps two stones together under water. The loudness of the sound

441

442 *PHYSICS.*

is painful. It may even be sufficient to injure the hearing at a distance at which the sound produced by rapping the same stones together in air would be only faintly heard.

EXERCISE 47.

VELOCITY OF SOUND IN OPEN AIR.

Apparatus: A pendulum beating seconds. A small spy-glass. A board. A hammer or stone.

One experimenter strikes the board sharply just when the pendulum-bob passes the middle point of its arc. Another places himself at the start about 900 ft. distant from the first, in a line at right angles with the plane in which the pendulum swings, and then, looking through his glass, seeks to place himself at such a distance that the stroke made when the pendulum-bob passes through the middle point of its arc in one direction reaches his ear just when the bob passes through the middle point of its arc in the other direction. The distance which the sound travels in one second in one direction is thus roughly determined. As the wind may either help or hinder this movement, the conditions should afterward be reversed by having the experimenter with the glass strike the board, while the other stays at the pendulum and listens for the stroke, signalling to the striker to come nearer or go farther away until coincidence occurs as before. The mean of the two distances now found should be taken as the distance which sound would travel in one second in still air.

This method is not likely to give very accurate results. Its merit lies in its directness and simplicity. A still more direct and simple method is to place two persons a long distance apart, but in sight of each other, and have each in turn discharge a pistol or gun while the other notes, as well as he can, the number of seconds between the sight and the sound of the explosion.

In case this Exercise is impracticable, as it would be in most city schools, the velocity of sound in air in a tube may be calculated from the results of Exercises 48 and 49.

Changes of temperature greatly affect the velocity of transmission in air and other gases, the speed in air increasing about 60 cm. per second for each centigrade degree of increase in temperature. Change of pressure without

Fig. 4. Edwin Hall's emphasis on experiment is apparent in these pages from the 1897 revision of his *Text-Book of Physics.* The "Exercise" is from the "Harvard Descriptive List."

Eliot, in writing the introductory section, pointed out that the conferences on physical science, natural history, and geography all "dwell on laboratory work by the pupils as the best means of instruction"; teaching based exclusively on discursive, authoritarian textbooks was unanimously rejected in favor of experiential forms of education. This thrust toward the instructional laboratory was particularly evident in the report of the "Conference on Physics, Chemistry, and Astronomy"—a group of secondary school and college educators that included Alfred Gage. Gage and his colleagues specifically proposed that at least one-half of the time in secondary school physics be devoted to laboratory experiments, especially those quantitative in nature. Echoing Harvard's earlier guidelines and Gage's own pioneering text, the committee included a list of 57 formal experiments suitable for high school or academy pupils and recommended that students maintain a laboratory notebook which was to be submitted as part of a college entrance examination. For these science educators, no argument was needed to justify such laboratory work—it was self-evident that study based solely on a textbook was "of little, if any, value."

Hall's revision

During the closing decade of the 19th century, wholesale enthusiasm for inductively-structured, quantitative experimentation diminished slightly. In the preface to their 1892 text, Carhart and Chute remarked that "a few years ago it seemed necessary to urge upon teachers the adoption of laboratory methods to illustrate the text-books; it would now seem almost necessary to urge the use of a text-book to render intelligible the chaotic work of the laboratory." Writing in 1902 about the original Harvard "Descriptive List," Edwin Hall similarly admitted that the program and the textbooks conforming to it had all overemphasized "exacting laboratory work" and had neglected alternative instructional routes; he maintained, however, that this "defect" was largely the responsibility of influential chemists (that is, President Eliot and Professor Cooke) who "made the mistake of treating the physics just as they treated chemistry."

Recognizing this deficiency, Hall introduced into the 1897 "Harvard Catalogue" a liberalized requirement which indicated that quantitative laboratory instruction should be supplemented, for example, by general lectures, numerical problems, and demonstrations. In addition, he amended the "Descriptive List" to allow for a much broader selection of experiments. The pupil was now asked to complete only 35 exercises from a list that had been extended to 61 (Table II). Finally, the 1897 edition of Hall's *Text-Book of Physics* reflected this move toward greater flexibility and away from quantitative rigor.

By the time of this revision the Harvard course had attracted a new group of enthusiastic promoters who were hoping to encourage a wider national acceptance of the program. At least five equipment manufacturers in Boston and Chicago were offering "at a reasonable price" (Fig. 5) complete sets of apparatus for the experiments in the "Descriptive List," which the entrepreneurs had renamed the "National Physics Course." Consequently, Hall could announce that "several manufacturers of school apparatus keep the articles of the Harvard list in stock and know these articles by the numbers they bear in that list, a fact which facilitates ordering from them if one has the Harvard list at hand." As we shall see, quite a few educators had the list "at hand" by the turn of the century.

NOTE.

The following estimates of cost for apparatus and materials are only approximate. It is hardly possible to make an accurate estimate, as prices will vary from time to time and different dealers may have somewhat different grades of apparatus. The cheapest is not necessarily the best to buy.

FOR ALL EXERCISES OF THE FIRST PART.

Apparatus for a laboratory section of one	\$6.00
Apparatus for a laboratory section of twelve	72.00
Tables for a laboratory section of twelve	50.00

These tables serve also for the Second Part.

FOR ALL EXERCISES OF THE SECOND PART.

Apparatus for a laboratory section of one	\$90.00
Apparatus for a laboratory section of twelve	750.00

FOR THIRTY-FIVE EXERCISES OF AVERAGE EXPENSE.

Apparatus for a laboratory section of twelve	\$475.00
Tables for a laboratory section of twelve	50.00

Fig. 5. Although there are many similarities between physics instruction at the turn of the century and today, this note that Hall included in the 1897 edition of the "Descriptive List" indicates one major difference — the cost of equipping a laboratory. (In this note the "exercises of the first part" are those numbered 1 through 25 in Table II; the "exercises of the second part" are numbered 26 through 61.)

Spread of the Harvard plan

Hoping to elaborate upon the findings of its Committee of Ten, the National Educational Association in 1895 organized a Committee on College-Entrance Requirements. This Committee's official report, distributed in 1899, contained general recommendations on college prerequisites in physics—recommendations that originated with a special NEA subcommittee headed by Edwin Hall. Although granting that a secondary school program should include textbook instruction as well as quantitative laboratory work, Hall managed to have the basic components of his "Descriptive List" explicitly included in the NEA requirements. Physics teachers were once again encouraged to adopt the Harvard approach with Hall's revised list of 61 experiments.

That Hall was chairman of this nationally important subcommittee comes as no surprise; by the late 1890s he was a full professor at Harvard and an internationally known research physicist. And besides being a talented writer who frequently contributed pedagogic and popular articles to the many blossoming journals of this period, he was respected as discoverer of the "Hall Effect." Although perhaps over-reacting, William Thomson (later Lord Kelvin), writing in the 1880s, considered this effect to be "by far the greatest discovery that has been made in respect to the electrical properties of metals since the times of Faraday . . . ."

The establishment in 1899 of the College Entrance Examination Board (CEEB) insured even wider adoption of the revised Harvard course of experiments by the secondary schools of the nation. With Charles Eliot (and, to a lesser extent, Edwin Hall) involved in the CEEB, the Harvard position again prevailed. The statement on physics in the "Definition of Requirements" is almost identical to prior Harvard dictums and was in fact based on the earlier report of Hall's NEA subcommittee. If a student were applying to a college that chose to participate in the CEEB's system of uniform tests, then he was obliged to master the experiments in the "Descriptive List." In June of 1901, at a dozen or so prominent colleges and universities, 175 such students took the first CEEB examination in physics. It is safe to say that laboratory physics was becoming the norm—at least in those secondary schools offering college preparatory courses.

Criticism

By about 1900—approximately 20 years after Clarke, Wead, and Gage had first advocated laboratory instruction—it was clear that the approach was being adopted nationwide. It was also clear by about 1900 that there were serious faults with the approach. One of the earliest and most influential critics of secondary school experimental physics was the noted psychologist and educator, G. Stanley Hall. As part of his general attack on the Committee of Ten, G. S. Hall focused on the committee's deleterious effect on physics teaching. In the five years following the publication of the committee's report, he observed, the percentage of public high school pupils studying physics throughout the nation had declined from about 25 percent of all students to 20 percent. Regarding this large drop, he explained that "the trouble with physics is simply that it has failed to take account of the nature, needs, and interests of high school boys and girls"; laboratory experiments and textbooks were too quantitative and too concerned with precise measurements.

Although G. S. Hall objected to the rigor of laboratory instruction, he did not wish to eliminate all experimental work. This moderate stance was also taken by the many critics who succeeded and often echoed Hall in the first decade of the century as the ranks of professional physics teachers grew. For example, in 1905 James Stevens objected only to the emphasis on often meaningless exercises such as the measurement of the top of a table. Writing in 1906, Robert Millikan felt that the "tyranny of the university over the high-school" had unfortunately led to "a minute study of the mathematical and mechanical foundations upon which technical science is built, rather than an inspiring insight into the meaning of the physical world." Similarly, in 1906, James Woodhull decried the control of high school physics by college educators holding esoteric and elitist goals; he advocated a more qualitative program that built on "common sense" and dealt with "practical" applications of physics. And in 1909, Charles Mann maintained that the laboratory approach had been perverted when "the college pressure was applied in 1886." Before Harvard stepped in, Mann felt, educators like Charles Wead had been moving in the proper direction by advocating laboratory work "of a kind to possess significance to the pupils—done with home-made apparatus and 'kitchen utensils.' "

In response to these critics, both Charles Eliot and Edwin Hall defended the Harvard program as set out in the "Descriptive List" and reflected in the reports of the Committee of Ten, the NEA Committee on College-Entrance Requirements, and the College Entrance Examination Board. They felt that the Harvard plan, particularly as liberalized in 1897, was a significant improvement over traditional textbook physics. In rebuttal to G. Stanley Hall, Eliot insisted that quantitative experiments provided the best means of training the mental faculties and cultivating the "powers of observation"—goals he valued highly. Edwin Hall contended that the Harvard course itself allowed adequate room for qualitative and descriptive studies. In making this point, he wrote, "I have never said . . . that precision of measurement should be the chief object of such a course"; but he betrayed his quantitative leaning by adding, "There are many experiments . . . in which . . . errors of two, three, even five or ten percent, are allowable, if not unavoidable." Generally speaking, however, Edwin Hall readily acknowledged specific deficiencies in existing methods of laboratory instruction, and he spent much time grappling with such basic problems as the feasibility of alternative courses for pupils not intending to enter college.

Conclusion

Regardless of its deficiencies, the physics laboratory was, as Hall said in 1902, "permanently established for the better class of American secondary schools . . . ." Moreover, many elements of the Harvard program dominated; a survey involving 165 high schools conducted by the Central Association of Science and Mathematics Teachers in 1906 revealed that experiments from the "Descriptive List" were the overwhelming favorites of teachers around the country. Although Hall perhaps exaggerated when he proclaimed that instructors in the high schools and academies of the United States had "little or nothing to learn from the corresponding schools in France, Germany, or England," it is clear that by the opening years of the 20th century Americans had accepted a new approach to teaching—an approach that persists in many introductory physics courses to this day.

Furthermore, the process which led to this acceptance of laboratories had important implications for the institutional development of the American community of physics teachers. From Frank Clarke's advocacy of laboratory instruction in 1880 until Charles Mann's moderating criticism after the turn of the century, the laboratory issue had become a common concern of physics teachers around the country. Out of this concern emerged first, a fairly widespread consensus on the specific makeup of a secondary school physics course appropriate for college preparatory students, and second, the definition of a unique and essential role for physics teachers. Writing in his later years, Edwin Hall concluded that through the "great influence" of the laboratory requirement, "the school teaching of physics came to be a recognized and stable profession, instead of the casual employment of someone whose main interest lay elsewhere." The complex process that led in the late 19th century to the adoption of laboratory work in secondary school courses and in college admission standards thus provided a stimulus that helped move the national community of physics teachers toward institutional maturity and fuller professional status. In a few active decades, the modern physics teacher, as well as the modern introductory course, had emerged — both bear the imprint of Edwin Hall.

A selected bibliography

Percy W. Bridgman, "Biographical Memoir of Edwin H. Hall: 1855-1938." *NAS Biographical Memoirs*, 21 (1941), 73-86; bibliog., 87-94.

Frank W. Clarke, "A Report on the Teaching of Chemistry and Physics in the U.S." *Circulars of Information of the Bureau of Education*, No. 6 (1880).

College Entrance Examination Board of the Middle States and Maryland, *First Annual Report of the Secretary* (The Board, New York, 1901).

Edwin H. Hall, "Physics: 1869-1928," *The Development of Harvard University*. Edited by Samuel E. Morison (Harvard University, Cambridge, 1930), pp. 277-291.

Edwin H. Hall, "Physics Teaching at Harvard Fifty Years Ago," *The American Physics Teacher*, **6**, 17 (1938).

Edwin H. Hall and Alexander Smith, *The Teaching of Chemistry and Physics in the Secondary School.* (Longmans, Green, New York, 1902).

Edward A. Krug, *The Shaping of the American High School: 1880-1920* (University of Wisconsin Press, Madison, 1969).

Charles R. Mann, *The Teaching of Physics* (Macmillan, New York, 1912).

Albert E. Moyer, "The Emergence of the Laboratory Approach in the Teaching of Secondary-School Physics in Late Nineteenth-Century America." Master's paper, University of Wisconsin, 1974.

National Educational Association, *Report of Committee on College-Entrance Requirements* (The Association, Chicago, 1899).

National Educational Association, *Report of the Committee on Secondary School Studies* (U.S. Bureau of Education, Washington, 1893).

Charles K. Wead, "Aims and Methods of the Teaching of Physics," *Circulars of Information of the Bureau of Education*, No. 7 (1884).

David L. Webster, "Contributions of Edwin H. Hall to the Teaching of Physics," *The American Physics Teacher*, **6**, 14 (1938).

John F. Woodhull, "The Teaching of Physical Science," *Teachers College Record*, **11**, 5 (1910).

Early history of physics laboratories for students at the college level

Melba Phillips
351 West 24th Street, New York, New York 10011
(Received 2 June 1980; accepted 5 September 1980)

Student laboratories became an integral part of the study of physics around 1870, some forty years later than those in chemistry. The motivation was both cultural and utilitarian, and the situation that produced the innovation was the growth of manufacturing industry in the mid-19th century. The first published and most influential laboratory manual was that prepared by E. C. Pickering for use at the Massachusetts Institute of Technology.

The history of the physics laboratory in American high schools has been traced in the journals more than once[1,2] and its importance was recognized by the American Association of Physics Teachers in the presentation of the second Oersted Medal to E. H. Hall in 1937.[3] The introduction of student laboratories in colleges and universities that took place a very little earlier has received less attention.[4] The two developments were related to the rise of industrialization, both in Europe and in the United States, particularly in the northeastern states where industrial growth was more rapid than in other parts of the country. The first institution to *require* laboratory work as we know it for physics was the fledgling Massachusetts Institute of Technology—it would hardly be an exaggeration to say that the Institute was invented for the purpose of requiring laboratory work in science. The MIT lab was organized to be ready for full operation in the fall of 1869 "to instruct the third year class by laboratory work; and if an experience of one year shall be favorable, as I feel it must be, we can gradually enlarge our facilities and take in the lower classes. I am convinced that in time we shall revolutionize the instruction in physics as has been done in chemistry."[5]

The honor of being first to require physics laboratory work may be shared with King's College, London, for it was reported in 1871 that "Professors of physics at different universities have usually selected their best students to assist them in their private laboratories, to the mutual advantage of professor and student, but I believe that Professor Clifton was the first to propose, more than three years ago, that a course of training in a physical laboratory should form a part of the regular work of every student of physics. This system was adopted and at once put into action at King's College, on a very considerable scale for a college with no endowment whatever, and has been working now for nearly three years."[6] We shall see that there are other institutions with priority for some aspects of student laboratory work, but there seems no doubt that the first laboratory manual for student use to appear in print was prepared by Edward Charles Pickering.[7] Pickering had designed and initiated the MIT laboratory, and his book is dedicated to "Professor William Barton Rogers, the first to propose a physical laboratory." William B. Rogers was the founder and first president of MIT.

Before examining the developments at MIT we should note that experimental research and laboratories for the instruction of students came into vogue earlier in chemistry than in physics. Laboratory techniques were invented and to some extent transmitted by the alchemists down through the ages, but alchemy is on the whole antiscientific, and the science of chemistry did not arise until linkages first to medicine and then to mining and manufacture were made. Except for practical courses in medical schools it appears that the first laboratory in which the experimental methods of chemistry were taught systematically to students was at the University of Giessen, and was established by Justus von Liebig soon after 1824.[8] This laboratory soon became world famous, and the pattern for chemical education was in large part responsible for the development of the great chemical industry in Germany later in the 19th century. Lord Kelvin claimed[9] that the first chemistry labs for students were established at the University of Glasgow prior to 1831, but the German laboratories were much more influential. That of Liebig, in particular, served as an acknowledged model for several physics laboratories some forty years later.

In chemistry, and in physics as well, the demonstration lecture was already much in vogue. The Royal Institution in London was famous for its lectures, given by Davy, Faraday, and others, replete with elegant experiments. College instruction in both chemistry and physics involved demonstration; at the Sorbonne both Gay-Lussac and Dulong were masters of experimental demonstration, and the story of Oersted's discovery of electromagnetism during a lecture is too well known to bear repetition here.[10] There were also lecturers who traveled with a kit of demonstrations. Most of these remain unknown, but the first practical electromagnet was made and used by William Sturgeon,[11] who was to become a successful itinerant lecturer. Such lectures were also in demand in the United States,[12] a fact that influenced the style of Rensselaer Polytechnic Institute, founded in Troy, New York, in 1824.

Stephen van Rensselaer, the founder of RPI, put down as his principal object[13] "to qualify teachers for instructing the sons and daughters of farmers and mechanics, by lecture or otherwise, in the applications of experimental chemistry, philosophy, and natural history to agriculture, domestic economy, the arts, and manufactures." The method was spelled out explicitly for chemistry: "In giving the course in chemistry, the students are to be divided into sections, not exceeding five in each section. These are to be taught by seeing experiments and hearing lectures, according to the usual method. But they are to lecture and experiment by turns, under the immediate direction of a professor or a competent assistant. Thus by a term of labor, like apprentices to a trade, they are to become operative chemists." This procedure was carried out by Amos Eaton, who had been on the lecture circuit himself. The plan was no doubt also used to teach what we would call physics, for a list of laboratories made by a student in 1829 includes a "Phi-

 0002-9505/81/060522-06$00.50

losophy room which contains a fine set of philosophical apparatus." But the method proved too expensive to maintain, and the cirriculum was revised in 1829. Civil engineering was added, and for more than a quarter of a century RPI was the most important school (except possibly for West Point) in supplying engineers in demand for the conquest of the West.

Physical laboratories for students on what became a more conventional pattern were gradually introduced into German universities. That in Heidelberg was initiated in 1846 by Philipp von Jolly, inventor of the Jolly balance that has been used by generations of students to determine the specific gravity of solids and liquids. Coincidentally William Thomson (to become Lord Kelvin) began to develop physical laboratories for students at the University of Glasgow, but in both instances the laboratory work was optional, not a required part of the regular study of physics.[14]

1846 was also the date of a document known as "A Plan for a Polytechnic School in Boston" which has been called the embryo of the Massachusetts Institute of Technology.[15] It consists of two letters from William B. Rogers to his brother Henry, who had conceived the plan for setting up just such a school. The time seemed ripe. According to Dean Samuel C. Prescott, writing in *When MIT was "Boston Tech,"* "The industrial development of New England and the energy of Boston capitalists in finding new outlets for their enterprise had created a need for well-trained engineers. Various schemes for technical education to meet this demand were in the air at the time. . . ." Schools of science at Harvard and at Yale were both established in 1847, although neither was at first a very prestigious part of its university. An idea for a new type of school had been considered by the Rogers brothers for several years, but it took another two decades for the plan to become reality.

William Barton Rogers (1804–1882) and Henry Darwin Rogers (1808–1866) are sometimes treated together in biographical dictionaries and are classified as geologists.[16] They were born in Philadelphia and educated mainly at William and Mary College, where their father had become professor of natural philosophy and chemistry. Both taught at several colleges; both served as state geologists, William in Virginia and Henry in New Jersey and Pennsylvania; both were deeply involved in the organization of the Association of American Geologists (1840) and its transformation into the American Association for the Advancement of Science in 1848. In 1832 Henry had gone to London with Robert Dale Owen, son of Robert Owen, the manufacturer and reformer who was also one of the most important utopian socialists. Even earlier young Rogers had become profoundly interested in and influenced by the social and educational theories of Owen's circle. In London both young men studied chemistry, gave lectures on science, and made a start toward their professional careers as geologists. After his return to America Henry kept in close touch with his brother William, and together they began to develop plans for "organizing a more rigorous and practical kind of scientific education involving sustained laboratory work." After unsuccessful attempts to carry out such plans the younger brother went to Scotland in 1855 and became Regius Professor of Natural Philosophy and Geology at the University of Glasgow; it was left for William Rogers to pursue their ideas for a school in Massachusetts. (The other two Rogers brothers both studied medicine and chemistry, and wrote one of the earliest textbooks of chemistry in the United States.)

A charter for the Massachusetts Institute of Technology was obtained from the Massachusetts legislature in 1861, and the Institute was organized with William Rogers as president in 1862. But even with some help from the Morrill Land Grant provisions there were delays in raising the necessary capital, and its doors were opened to students (15 of them at the start) on February 10, 1865. Rogers himself was the first professor of physics,[17] a subject in which he had conducted research and published several papers. But his health was poor, and other duties were pressing; in 1867 he hired Edward Charles Pickering, fresh from the Lawrence School of Science at Harvard, to take charge of the physics department. Thus it was Pickering who carried out the ideas for a student laboratory in physics, and who prepared the first laboratory manual. E. C. Pickering left MIT for the Harvard Observatory in 1877, and is chiefly remembered for notable contributions to astronomy proper; to physicists he is known for his discovery, in stellar spectra, of the series of lines due to singly ionized helium, so very similar to the Balmer series in the hydrogen spectrum. The similarity was not understood until the emergence of the Bohr theory of the atom, much later.

For sources of experiments Pickering used scientific journals, textbooks, and reference works, as well as drawing on his own considerable ingenuity. He prepared instructions for student use, at first available only in single copies, but he soon became convinced that access to these instructions prior to entering the laboratory would be helpful to the students. Other institutions were showing interest in the student laboratory, and the preface to *Elements of Practical Manipulation*, published in 1873, begins: "The rapid spread of the Laboratory System of teaching Physics, both in this country and abroad, seems to render imperative the demand for a special textbook, to be used by the student. To meet this want the present work has been prepared, based on the experience gained in the Massachusetts Institute of Technology during the past four years." He may have been a bit overoptimistic about the growth of the idea, but Pickering's manual was extremely influential. He described its plan further:

"The preliminary chapter is devoted to general methods of investigation, and the more common applications of the mathematics to the discussion of results. The graphical method does not seem to have attracted the attention it deserves; it is accordingly compared here with the analytical method. Some new developments of it are moreover inserted. It is of fundamental importance that the student should clearly understand how to deal with his observations, and reduce them, and that he should be familiar with the various kinds of errors present in all physical experiments. A short description is also given of the various methods of measuring distances, time and weights, which, in fact, form the basis of all physical investigation. This chapter is intended as the ground-work of a short course of lectures, given to the students before they begin their work in the laboratory. It should be so far extended by the instructor, as to render them familiar with the general principles on which all physical instruments are constructed, thus greatly aiding them when they have occasion to devise apparatus for their own work.

"The remainder of the volume is devoted to a series of

experiments which it is intended that the student shall perform in the laboratory. Each experiment is divided into two parts; the first called *Apparatus*, giving a description of the instruments required, and designed to aid the instructor in preparing the laboratory for the class. The student should read this over, and with it the second part, entitled *Experiment*, which explains in detail what he is to do.

"Perhaps the greatest advantage to be derived from a course of physical manipulation, is the means it affords of teaching a student to think for himself. This should be encouraged by allowing him to carry out any ideas that may occur to him, and so far as possible devise and construct with his own hands, the apparatus needed. Many such investigations are suggested in connection with some of the experiments. . . . To aid in this work, a room adjoining the laboratory should be fitted up with a lathe and tools for working in metals and wood, as most excellent results may sometimes be attained at very small expense, by apparatus thus constructed by students.

"The method of conducting a Physical Laboratory, for which this book is especially designed, and which has been in daily use with entire success at the Institute, is as follows. Each experiment is assigned to a table, on which the necessary apparatus is kept, and where it is always used. A board called an indicator is hung on the wall of the room, and carries two sets of cards opposite each other, one bearing the names of the experiments, the other those of the students. When the class enters the laboratory, each member goes to the indicator, sees what experiment is assigned to him, then to the proper table where he finds the instruments required, and by the aid of the book performs the experiment.

"Any additional directions needed are written on a card also placed on the table. As soon as the experiment is completed, he reports the results to the instructor, who furnishes him with a piece of paper divided into squares if a curve is to be constructed, or with a blank to be filled out, when single measurements only have been taken. In either case a blank form is supplied, as a copy. New work is then assigned to him by merely moving his card opposite any unoccupied experiment. By following this plan an instructor can readily superintend classes of about twenty at a time, and is free to pass continually from one to another, answering questions and seeing that no mistakes are made. He can also select such experiments as are suited to the requirements or ability of each student, the order in which they are performed being of little importance, as the class is supposed to have previously attained a moderate familiarty with the general principles of physics. Moreover, the apparatus never being moved, the danger of injury or breakage is thus greatly lessened and much time is saved. To avoid delay, the number of experiments ready at any time should be greater than that of the students, and the easier ones should be gradually replaced by those of greater difficulty.

"English weights and measures are occasionally used as well as French to familiarize the student with both systems, as in many of the practical applications of physics the general prevalence of the foot and pound as units seems to render premature the exclusive introduction of the metric system. The second volume of this work, including Heat, Electricity, a list of books of reference, and other matters of general interest to the physicist, will be issued at as early a date as possible."

The "method of conducting a physical laboratory" outlined here is somewhat reminiscent of much later developments, the Library of Experiments instituted at the University of Colorado by Oppenheimer and Correll,[18] for one, but there is no evidence the students worked in teams or even in pairs. The individual experiments vary in length, but no set period of hours is suggested.

In addition to an introductory discussion of general methods of physical investigation, Volume I of Pickering's manual contains experiments teaching various techniques, then sets of experiments arranged by subject matter—mechanics of solids, of liquids and gases, then sound and light in that sequence. Volume II (published in 1876) begins with electricity, proceeds to heat, then veers toward applications: mechanical engineering and meteorology. The latter, by the way, includes terrestrial magnetism and "electricity of the air." There is also a section on astronomy: "Surely nothing could be more valuable to the civil engineer or explorer than to be able to determine his latitude, longitude, and time, by the sextant or transit." Moreover, Pickering argued, every physicist should have some acquaintance with astronomy and its methods, especially the determination of errors and the application of corrections. The final experimental section is on Lantern Projection, including such sophisticated uses as the projection of spectra. Volume II ends with some thirty pages of tables, a list of references, and more general discussion of laboratories. Much importance is attached throughout to graphical methods. Graphs were to be submitted for approval of the instructor before being pasted into the student notebooks. Notebooks were evidently required, not separate laboratory reports.

The almost simultaneous emergence of the academic research laboratory is another story (although not unrelated), but Pickering listed a hundred additional experiments, with this introduction:

"One of the greatest advantages of a Physical Laboratory would, however, be lost if the work should be confined to what has been already described. The highest aim of every physicist should be to direct, not only his own efforts, but those of his students, toward original investigations or the determination of new facts and laws. Without this he is liable to become a mere machine, disseminating knowledge, but never advancing it. The whole aim of this book has been in this direction, and without it, we may educate followers, but never leaders in science. . . ." The list of 100 was meant to be only suggestive, and could be extended almost indefinitely: "In fact, the current numbers of the scientific periodicals, especially Poggendorff's *Annalen* and the *Comptes Rendus*, are full of them." It was noted that supervising such student work would by no means be easy for the instructor, especially when has a large number of students to direct.

Pickering had reported on his new laboratory in *Nature*, January 26, 1871.[19] He was modest about priority: "It is well known that chemistry can be taught far better by a laboratory in which the student performs the various experiments, than by any system of lecture. Now, although for many years physicists have been in the habit of instructing their special students and assistants in this way, yet it is only recently that the same plan has been tried with large classes in physics. One of the first institutions to attempt this method, in America at least, was the Massachusetts Institute of Technology of Boston; and as I find many colleges here establishing physical laboratories, I trust that our experience will prove of some interest." After describing the plan he remarked: "There are now in America at least four similar laboratories in operation or preparation, and the chances are that in a few years this number will be greatly increased." He did not name the four institutions.

The State University of Iowa in Iowa City could have qualified as one of them.[20] The physics laboratory there was designed by Gustavus Hinrichs; he credited as his model Liebig's plan of laboratory work at Giessen. In the fall of 1871 he reported that "more than 200 students have experimented within six months" at the university, and we have another account two years later. Rossiter W. Raymond, U.S. Commissioner of Mining, on visiting the university in 1873, wrote "Professor Hinrichs was not the first, though one of the first, to insist that physics as well as chemistry should be studied in the laboratory, and that every law, principle, and mathematical relation should be verified by the actual observation of the pupils. But he is the first to show how this can be done conveniently and efficiently, with simple and inexpensive apparatus, and for large numbers of students." Reviews of a text by Hinrichs and Nos. 1 and 2 of a laboratory manual in magazine format for the schools appeared in *Nature*.[21] He thought physics should be taught at three levels, including the lower schools, high schools, and college. Unfortunately, physics at Iowa did not have a record of continuous growth; Hinrichs's activities came into disfavor, and were revived only in 1877 with a change in university administration.

In 1869 the acting president of MIT wrote "I find that the Cornell University Catalogue states that a physical laboratory is in early contemplation." It was also the recollection of Edward L. Nichols (founder of the *Physical Review*) that student laboratory study in physics "was begun for the first time in this country at MIT and Cornell almost simultaneously."[22] But W. A. Anthony, who created the physics department at Cornell, did not arrive there until 1872, and his letter detailing plans and requesting money for a student laboratory is dated September 1873.[23] Evidently laboratory work actually arrived at Cornell a little later than at Iowa and at MIT.

Land Grant Schools (such as Cornell) and a number of recently established state universities, in addition to technical schools, made provisions for student laboratories earlier than the long-established colleges and universities. The older instituions recognized the need for laboratory work, but proposed to satisfy it at the secondary level. It is sometimes forgotten that Hall's famous list of forty experiments, published in 1887, was set forth as an admission requirement for students entering Harvard, not as being taught there.

Although other universities did not institute such a rigid requirement Hall's list had a great impact on secondary school physics. It was widely circulated; its prestige was high and so were the standards implied for accuracy and precision. The social and educational climate favored the introduction of laboratory for both utilitarian and cultural reasons, and the plan caught on readily. But enthusiasm carried the movement too far. The laboratory became a fad, a thing in itself, with no theoretical context. Within a few years is was necessary to modify such requirements so as to give proper place to the textbook and to qualitative demonstrations, as well as to simplify many of the experiments. Hall had gleaned them from Pickering and other college texts, and he himself recognized that not all were suitable at the high school level. Meanwhile it became increasingly clear to all that physics laboratory work could not be left entirely in the hands of the secondary schools, although some of them were very advanced.

The status of physics in the high schools was and remained vastly uneven, but that of college physics was even more chaotic. In 1878 the United States Commissioner of Education carried out a survey of programs in physics and chemistry at all levels. The data were analyzed and summarized by Frank W. Clarke, then professor of chemistry and physics at the University of Cincinnati.[24] In grouping together universities, colleges, and schools of science, Clarke remarked that it was difficult to select useful dividing lines despite extremely wide variations. In addition to statistical tables listing all the institutions surveyed he gave brief descriptions of programs in those institutions in which student laboratory work was carried as far as qualitative analysis. These add up to less than 150 nationwide, of which some thirty-five included laboratory instruction in physics. In the colleges listed only in the statistical tables physics and chemistry occupied much the same position as in the ordinary high schools. In fact, "many high schools are actually doing more and better work with these sciences than is done in a very considerable number of colleges bearing good reputations." (Note that this was nearly a decade before the appearance of Hall's list.) Clarke attributed the subordinate position of scientific studies to "the old-fashioned plan of a fixed curriculum." With all students required to follow the same course of study no more than a smattering of science could be included. The colleges doing the best work, "with a few honorable exceptions," had adopted the elective system.

Before the end of the century the situation had changed remarkably. All universities had instituted physics laboratories, and so had most colleges; the exceptions were small sectarian schools that retained rigid curricula. But problems multiplied as well. As the number of students increased, so did the difficulties of arranging laboratory work for large classes. For Pickering a large class numbered twenty; there were hundreds of physics students in large schools by the end of the century, and today there are thousands. One attempt to meet the problem was to let all the pupils perform the same experiment at the same time, but this is expensive if the apparatus is of good quality and unless the instructions are amended from year to year students tend to copy from old reports easily accessible to them.[25] When Cajori wrote in 1898 he reported that a combination of the two methods was usual in most colleges and universities, but that many difficulties persisted. We are well aware that three-quarters of a century later they have been further magnified.

The question has sometimes been raised as to why physics laboratories for students were introduced so late, at least forty years later than those in chemistry. There are several reasons. Physics was already highly theoretical, whereas chemistry was largely an empirical science for much of the 19th century. Newtonian mechanics was not always clearly distinguished from mathematics, and subjects like heat and electricity were regarded as branches of chemistry. Black, Cavendish, Davy, even Faraday, considered themselves chemists. Joule, Kelvin, Maxwell, and others made these subjects not so much a part of physics as a part of natural philosophy. The very word physics was a late comer in English, having been used somewhat earlier in both German and French.[26] Chemistry as a profession was also developed, especially in Germany, whereas physics remained a rich man's avocation or narrowly academic. The reasons cited by Cajori for the lag in physics relative to chemistry laboratories were the greater apparent relevance of chemistry to practical life (e.g., metallurgy) and the fact that physical apparatus is much more costly than that for chemistry.

There is still another reason for the tardy advent of laboratory education: inertia. We have seen that student laboratories were slow to come in well-established American colleges and universities—this was even more true in England. In reply to a letter of congratulation from Lord Rayleigh on his professorship at Cambridge (1871), Maxwell wrote: "I hope you will be in Cambridge occasionally yourself for it will need a good deal of effort to make Exp. Physics bite into our university system which is so continuous and complete without it."[27] Only after Rayleigh took the professorship following Maxwell's death in 1879 was student laboratory teaching introduced into the Cavendish Laboratory. And for class teaching in experimental physics Rayleigh adopted the methods Pickering had established at MIT. Before that, all laboratory teaching had been carried out on an individual basis. R. T. Glazebrook, who was largely responsible for adapting the MIT system to the Cavendish, said that most experimental physicists in Britain as well as in America were brought up on this system throughout the remainder of the century.[28]

However slow the start, the 19th century did see the development of physics laboratories for students and for research as well, often correlated with each other. The need for educational laboratories had been foreseen by earlier reformers: Comenius (1592–1670) is quoted as saying "Men must be instructed in wisdom so far as possible, not from books, but from the heavens, the earth, the oaks and the beeches—that is, they must learn and investigate the things themselves, and not merely the observations and testimonies of other persons concerning things." And further: "Who is there who teaches physics by observation and experiment, instead of by reading an Aristotelian or other textbook?"[29] This view was expanded and emphasized by social and educational reformers in the 19th century. In fact, the role of the laboratory was sometimes idealized. In retrospect James Phinney Munroe could say in eulogizing William Barton Rogers that his founding of MIT was but an insignificant part of what he did for education. "His greater service was in showing that the laboratory method is the basic means of all true teaching, and in forcing the colleges, the high schools, the whole teaching hierarchy, slowly to accept this truth. The change has come about so gradually as to make it hard to realize that, whether in the primary school or in the university the attitude of the teacher toward his pupil, of the pupil toward his work, of the public toward the means and ends of education, is enormously different from that of forty years ago. Then education was receptive, today it is creative; then the pupil was to be instructed, now he is to be developed; then the important element was the lesson learned, today it is the student learning. What the seers in education had been preaching for centuries is meeting general acceptance only in the last fifty years. That this right view is so rapidly prevailing is because of the development, in the broad meaning of the term, of the teaching laboratory."[30]

Would it were true! We owe a great debt to Rogers, Pickering, and others, and can agree wholeheartedly with those who argues, in the words of F. W. Clarke's 1880 report, "much teaching of science preliminary to laboratory practice is like lecture upon swimming before the pupil enters the water." But it is not in the nature of things that laboratories alone can solve all problems in science teaching, and difficulties with the laboratories themselves multiplied as did their number. Physics teachers still struggle with great ingenuity, sometimes make notable advances, and expect to continue doing so. In this century it has become increasingly evident that there are no final answers, either in education or in science itself. But there are always better answers to be found, and discovering them is the continuing challenge.

[1]Sidney Rosen, Am. J. Phys. **22**, 194 (1954).

[2]A. E. Moyer, Phys. Teach. **14**, 96 (1976). (Hall's list of required experiments for entering Harvard is given in this reference as well as Ref. 1.)

[3]F. K. Richtmyer and E. H. Hall, Am. J. Phys. **6**, 14 (1938). These remarks have been reprinted in *On Teaching Physics* (AAPT, Stony Brook, NY, 1979).

[4]F. Cajori, *A History of Physics including the Evolution of Physical Laboratories* (Macmillan, New York, 1899) is the best general reference. See also F. P. Whitman, Science **8** (NS), 201 (1898).

[5]Mrs. W. B. Rogers, *Life and Letters of William Barton Rogers* (Boston, 1896), Vol. II, p. 287.

[6]W. G. Adams, Nature **3**, 323 (1871). This communication was in response to one from E. C. Pickering, cf. Ref 19. Robert Bellamy Clifton planned the Clarendon Laboratory at Oxford, said to be the first in England especially built and designed for the study of experimental physics (1868–70), but he was not able to establish a student laboratory at Oxford so early.

[7]E. C. Pickering, *Elements of Physical Manipulation* (Hurd and Houghton, 1873), Vol. I; *ibid.* (1876), Vol. II.

[8]Ira Remsen, Nature **49**, 531 (1894). Ira Remsen in an address at the opening of Kent Laboratory at the University of Chicago related the story of Liebig's laboratory.

[9]Lord Kelvin, Nature **31**, 409 (1885). An address at the opening of the physical laboratory at University College, Bangor, Wales.

[10]J. Rud Nielsen, Am J. Phys. **7**, 10 (1939).

[11]Sylvanus P. Thompson, *The Electromagnet* (Spon, London, 1891). Appendix A, on William Sturgeon, includes a memoir by James P. Joule.

[12]Some lecturers were not financially successful. James P. Gray, in *The University of Minnesota 1851–1951*, says that at one point "Minnesota's apparatus for [physics] instruction consisted of a few odd pieces of equipment that had been bought of a traveling lecturer in 'Natural Philosophy' who had visited St. Anthony in the 1870s and became stranded."

[13]P. C. Ricketts, *History of Rennselaer Polytechnic Institute* (Wiley, New York, 1914, 1934).

[14]Compare Ref. 4, p. 292.

[15]Samuel C. Prescott, *When MIT was "Boston Tech"* (Technology, Cambridge, 1954).

[16]For example, *Dictionary of Scientific Biography* (Scribners, New York, 1975).

[17]H. M. Goodwin, Technol. Rev. **35**, 287 (May 1933).

[18]F. Oppenheimer and M. Correll, Am. J. Phys. **32**, 220 (1964).

[19]E. C. Pickering, Nature **3**, 241 (1871).

[20]James P. Wells, From Natural Philosophy to Physics and Astronomy, Annals of the State University of Iowa, unpublished copy in the AIP Center for the History of Physics (1979).

[21]Nature **4**, 421 (1871).
[22]Letter from Ernest Merritt, initial co-editor of *The Physical Review*. Cornell University archives, microfilm copy available from AIP (unpublished).
[23]Morris Bishop, *A History of Cornell* (Cornell University, Ithaca, 1962).
[24]F. W. Clarke, *A Report on the teaching of chemistry and physics in the United States* (U.S. GPO, Washington, DC, 1880).
[25]C. R. Mann, *A Study of Engineering Education* (Carnegie Foundation for the Advancement of Learning, New York, 1918).
[26]D. S. L. Cardwell, in *Scientific Change*, edited by A. C. Crombie (Heinemann, London, 1963).
[27]Lord Rayleigh, *John William Strutt, Third Baron Rayleigh* (Arnold, London, 1924).
[28]J. G. Crowther, *The Cavendish Laboratory 1874–1974* (Science History, London and New York, 1974).
[29]W. H. Welch, in *The Electrician* (London, 1896), Vol. 37, p. 172. Comenius is quoted by Cajori.
[30]J. P. Munroe, *William Barton Rogers, founder of the Massachusetts Institute of Technology* (Ellis, Boston, 1904). Also Technol. Rev. **6**, 501 (1904).

Laboratories and the rise of the physics profession in the nineteenth century

Melba Phillips
351 West 24th Street, New York, New York 10011

(Received 16 August 1982; accepted for publication 6 December 1982)

There were no established research laboratories before the nineteenth century. Galileo had a small shop, Hooke was a curator of instruments, Boyle worked at home, Newton in his college rooms. Scientists were often amateurs with their own laboratory facilities. Even in universities they could obtain prominence as individuals, but took students only privately, as assistants or apprentices. There was no profession of physics to which one could aspire as a career. The tradition of modern research laboratories and the research school can be traced to Liebig in chemistry, and the extension to physics can be attributed in large part to Gustav Magnus of the University of Berlin, who also invented the physics colloquium. German universities continued to lead the way throughout the second half of the nineteenth century, but by 1900 other industrialized countries (especially Britain, France, and the United States) had developed research schools leading to careers in physics. Research laboratories were also being established by industry and officially by governments.

Laboratories did not play a part in either elementary or advanced education until well into the 19th century.[1,2] In fact laboratories in the modern sense did not exist before 1800. Almost without exception earlier research scientists were amateurs, who gained prominence as individuals with private facilities. Pupils studied with masters as disciples or apprentices but there were no schools in which students could study the physical sciences to become practitioners. In neither physics nor chemistry was there a profession, a group recognized as experts in a common discipline. The development of professionalism occurred somewhat earlier in chemistry than in physics,[3] but by the end of the century each was a discipline in its own right, with schools and an increasing variety of opportunities for continuing careers. Research laboratories played a key role in this development, which depended on changes taking place in several countries. The most significant features came to fruition first in Germany, and appeared rather later in English-speaking countries. France played a seminal but somewhat static role in the evolution of laboratories and of professionalism. The roots and the strands of the development are diverse, but the main features can be traced.

An underlying tradition of experimentation existed well before 1800. Galileo is credited with giving us experimental physics, although he did not have what could be called a laboratory. While in Padua (1592–1610) he did establish a small shop, and employed a skilled workman who made his home with the Galilei family.[4] There he designed and manufactured mathematical apparatus, notably "geometric and military compasses," that were sold to students and others and added to the family income. After the appearance of William Gilbert's book (1600) magnets of various shapes were produced. The most important instruments to come from his shop were Galileo's telescopes. It is not clear just what role the shop played in the development of Galileo's mechanics; *The Two New Sciences*, written more than a quarter of a century later, is thought to be a reworking and expansion of results obtained during the Paduan period and communicated only by letters to friends and pupils at the time. The book begins with a well-known reference to the arsenal at Venice—this arsenal, by the way, was a shipyard producing commerical as well as naval vessels, not primarily armaments. But Galileo had not been in Venice for years when the reference was made.

Later in the 17th century the center of physical science had moved north and west. Robert Boyle, the most famous scientist of his generation, had a succession of laboratories at home; his most gifted and famous assistant was Robert Hooke. Newton used his college rooms for experimentation, but it would not have occurred to him that the university would furnish a room for the purpose of experimentation. Robert Hooke might be called a professional scientist of the 17th century.[5] At the suggestion of Boyle he was made curator of instruments of the Royal Society in 1662. He designed apparatus and did notable research in the laboratory he developed there, but his first duty was to furnish the Society with "three or four considerable experiments" at the weekly meetings. Hooke was with the Royal Society for forty years, but the laboratory was dismantled early in the 18th century, when Newton had become president. Regardless of his splendid experiments in optics Newton was essentially a mathematical physicist, and both his personal taste and his antipathy to Hooke kept him from preventing the deterioration of the Royal Society Laboratory.

The laboratory as a special room or building for experiments arose in connection with alchemy, but the tradition did not hold for most of the pioneers of modern chemistry.[6] There were exceptions. That universal genius Lomonosov persuaded the Russian Academy of Sciences to build and equip a good laboratory in St. Petersburg (actual construction 1748) where he studied light and electricity as well as chemistry and metallurgy. But he founded no school, and his troubles with other professors in the Academy limited the duration of the activity to about a dozen years. Unfortunately, his influence did not extend much beyond Russia. The more influential Lavoisier[7] had a fine laboratory in the arsenal in Paris, where he also lived, from 1776 to 1792. He investigated much the same subjects as Lomonosov, but his primary duty there was to make sure of the quantity and quality of gunpowder. His visitors to the laboratory included Benjamin Franklin and James Watt; a description of it was recorded by the famous traveler Arthur Young.[8] Later (from 1818) Gay-Lussac also had a laboratory in the Paris arsenal. It was there, as we shall see, that Liebig was given an opportunity to try his hand at serious experimentation in chemistry.

In the 18th century Edinburgh had been among the best scientific schools, but no university anywhere prepared students to become scientists as such. Science was done either by amateurs or by professional people like physicians and engineers, even churchmen, who pursued it in their spare time. Science was largely an avocation. There were noted professors, such as Joseph Black, but Black was trained as a physician and continued his practice even as a full-time professor of chemistry. His discoveries on heat were demonstrated in lectures but were made generally known only when the lectures were published after his death.[9] His research experimentation, once he was a professor, seems to have been almost indistinguishable from his preparation of lecture demonstrations. Meanwhile England had produced Joseph Priestley and Henry Cavendish, both of whom worked with his own private laboratory facilities. But after Newton the pendulum had swung back to the continent in both physics and chemistry.

During the early part of the 19th century France possessed the most brilliant galaxy of scientists the world had ever seen, many of them in experimental physics.[10] One need only think of Ampeŕe, Biot, Savart, Arago, Fresnel, Carnot, Dulong, Petit, in addition to the more mathematical Fourier, Lagrange, Laplace, Poisson—the list goes on. Arago lived and worked in the Paris Observatory, and there was usually limited space in the colleges available to professors for their own work. The Ecole Polytechnique, founded in 1794, centainly had laboratory facilities, apparatus acquired by revolutionary confiscation from the laboratories of wealthy amateurs at the beginning, and later from Napoleon's war booty. (In the last category one hears of alum and great stores of mercury, both much prized.) A splendid Voltaic pile was constructed at one point, but a contemporary visitor reported that Gay-Lussac had to wear wooden clogs because of dampness in his workroom there. Many of the early developments in electrodynamics were made in France, but no original records have survived and there is no way of knowing exactly what was done, or where. There is in fact evidence that some of the experiments reported were never carried out at all.[11] There was neither opportunity nor incentive for advanced students to become research scientists anywhere in France. Professors were poorly paid, took several jobs, and there was no place for young men. Almost all scientific activity was concentrated in Paris and positions elsewhere were undesirable even when they existed. But lectures in the Paris colleges were often beautifully illustrated with demonstrations; Dumas and Gay-Lussac were both famous lecturers.

The tradition of both research and student laboratories of modern times can be traced to France through Justus von Liebig.[12] Liebig was frustrated as a student in Germany by the lack of opportunity for "practical work" in chemistry, although he obtained a good background there. He went to study in Paris, was much impressed with Gay-Lussac's lectures, and was fortunate to come to the attention of Alexander von Humboldt, who was a good friend of both Arago and Gay-Lussac. It was in this way that he gained access to the latter's arsenal laboratory, more or less as a young colleague. At that time no school in Paris had a laboratory that could accomodate a student. In 1824, when Liebig was only 21, Humboldt arranged for him a chair at the small University of Giessen; he remained there for 32 years. His laboratory opened in 1826. Until 1839 Liebig had no private laboratory room of his own, but his school was becoming famous throughout the world.

An autobiographical sketch found among Liebig's papers after his death describes his method.[13] "Actual teaching in the laboratory, of which practical assistants took charge, was only for the beginners; the progress of my special students depended on themselves. I gave the task and supervised the carrying out of it, as the radii of a circle have all their common center. There was no actual instruction; I received from each individual every morning a report upon what he had done on the previous day, as well as his views of what he was engaged upon. I approved or made my criticisms. Every one was obliged to follow his own course. In the association and constant interchange with each other, and by each participating in the work of all, every one learned from the others. Twice a week, in winter, I gave a sort of review of the most important questions of the day; it was mainly a report of my own work and their work combined with the researches of other chemists." These methods are in marked contrast to the accepted practice of private laboratories. That of the famous Swedish chemist Berzelius was an adjunct to his kitchen. Berzelius worked mostly alone; he was willing to accept one student each year, but he was often visited, sometimes for rather extended stays, by other professional chemists. Liebig was certainly the first to set up a successful research school for the education of professional scientists.

The idea of setting up what might be called a graduate school in chemistry also arose in Scotland. As early as 1817 Thomas Thomson of the University of Glasgow wrote that he hoped "to erect a laboratory upon a proper scale to breed up a set of young practical chemists."[14] Exactly what he meant by "practical chemists" is hard to say, but chemistry was at that time almost entirely empirical. Beginning in the 1820s he worked at this project, but with very little success, mainly because he had no backing. Liebig's success in Germany was due to public support, furnished by the State of Hesse-Darmstadt for the university at Giessen. He came to have a laboratory steward, an assistant paid by the institution, and money for supplies, none of which was furnished to Thomson.

The scientific climate in England favored such establishments even less. There were training schools for mechanics and artisans, but not in basic science. Prince Albert was instrumental in setting up a Royal College of Chemistry with advice from Liebig in 1845; A. W. Hofmann, a protege of Liebig's, was made director.[15] British landowners and industrialists were interested. Hofmann actually founded a very promising school of chemistry in London and instituted a research program specializing (very appropriately for England) in compounds derived from coal tar. William Henry Perkin, inventor of the first synthetic dye, was a student of his. But although the school was influential it did not prosper as Hofmann hoped; there was little interest from either the industrialists or the government in supporting research projects which might or might not be practical. The college was later consolidated with the Royal School of Mines, and Hofmann returned to Germany after nearly twenty years in England. There he became a professor at the University of Berlin, where he was free to pursue fundamental researches that were to become important to the chemical industry in that country. The universities in Britain were not yet seriously involved in the growth of experimental science.

The facilities of the Royal Institution date from 1800, and its laboratories, made famous by Davy and Faraday, began to break new ground in physics research. Rumford's object in founding the Institution was to alleviate the problems of the poor by the promotion of applied science, although the emphasis soon shifted to pure scientific research.[16] Davy's chief responsibility was lecturing, and the success of the Institution was due in large part to Davy's great success as a lecturer. Davy was a chemist, to be sure, but so was Faraday. In fact, Faraday objected strenuously to the word *physicist* when it was introduced into the English language. Physics did not constitute a distinct specialty, especially in Britain.

The word *science* came into English usage in the Middle Ages as a French importation, synonomous with *knowledge*. It was brought into prominence with its modern meaning with the creation of the British Association for the Advancement of Science in 1831. The name *scientist* was proposed by William Whewell in an unsigned book review published in 1834, and more formally in 1840.[17] "...and thus science, even mere physical science, loses all traces of unity. A curious illustration of this result may be observed in the want of any name by which we can designate the students of the knowledge of the material world collectively. We are informed that this difficulty was felt very oppressively by the members of the British Association for the Advancement of Science at their meetings in York, Oxford, and Cambridge in the last three summers. There was no general term by which these gentlemen could describe themselves with reference to their pursuits." And, later, "As we cannot use *physicien* [the French word] for a cultivator of physics, I have called him a *physicist*. We need very much a name to describe a cultivator of science in general, I should incline to call him a *scientist*." Faraday, who had consulted Whewell in naming ion, cathode, diamagnetism, etc., approved *scientist*, but rejected *physicist*. Kelvin objected to *physicist* as late as 1890. Chemist and chemistry were old established words, derived from the even older *alchemy*, but most fields of physics were included in the category "natural philosphy."

The absence of what might be called "separate schools of science" is indicated in the purview of the four "committees" set up at the founding of the British Association for the Advancement of Science in 1831.[18] They were (i) pure mathematics, mechanics, hydrostatics, hydraulics, plane and practical astronomy, meteorology, [terrestrial] magnetism, philosophy of heat, light, and sound; (ii) chemistry, mineralogy, electricity, magnetism; (iii) geology and geography; and (iv) zoology, botany, physiology, anatomy. It is easy to see that Faraday would belong in category (ii), but where would Joseph Black have felt at home if he had lived into the 19th century? It may be that physics and chemistry are distinguished only by differences in emphasis, but some differences become crucial as the body of scientific knowledge grows beyond the possibility of one person being expert in many fields.[19] We shall return to the situation among English-speaking scientists, but the path to professionalism began mainly in Germany. In German universities there were professors of physics as well as professors of chemistry. It was to such a professor at the University of Berlin that we owe the extension of the idea of a research school to physics as a field in itself.

Heinrich Gustav Magnus[20] was born in 1802, the year before Liebig, but his university career got under way a year later than Liebig's. After taking his doctorate in chemistry at the University of Berlin in 1827 he spent time with Berzelius and with Dulong and Gay-Lussac in Paris. He taught at a technical school for a few years before joining the University of Berlin in 1834; he was made professor of physics and technology in 1845. Soon after 1840 he was already taking research students, and by 1843 he had initiated the first physics colloquium, which became almost as well known as his laboratory. Among his most famous research students were Helmholtz and Tyndall. Students and visitors came from all over the world to the school he built up—J. Willard Gibbs attended his lectures during 1867–68, for example.[21] His laboratory rooms were at first in his dwelling, which was furnished by the university; a spacious new physics laboratory was finally built in 1863. The seminar had been introduced at German universities as early as 1830 to enable students of history and the social sciences to become acquianted with methods of research in those fields, but the physics colloquium invented by Magnus was different: students reported on recent scientific advances, and visitors occasionally spoke. Topics of interest were decided on at the beginning of the term, and regular participants undertook the responsibility for preparing on different subjects. Discussion followed the presentations. The list of participants for the 27 years 1843–1870 shows 268 different signatures, including Clausius, Kirchhoff, Kundt, Wiedermann, and many others who were or who became famous. Two basic aspects of professionalized science are said to be laboratory teaching and the research school; the physics research laboratory and the colloquium–seminar introduced by Magnus are essential parts of the research school in physics. Other German universities built up similar schools, a pattern which was to be copied abroad.

Meanwhile the growth of science in British universities had got under way, although the role of the German model was not conspicuous at the start. Laboratories were introduced first in Scotland, even though the efforts of Thomas Thomson had not been very productive. Soon after William Thomson (to become Kelvin) was made professor of physics at the University of Glasgow (1846) he made a place for his own experiments in an abandoned wine cellar near his lecture room, and later added an unused examination room.[22] His initial reason for inviting students there was that "the labor of observing proved too heavy, much of it could scarcely be carried on without two or more persons working together." The laboratory was formally recognized as part of the university in 1866, but with all his enormous fame and his very strong personality Thomson did not found a great research laboratory at the University. His tremendous reputation reflected a wide range of scientific achievement, but it is clear in retrospect that he was most interested in engineering. His greatest pride was in the trans-Atlantic cable. While remaining a professor of physics he became a wealthy man as a successful industrialist, and his students were often put to work on problems of industrial design. He was sought after by both Oxford and Cambridge—it is probably fortunate for the latter that he did not accept the Chair of Experimental Physics associated with what became the Cavendish Laboratory. The great Cavendish tradition grew out of the leadership of even better physicists: Maxwell, Rayleigh, and J. J. Thomson.

Student laboratories were set up in University College, London (1866) and Kings College, London (1868)[23] but laboratory work was not at first integrated into the regular study of physics, and there was no general plan for involving students in research. The first university building in England devoted to physics was the Clarendon Laboratory founded by R. B. Clifton at Oxford, begun in 1868 and completed in 1870. Clifton did not receive great support for making use of and expanding the facility. Meanwhile a plan for founding a professorship and laboratory at Cambridge to promote the teaching and study of experimental physics was given funding by the Duke of Devonshire.[24] Maxwell became the first holder of the chair in 1871, and retained the position until his death in 1879. The Cavendish Laboratory was ready by 1874, but students were few, discouraged for the most part by the Cambridge examination system. The first person to break this impasse was Arthur Schuster, who had studied in Manchester and in more than one German university. He refused to sit for the examinations because he had a degree elsewhere; the Demonstrator at the Cavendish got him admitted to one of the colleges, though in no recognized category, and he became the first graduate research student in the new laboratory (1876). Schuster became a professor at Manchester, where he planned and built the Physical Laboratory at that university.

Owens College, the forerunner of Manchester University, was founded in 1851, with a staff of professors that included Balfour Stewart in physics and Osbourne Reynolds in engineering.[25] Soon it became the best England could offer in chemistry; H. E. Roscoe, trained under Bunsen, was professor of chemistry, and the laboratories he designed in 1868 followed the German pattern. J. J. Thomson, who entered Owens at the age of 14 to study engineering and spent five years there, put down vivid recollections of physical facilities that were in his time meager and primitive but with which significant results were obtained. Schuster returned to Owens as professor of mathematics in 1881 and succeeded Stewart in physics in 1888. It was he who made the University of Manchester one of the greatest physics schools in the world. A characteristic act was his resignation in favor of Rutherford, who had been at McGill University in Montreal. This offer was made in 1906, when Schuster was by no means at retirement age. Rutherford continued this perceptive tradition of recognizing merit: a famous example was his encouraging welcome to Bohr who had not felt comfortable in Cambridge. But this was later, after other "Red Brick" universities had been created in the British provinces. By the end of the century the University of London had also begun to establish good facilities for scientific research.

In contrast to its very impressive record of inventions, the United States was backward in science itself. The most outstanding experimental physicist in America during the first half of the nineteenth century was Joseph Henry, who was largely self-educated.[26] His mathematics training at the Albany Academy included calculus, which was among the courses he later taught there, but his electrical experiments grew out of his own reading on the subject. At least in part his laboratory was a spare classroom, liable to be needed for regular classes. Even at Princeton Henry founded no school to carry on what might have become a tradition in physics research. Emphasis in America was focussed almost exclusively on applications, or purely empirical inventions. Some engineering and technical schools had been set up. Rensselaer Polytechnic Institute, dating from 1824, became a chief source of civil engineers in demand for the conquest of the west. In 1847 both Harvard and Yale established schools of science, the Lawrence and Sheffield schools, respectively, but neither was a prestigious part of its university. The document called the embryo of the Massachusetts Institute of Technology, "A Plan for a Polytechnic School in Boston," dates from 1846, but MIT did not get under way until the 1860s. This "new kind of school" was a pioneer in improving physics education, particularly in laboratory work, but its emphasis was on engineering and applications until well into the twentieth century.[27]

Two contributors to physics who flourished around 1870 are usually overlooked today: the Drapers, John William and his son Henry, were pioneers in the photography and analysis of solar spectra.[28] Both taught at what is now the Medical School of New York University, and used their own funds for what was for them professionally an extracurricular activity. Not properly recognized for their accomplishments they were nevertheless in the mainstream of physics research, although neither of them ever taught the subject. The first truly professional experimental physicist in America was Henry Rowland. The recognition of Rowland is associated with the founding of the first research school in the country, the Johns Hopkins University.[29]

The Ph.D. degree had been introduced to America at Yale (J. Willard Gibbs was the first recipient, in 1863) but Yale had no research school until the 1880's. Rowland's first papers, submitted from Rensselaer Polytechnic Institute, were rejected by *Silliman's Journal*, the foremost scientific journal in America. He had the temerity to send the second paper to Maxwell, who had it published in the *Philosophical Magazine* and encouraged the young American to follow up his work. About this time (1873) Daniel Coit Gilman, after trying and failing to introduce research science at the new University of California, was beginning to plan a new university in Baltimore, with money supplied by Johns Hopkins. This was the first American graduate school, complete with laboratories, seminars, and dissertations, one patterned consciously on those in Germany. Gilman combed the country for scientific talent, procured Rowland, and sent him off to Europe, along with Ira Remsen, his future professor of chemistry, in preparation for the tasks ahead. Rowland spent time in Berlin with Helmholtz (who had succeeded Magnus at the university there), after visiting scientific establishments in Britain. Johns Hopkins University opened its doors to students (men only) in 1876. Rowland had nearly $6500, then a lavish sum, for equipment, and began to plan the construction of a physics building that was to be the largest structure at the university for years. It was the first and one of the very few schools in which graduate study began at the start. The University of Chicago also began with both a graduate school and an undergraduate college, in 1890.

The idea of a research school did not spread very rapidly in the United States. The Jefferson Physical Laboratory at Harvard and the Sloan Laboratory at Yale were built in the 1880s, but Harvard did not establish a graudate school until 1890. Columbia introduced graduate studies as early as 1880, but the actual situation in physics at Columbia at that time has been vividly described by Michael Pupin,[30] who was a college student there 1879–83. There was no physics laboratory, and very little instruction in theory. Pupin had learned more physics from reading Tyndall's popular lectures (given in 1872, published in '73) than was available in college lectures, and when he arrived in Cambridge in 1883 to study with Maxwell he had no idea the great man had died four years earlier. Pupin did learn mathematics in Cambridge, but he was embarrassed at the idea of studying with the new professor (J. J. Thomson) who was only two years older than himself, and he went on to Berlin. When he returned to Columbia in 1889 to become "Teacher of Mathematical Physics in the Department of Electrical Engineering" he had at his disposal "a dynamo, a motor, an alternator, and a few crude measuring instruments only, all intended to be used every day for the instruction of electrical engineering students...." He was expected to lecture three or four hours each morning and to oversee student laboratories in the afternoon. Pupin did manage to do research, although his early work was for the most part anticipated by J. J. Thomson.

But by that time there was no lack of competent physicists in America, most trained in Europe but beginning to come from Johns Hopkins, as did E. H. Hall of Harvard, a Johns Hopkins Ph.D. of 1880. Research by professors was not highly valued at most schools, however. One exception should be noted: Clark University, in Worchester, Massachusetts, was founded in 1887, with a truly academic graduate school planned from the start and under way by 1889. But there were financial troubles almost from the very beginning, and the very advanced plans were curtailed. Some of its most distinguished professors, including A. A. Michelson, were hired away by William Rainey Harper[31] for the new University of Chicago in 1891, in what became known as "Harper's Raid." But Clark University continued to be influential; its best known physicist was Arthur Gordon Webster, who was instrumental in founding The American Physical Society in 1899.

In hindsight Rowland's "Plea for Pure Science," the title of his address to the American Association for the Advancement of Science in 1883 and the theme of his American Physical Society presidential address in 1899, seems extreme: we should not "waste the intellect of the country in the pursuit of the so-called practical science which ministers to our physical needs...."[32] Rowland was in fact trained as an engineer, and was a well-paid consultant for the Niagara Power and Construction Company. He was trying to raise the level and prestige of research single-mindedly in these statements. This emphasis on pure science for universities was overly emphatic in some opinion. At a meeting of the AAAS in 1890, T. C. Mendenhall[33] deplored the "unfortunate and perhaps growing tendency among scientific men to despise the useful and the practical in science." He pointed out that "the arrogance of genius is not less disagreeable than that of riches." Actually both practical and pure research were fairly well established in American universities before 1900, and both government and industry were beginning to develop research laboratories, as we shall see, although the United States did not compare favorably with Germany or Great Britain at that time.

France continued to produce great scientists. There were inspiring professors, but little or no encouragement was given students who had no connections at high levels. In no other country, by the way, has the tradition for families of scientists been so marked; one thinks of Bequerel, Curie, Langevin, Perrin, Brillouin, even de Broglie. There was very little demand for physicists outside Paris, whether in universities or otherwise. A famous plea for a government laboratory made by Pasteur in 1868 met with some favorable response that hardly extended to physics. The universities in Paris continued to receive public support, but industry fell well behind that emerging in Germany, and the market for trained physicists in France remained very limited.

The growth of research laboratories in universities and the development of research schools were crucial factors in the professionalization of science; until the latter half of the 19th century it was almost impossible to earn a living doing science in any country. The existence of the profession depended on social demand, to be sure, the vastly increased need for scientific services. This need had existed, had even been recognized, for many many years, and there was historical precedent for governmental support in a related science: Global exploration and later mercantilism depended vitally on astronomy, and the first nationally supported governmental observatories were founded in the same decade by both France and England.[34] The Observatoire de Paris was completed in 1672, with Cassini as director, and the Royal Observatory of Greenwich was established by Charles II in 1675. John Flamsteed, appointed "Our Astronomical Observer," had to furnish all the instruments himself; he had a salary of 100 pounds per year, but since he was out of pocket for all apparatus and expenses it is not surprising that he wanted to publish his famous star catalog in his own way. This was the subject of a long and painful dispute with Newton. In both observatories, and other less official ones, important observations were made which could never be of direct interest to navigation, and other researches were undertaken. Ole Roemer's deduction of the finiteness of the velocity of light from Cassini's observations of Jupiter's moons dates from 1675 at the Paris observatory, for example, and we have noted that Arago lived and worked there on a variety of problems early in the 19th century.

National laboratories for more general research beyond the resources of universities developed only toward the end of the 19th century.[35] The Royal Institution in London never depended on public funds, and the Conservatoire des Arts et Metiers, founded in Paris in 1793, had been primarily a sort of museum, although a physics laboratory was established in 1829. Universities in France did receive public money, as we have noted, and the French model was an inspiration for ideas of Joel Barlow and Thomas Jefferson in the United States, ideas that were not realized. John Quincy Adams developed much more comprehensive plans, but they too came to nought. The Smithsonian Institution, with its own endowment but federal trusteeship, was finally established in 1846; under Joseph Henry it initiated various scientific projects, mostly carried out by other agencies, and it made original research more respectable in the United States. But its endowment dwindled in importance as demands increased, and it was never a governmental laboratory.

In building a true national laboratory Germany again took the lead. The various states had long supported science in universities, and after German unification in 1871 the country took full advantage of its scientific capabilities. Prussia and not the more democratic West dominated policy for the Empire, but Bismarck and his conservatives saw the value of science in making Germany preeminent industrially, as indeed it became. The Imperial PhysicoTechnical Institute in Charlottenburg, known as the Reichsanstalt, was established in 1884, a facility much envied in other countries. In 1888 Helmholtz became its director. It was supported in part by industrialists (Werner Siemens donated to its founding) but largely by funds voted by the Reichstag. There were departments for pure research but problems useful to industry were also undertaken. Even industries from abroad applied for services there.

The fixing of standards has been a problem for governments since ancient times, but qualitatively new and urgent demands accompanied the rise of the electrical industry. An international conference during the Columbian Exposition at Chicago in 1893 adopted standard electrical units, and gave new impetus to national institutions for maintaining them, among other duties. In Great Britain the Kew Observatory had been made responsible for standards in 1871, and evolved into the National Physical Laboratory in 1899. In the United States the Office of Weights and Measures was by no means adequate for the increased load, and plans for the National Bureau of Standards were undertaken in 1897. The result was not "a copy of the Reichsanstalt, but a standardizing bureau adapted to American science and American manufacture," according to the instigator of the project, Henry S. Pritchett. Despite the limitations implied by this statement, the Bureau established in 1901 did expand its activities into original research, since problems of physical standards clearly depend on many areas of physics and chemistry. But it did not become a national laboratory; such governmental research as was done was divided between many bureaus. No centralized rational organization of government science ever developed in America. Moreover there was a well-established assumption that basic research belonged to the universities, and only applied research should be left to the government. Nevertheless the various government bureaus gave employment to the largest fraction of scientific personnel trained at the universities. The schools themselves were also expanding, and a new demand for physical scientists was beginning in the larger industries.

Industrial research laboratories had begun to evolve in the latter half of the 19th century. It had long been necessary for factories to maintain shops for testing products and for quality control, but problems were not taken up unless they were directly related to the products manufactured or to be manufactured. The first industrial laboratories that undertook basic research were those in the chemical industry. This happened in Germany as early as 1870, the research being concerned largely with dyes.

In America Edison was one of the more scientific inventors, and established a laboratory in Menlo Park in 1876, but he was not a part of the scientific community. In fact, he disdained it, and took pride in decisions not to undertake inventions for which there was no definite market.[36] Nevertheless he employed one theoretical physicist (Francis R. Upton, who had spent a year with Helmholtz in Berlin) and remarked that it was just as well "to have one mathematical fellow around, in case we have to calculate something out."

Meanwhile the the role of professional applied science was growing.[37] In 1886 Authur D. Little set up a consulting firm in the Boston area, offering research and development services on a gradually broadening scale.

Some industries contributed to associations in order to set up facilities for applied science services. Individual industries were also undertaking research in increasingly wider areas. George Westinghouse founded his electric company in 1886, but he had already in 1885 had W. Stanley "make, set up, and demonstrate the advantage of the alternating current system, the patents for which he had obtained in England." His original plant in Pittsburgh had an associated small laboratory, which developed gradually so as to undertake various problems of concern to the electrical industry. What was to become the largest industrial laboratory in the world, Bell Telephone Laboratories, had its origin in a number of fragmented efforts, small and widely scattered; Bell hired his first engineer–physicist, H. V. Hayes, in 1885 as chief of the Mechanical Department. The amalgamation took place only in 1925. The first industrial laboratory in America dedicated to fundamental research was the General Electric Research Laboratory, founded in 1900. Industrial laboratories were slow in getting under way in Britain, owing largely to the bias against "trade" as vulgar and degrading; some sons and grandsons in manufacturing families were apt to escape industry and commerce, especially if they were sent to the desirable "public" schools.[38] In contrast to the German scene industrial research laboratories in chemistry arose very late in both Britain and America. The du Pont who came to America and was encouraged by Jefferson to start a powder mill in 1802 had been a student of Lavoisier, but the du Pont Company organized its first research laboratory in 1902.

By 1900 research laboratories had been established in the major universities of the industrialized countries, and the importance of industrial research in both physics and chemistry was becoming universally evident. Government laboratories lagged except for the Reichsanstalt; their phenomenal extension in the 20th century awaited growing involvement of science with the military, clearly a government affair. The emergence of German leadership in science during the second half of the 19th century arose from the excellence of its universities, but other countries, chiefly although not exclusively those included in this account, were becoming competitive.

[1]F. Cajori, *A History of Physics including the Evolution of Physical Laboratories* (Macmillian, New York, 1899).

[2]Melba Phillips, Am. J. Phys. **49**, 522 (1981).

[3]H. M. Leicester, *The Historical Background of Chemistry* (Wiley, New York, 1956; also Dover reprint, 1971), especially Chap. XXII.

[4]Ludovico Geymonat, *Galileo Galileo* (published in Italian 1957, translated by Stillman Drake; McGraw-Hill, New York, 1965).

[5]Margaret Espinasse, *Robert Hooke* (University of California, Berkeley, 1962). See also J. G. Crowther, *Founders of British Science* (Cresset, London, 1960).

[6]Aaron J. Ihde, *The Development of Modern Chemistry* (Harper and Row, New York, 1964).

[7]Douglas McKie, *Antoine Lavoisier, Scientist, Economist, Social Reformer* (Shuman, New York, 1952).

[8]Arthur Young, *Travels in France during the years 1787, 1788, and 1789* (Doubleday, New York, 1969), p. 72.

[9]J. G. Crowther, *Scientists of the Industrial Revolution* (Cresset, London, 1962).

[10]Maurice P. Crosland, *The Society of Arcueil; a view of France science at the time of Napoleon* (Harvard University, Cambridge, MA, 1967).

[11]R. A. R. Tricker, *Early Electrodynamics* (Pergamon, Oxford, 1965). Ampère, for example, was clearly both a gifted experimenter and a talented theorist, but Tricker has pointed out that he reported experiments which he frankly admitted had not been performed: "I ought to say in finishing this memoire, that I have not yet had the time to construct the instruments shown in Fig. 4 of the first plate... . The experiments for which they were designed have not yet been performed; ... ." The assumed results were an essential part of his argument.

[12]See, for example, Ihde, Ref. 6, or Leicester, Ref. 3.

[13]Justus Liebig, "Autobiographical Sketch," translated by J. Campbell Brown. Smithsonian Institution, Annual Report 1890/91, p. 257. First printed in *The Chemical News*, London, June 5 and 12, 1891.

[14]J. B. Morrell, Ambix **19**, 1–47 (1972).

[15]See Ihde, Ref. 6.

[16]D. S. Cardwell, *The Organization of Science in England* (Heinemann, London, 1957).

[17]Sydney Ross, Ann. Sci. **18**, 65 (1962).

[18]The British Association for the Advancement of Science, A Retrospect 1931–1931, O. J. R. Howarch, Sec'y. London, 1931.

[19]John Theodore Merz, *A History of European Thought in the Nineteenth Century*, (Blackwood, London, 1904–1912. Dover reprint 1965), Vol. II. Merz has pointed out that the nonexistence of English words to distinguish the various specialties may have reflected a fruitful aspect of British scientific thought: "In England alone the name of natural philosophy still obtained, and in the absence of separate schools of science, such as existed abroad, suggested, at least to the self-taught amateur or to the practical man, the existence of a uniting bond between all natural studies. It is significant that the term under which we now comprise...all natural agencies, the term Energy, was first distinctly used in this sense by Dr. Thomas Young. ...Young himself was a medical man, as were Robert Mayer and Helmholtz after him. Practical men such as Watt felt the necessity of measuring not so much forces (in the Newtonian sense) as the action of forces, and introduced the term power... . Uninfluenced by the theoretical views which were developed and firmly held by the school of which Laplace was the most distinguished representative, natural philosophers like Black, Rumford, and Davy had approached the study of those phenomena where heat and chemical change are the prominent feature." Merz goes on to say that on the continent "...and especially by his great influence as a teacher, Liebig himself did much to bring about an alliance of the separate sciences and a connection between practical pursuits and abstract research, and to draw attention to the interdependence of the various forces of nature."

[20]A. W. Hofmann, Ber. Deutsch. Chem. Ges. **3**, 993 (1870). Reprint pamphlet Berlin 1871.

[21]Lynde Phelps Wheeler, *Josiah Willard Gibbs* (Yale University, New Haven, CN, 1962).

[22]Andrew Gray, *Lord Kelvin* (Dent, London and Dutton, New York, 1974). The quotation is from Kelvin's address at the opening of the physical laboratory at University College, Bangor, Wales [Nature **31**, 409 (1885)].

[23]Romualdos Sviedrys, Hist. Stud. Phys. Sci. **7**, 405 (1976).

[24]J. G. Crowther, *The Cavendish Laboratory 1874–1974* (Science History, London, 1974).

[25]J. G. Crowther, Bull. Inst. Phys., 294 (November 1960).

[26]Thomas Coulson, *Joseph Henry, His Life and Work* (Princeton University, Princeton, NJ, 1950).

[27]See Ref. 2.

[28]*Science in Nineteenth Century America, A Documentary History*, edited by Nathan Reingold (Hill and Wang, New York, 1964).

[29]John C. French, *A History of the University Founded by Johns Hopkins* (Johns Hopkins, Baltimore 1941).

[30]Michel Pupin, *From Immigrant to Inventor* (Scribner's, New York, 1922).

[31]Milton Mayer, *Young Main in a Hurry, the story of William Rainey Harper* (University of Chicago Alumni Association, Chicago, 1957); or Richard J. Storr, *Harper's University, the Beginnings* (University of Chicago, Chicago, 1966).

[32]Henry A. Rowland, Pop. Sci. Mon. **XXIV** (November 1883) and also Bull. Am. Phys. Soc. **1**, 1 (1899). Reprinted in *Selected Papers of Great American Physicists*, edited by Spencer Weart (American Institute of Physics, New York, 1976).

[33]T. C. Mendenhall, Proc. AAAS (1890).

[34]R. J. Forbes and E. J. Dijksterhuis, *A History of Science and Technology, The Eighteenth and Nineteenth Centuries* (Cox and Wyman, London, 1963; also Penguin, Baltimore).

[35]A. Hunter Dupree, *Science in the Federal Government* (Harvard University, Cambridge, MA, 1957; reprinted, Harper Torchbook edition, Harper & Row, New York, 1964).

[36]Matthew Josephson, *Edison* (McGraw-Hill, New York, 1959).

[37]C. E. Kenneth Mees and John A. Leermakers, *The Organization of Industrial Scientific Research* (McGraw-Hill, New York, 1950).

[38]*A History of Technology, Volume V, The Late Nineteenth Century*, edited by Charles Singer, E. J. Holmyard, A. R. Hall, and Trevor I. Williams (Oxford University, New York, 1958).

Robert A. Millikan, physics teacher

Alfred Romer

To remember a man

We all know the name of Robert A. Millikan. He measured the charge of the electron by a method so simple and direct that it is still described in our textbooks and duplicated in our laboratories, even though as a student experiment it is something of a killer. It pleases us to remember names in the sciences and to link each name with an achievement. We celebrate Ohm and his law, Fresnel and his diffraction patterns, Pauli and his principle, Millikan and his oil drop. All that gets lost is the personality.

Millikan was a personality in the narrow sense of that word. At Cal Tech in the 1930s he bossed his research students, directed the physics department, and ran the college with such evident authority that he was known always as The Chief. Ten years later when people saw the crowd-packed, war-time Radiation Laboratory at M.I.T. they said in awe, "Not even Millikan could manage all those physicists."

This is not quite fair, however. Millikan was a good deal more than a supreme commander. As graduate students, we took him for granted, not realizing what a feat it was for the top administrator of our institution to maintain a personal research program. He came to the Tuesday and Thursday research conferences (as the physics colloquium was called), peppering his comments with the same italics we could read in his printed papers, and we never wondered where he found the time. He was The Chief, but with a special quality which I once heard described by a member of his faculty: "When Millikan says, 'Let's compromise,' he doesn't mean, 'We'll do it my way.' He meets you somewhere in between."

Years later my department head at St. Lawrence, John F. Smith, gave me a glimpse of Millikan at the University of Chicago in 1919-20. He was not the boss then, but rather a working faculty member to whom a graduate student could turn for sympathetic help and advice.

In this article, I want to go back to a still earlier time, before the days of the busy executive, before the cosmic-ray investigations, before the spectra of highly-ionized atoms, before the Nobel Prize work on the photoelectric effect and the establishment of *e*. I should like to tell how Millikan became a physicist, and what he did for the teaching of physics by the textbooks which he wrote.

On the way into physics

Like many physicists of his day, Robert Andrews Millikan was a country boy.[1] He was born and brought up in small towns of the Mississippi River country along the Illinois-Iowa borders, the second child of Silas Franklin Millikan and Mary Jane Andrews. His father was

Alfred Romer *has taught physics at both edges of the country (Whittier College, Vassar College, and St. Lawrence University), he has run summer institutes for PSSC teachers, has served on the editorial boards of both* **The Physics Teacher** *and the* **American Journal of Physics**, *and was for five years Secretary of the AAPT. He has worked in the history of physics, concentrating on the period between 1895 and 1920, from which his book* **The Restless Atom** *is drawn. He was irrevocably marked for physics by a lecture on cosmic rays which he heard Millikan deliver in the fall of 1928, and he knew him at closer, although by no means intimate range, during his graduate student days at Cal Tech from 1929 to 1933. (St. Lawrence University, Canton, New York, 13617)*

Robert Andrews Millikan

Robert Andrews Millikan was born on March 22, 1868 in Morrison, Illinois, and had his schooling in Maquoketa, Iowa, 40 miles to the west. He received his A.B. from Oberlin College (where he began his teaching of physics) in 1891 and his A.M. in 1893. He took his Ph.D. at Columbia University in 1895 and after a year's study in Europe joined the physics department of the University of Chicago with the rank of Assistant. He rose step by step to become Professor in 1910 and remained there until 1921, with the exception of 1917-18 when he served as Vice-chairman of the National Research Council and took a leading part in organizing the scientific community for wartime research. In 1921 he was called to be the chief executive officer of the California Institute of Technology, a post which he assumed not as President but with the more circumscribed title and authority of Chairman of the Executive Council; and to that he added the duties of Director of the Norman Bridge Laboratory of Physics. In 1946 he was succeeded by Lee A. DuBridge who took the title of President. He died in Pasadena on December 19, 1953.

Millikan is best remembered for his oil-drop experiment by which he showed that electric charges come in discrete units and measured the value of that unit. He also verified experimentally the Einstein equation for the photoelectric effect. For these two investigations, he was awarded the Nobel Prize in Physics for 1923. In addition, he initiated important work in ultraviolet spectroscopy and with the cosmic rays. Since he was a superb research director, a great deal of this was carried out and elaborated by his graduate students and associates.

a Congregational minister who moved from church to church and settled when the boy was seven in Maquoketa, Iowa. Although it was not a farm life, there were still imperative chores for the three Millikan boys: two cows to be milked, a vast truck garden to be kept in order, cordwood to be sawed into stove lengths. In spare time there was fishing in the Maquoketa River, there were once three colts to be broken for a neighbor, and for indoor amusement there was a family gymnasium rigged up in the barn.

Expectations are high for a minister's son. Robert Millikan was one of only two boys in the class of fifteen which graduated from the Maquoketa High School in 1885. His father's uncle, Peter Pindar Pease had been the first colonist in Oberlin and one of the founders of Oberlin College.[2] Robert would follow his father and his older brother Allan there, but college required money. To earn money he learned shorthand and took a year's job as court reporter. It was not until 1886 that he entered Oberlin in the Preparatory Department. This was usual. High school graduates of those days were rarely ready for college and colleges operated their own preparatory departments to close the gap. Thanks to his experience in the family barn, the 18-year-old sub-freshman secured the post of Acting Director of the college gymnasium, with a salary which would more than cover his college expenses.

Two years later, when he reached sophomore standing, he was introduced to physics. It was a numbing experience in which what would seem the best possible circum-

Fig. 1. The gymnasium at Oberlin. This is known as the Second Gymnasium. It occupied a one-story frame building about 75 by 25 feet and was used from 1873 to 1900 when it was moved to make room for its successor, Warner Gymnasium.

stances produced the worst possible results. The course was taught from a leading textbook of its day, the *Elementary Text-book of Physics* by William A. Anthony of Cornell and Cyrus F. Brackett of the College of New Jersey[3] (which in another decade would change its name to Princeton University). It was taught by one of the most revered members of the Oberlin faculty, Professor Charles H. Churchill, who had taught mathematics, astronomy, and physics there for 30 years.[4] We can guess what went wrong. Anthony and Brackett's text has many of the virtues which commend it to a teacher. It is clear, concise, complete, and correct; but it is also dull as ditchwater. There is little doubt that Churchill taught it in the style for which it was written, by lecture and recitation, and that he required no more in recitation than that the students repeat back its subject matter: the definitions, the laws, perhaps some derivations. To make matters worse, Millikan took the short course in the winter term which was limited to mechanics only.[5]

When a few months later he was offered a job teaching physics, his considered reply was that he did not know any physics. The occasion was the establishment of a physics course in the preparatory department; the offer came from its Assistant Principal, John F. Peck, who taught Greek in the college and had already sampled Millikan's intellect. "Anyone who can do well in my Greek," he said, "can teach physics," and Robert Millikan gave in.

He located a promising high school text and took it home to Wichita, where his father had recently moved, to study over the summer. The book was *Elementary Physics* by Elroy McKendree Avery.[6] Avery was a former Cleveland school teacher and a future Ohio politician. He was also a prolific author who ranged from science textbooks to local and national history. Millikan was attracted by Avery's generous provision of problems and worked his way through them all. In the fall, when he put the book in the students' hands, he was pleased to discover that physics

Fig. 2. John F. Peck who taught Greek in Oberlin College and at the time of this picture was Assistant Principal of the Preparatory Department. The picture dates from before 1893 when he became Principal, and so must represent him as Millikan knew him.

was a subject "which taught itself."

We all know how far from the truth that is. Nevertheless, once in a blue moon when the material is fresh and the instructor still elated from having mastered it, a class can go soaring. For that brief time there is indeed no more to teaching than holding the book and keeping up with the class.

Turning professional

Evidently things looked as satisfactory from the outside as they felt within. After his graduation in 1891 when jobs were scarce, Oberlin was willing to keep Millikan as Acting Director of the gymnasium and as teacher of preparatory physics with the formal rank of Tutor. For that he would receive what he later called "the extravagant salary of $600 per annum," enough to leave a comfortable surplus toward the education of his younger brother and sisters. Not to be idle, he dug his way through the 672 pages of Silvanus P. Thompson's *Dynamo-electric Machinery,*[7] then persuaded the college to accept that effort as the basis for a master's degree.

Probably the only mistake one can make with gifted students is to be gentle. Millikan had come up swimming when Peck had pushed him into the creek. Now he was tossed into deep water. Columbia University was advertising its graduate fellowships, including one in physics. Millikan's friends on the Oberlin faculty seized this as an ideal opportunity and applied for it in Millikan's name but without bothering to tell him. They were successful, and in keeping with the impersonality of it all, Millikan learned of the award by reading it in the newspaper.

Consequently the fall of 1893 found him at Columbia, which was still on Madison Avenue between 49th and 50th Streets, as the one graduate fellow and the one graduate student in physics. There he completed his Ph.D. in two years with a major in physics and minors in chemistry and electrical engineering. Although he lost his fellowship to campus politics, he made do in his second year with tutoring, and to occupy his time, he spent the summer between the two at the University of Chicago.

At Columbia he took courses from Michael Pupin, at Chicago he studied under A.A. Michelson, and both impressed him greatly. Both had European experience, both had worked in Helmholtz's laboratory in Berlin. Michelson had an established reputation as an experimenter, Pupin had been drilled in mathematical physics at Cambridge. What attracted Millikan was that both were at home in physics. Pupin could lecture without notes, confident that his mathematical dexterity could pull him through. Michelson seemed to have all of optics at his fingertips. In return, Millikan impressed them.

During the spring of 1895, Pupin began to insist that Millikan too must study in Berlin. Millikan had more need of a job; but when the market remained poor and the University of Vermont turned him down for another candidate, he consented to go abroad and to finance the trip with a $300 loan from Pupin, if he might be allowed to pay interest.

Columbia has always had an early calendar. Millikan sailed from New York in May and spent June and July in Jena warming up his German with various university lectures. In August and September he went on a bicycle tour from Dresden to Rome, Naples, and Paris, then a straight run of seven hundred miles in seven days to Berlin. There he spent the fall semester. Helmholtz had died the year before, but he could take courses from Planck and Warburg in theoretical physics, Neesen in thermodynamics, Schwartz in mathematics, and hear Rubens and Kaufmann in the colloquium. In the spring he moved to Göttingen where he took courses under Voight and Felix Klein and did an experimental study in Nernst's laboratory on the dispersion of electromagnetic waves in an artificial medium prepared by suspending droplets of benzene in water.

Then suddenly there were jobs. First came an offer from Oberlin of an appointment at $1600 a year, and when those negotiations were all but complete an unexpected

Fig. 3. The Columbia College building at the corner of Madison Avenue and 49th Street.

Fig. 4. Michael Pupin at a somewhat later age than when he sent Millikan to Europe.

cable arrived from Michelson offering a post at Chicago at just half that salary with the bottom-rung rank of Assistant. It was time to move, and Millikan left for the United States, pledging his trunk in England to the captain of the ship against payment of his fare in New York. Once there he went about disentangling himself from Oberlin and bargaining Chicago up to $900 even while he was accepting the job.

At Oberlin, it is likely that Professor Churchill was about to retire. He did indeed retire a year later at the age of seventy-three, turning over his physics and astronomy to

Fig. 5. Senior members of the staff of the Ryerson Laboratory at the University of Chicago. The picture was taken by Miss Crowe at 1:30 P.M. on June 12, 1908. Carl Kinsley, who had been two years behind Millikan at Oberlin, joined the Chicago physics department in 1902. Photo courtesy A.P. Niels Bohr Library.

C.E. St. John,[8] who had been teaching at the University of Michigan, who had a Harvard Ph.D., and who had been in Berlin the year before Millikan.[9] In the 1930s, St. John would be pointed out to Cal Tech students as a man who having retired after a lifetime of teaching could now do research as an astronomer at Mount Wilson.

Although no one could have foreseen the future more than very dimly, the difference in Millikan's and St. John's careers shows pointedly why Millikan rejected Oberlin for Chicago. It was the choice between the world where physics was taught and the world where physics was practised. Part of the bargain he drove with Michelson and with President Harper was that his teaching load would be limited to six hours a day. For a young man with country upbringing, that would leave another six hours free for research.

Four textbooks

The choice was definite, the planning was provident, it was only events which did not cooperate. For the next dozen years, Millikan's reputation, both within the University and outside, depended more on his accomplishments as a teacher than on his research. That reputation rested on a group of textbooks which were widely used when they were new and which went on being useful for a matter of four decades. Each held its market in its own way, some unaltered over the years, some steadily revised and rewritten. As their success became evident, some of them gave rise to adaptations to meet the needs of different student audiences. The complete set of originals, revisions, and adaptations is listed in the Table with their dates, coauthors, and changes of title.

When Millikan arrived in Chicago in the fall of 1896, he entered what was still a new department. Michelson had come as chairman only in 1892, the Ryerson Laboratory had been opened in 1894, and the departmental program was still developing. Samuel W. Stratton was lecturing to the freshmen and Millikan was put in charge of the laboratory for that course. For Michelson, physics was the science of precise measurements, and he insisted on experiments which would make that point. Millikan (with four years of teaching behind him) had strong opinions about the importance of the laboratory and the close degree to which it should be tied to the class work. To realize all these ambitions required strenuous effort in the invention and redesign of apparatus, a task which Stratton and

TABLE I

Millikan's Texts and Laboratory Manuals

1. *The Theory of Optics* by Paul Drude, trans. by C. Riborg Mann and Robert A. Millikan (Longmans, New York, 1902).
2. *College Course of Laboratory Experiments in General Physics* by S.W. Stratton and Robert A. Millikan (Chicago, 1898).
3. *Mechanics, Molecular Physics, and Heat* by Robert Andrews Millikan (Scott, Foresman, Chicago, 1902; Ginn, Boston, 1903, 1931).
   Mechanics, Molecular Physics, Heat, and Sound by Robert Andrews Millikan, Duane Roller and Earnest Charles Watson (Ginn, Boston, 1937).
   This is currently available in the MIT Paperback Series (MIT Press, Cambridge, 1965).
4. *A Short University Course in Electricity, Sound, and Light* by Robert Andrews Millikan and John Mills (Ginn, Boston, 1908, 1935).
   Optics by Robert A. Millikan, Duane Roller, and Earnest C. Watson (Calif. Inst. of Technology, Pasadena, 1943).
5. *A First Course in Physics* by Robert Andrews Millikan and Henry Gordon Gale (Ginn, Boston, 1906; Commercial Press, Shanghai, 1911).
   A First Course in Physics, rev. ed. by Robert Andrews Millikan and Henry Gordon Gale (Ginn, Boston, 1913).
   Practical Physics by Robert Andrews Millikan, Henry Gordon Gale, and Willard R. Pyle (Ginn, Boston, 1920, 1923).
   Elements of Physics by Robert Andrews Millikan, Henry Gordon Gale, and Willard R. Pyle (Ginn, Boston, 1927).
   New Elementary Physics by Robert Andrews Millikan, Henry Gordon Gale, and James P. Coyle (Ginn, Boston, 1936, 1941, 1944).
6. *A Laboratory Course in Physics for Secondary Schools* by Robert Andrews Millikan and Henry Gordon Gale (Ginn, Boston, 1906).
   A First Course in Laboratory Physics for Secondary Schools by Robert Andrews Millikan, Henry Gordon Gale, and Edwin Sherwood Bishop (Ginn, Boston, 1914).
   Exercises in Laboratory Physics for Secondary Schools by Robert Andrews Millikan, Henry Gordon Gale, and Ira Cleveland Davis (Ginn, Boston, 1925, 1945).
7. *A First Course in Physics for Colleges* by Robert Andrews Millikan, Henry Gordon Gale, and Charles William Edwards (Ginn, Boston, 1928).
   A First Course in Physics for Colleges, rev. ed. by Robert Andrews Millikan, Henry Gordon Gale, and Charles William Edwards (Ginn, Boston, 1938).
8. *A Manual of Experiments to Accompany "A First Course in Physics for Colleges"* by Robert Andrews Millikan, Henry Gordon Gale, and Charles William Edwards (Ginn, Boston, 1930).
9. *Physics* by Glenn Moody Hobbs and Robert Andrews Millikan (Amer. School of Correspondence, Chicago, 1908).
10. *Practical Lessons in Electricity* by Robert A. Millikan, Francis B. Crocker, and John Mills (Amer. Technical Soc., Chicago, 1914).
    Elements of Electricity by Robert A. Millikan and E.S. Bishop (Amer. Technical Soc., Chicago, 1917, 1918, 1920).
11. *Teachers Manual for Practical Physics* by Robert Andrews Millikan, Henry Gordon Gale, and Willard R. Pyle (Ginn, Boston, 1920).
12. *A First Course in Physics: A Review* by Robert Andrews Millikan (Ginn, Boston, 1922).

Millikan shared, and which produced presently a preliminary textbook, the *College Course of Laboratory Experiments in General Physics*.

Stratton was an officer in the Naval Militia. In May of 1898, he was called to active duty in the Spanish-American War and Millikan took over the entire freshman course, which he kept after Stratton's return. In 1901, Stratton left Chicago permanently to become the first Director of the National Bureau of Standards. In 1923 he became President of the Massachusetts Institute of Technology, an office which he held until his death in 1931.[10]

Given a free hand, Millikan developed the course according to his own ideas. He abandoned the formal lecture, relying on recitation and laboratory to cover the ground. He dropped most of the usual descriptive material, producing a lean, analytical course which concentrated on physical principles and which developed the theory with all the mathematics at his students' command. The laboratory was far from a separate enterprise, but was so much of the course that each chapter of the text culminated in an experiment. To one reared on conventional books, *Mechanics, Molecular Physics, and Heat*, the text published in 1903 to cover the first portion of the course, looks like a glorified laboratory manual, but such a judgment does it less than justice. Its topics were not chosen to fit the experiments on hand. The text set out the necessary material and the experiments were invented to support it.

As for the text, the description I have given sounds curiously apt for the book by Anthony and Brackett which Millikan had found intolerable. I have referred to the one as *lean, analytical*, and *concentrated on physical principles*, to the other as *clear, concise, complete, and correct*. Actually they differ quite remarkably, and differ in ways which are explicitly a matter of emphasis and implicitly a matter of personal style. Anthony and Brackett's book flows evenly from topic to topic, never hurrying, never lingering, never in a way raising its voice. Millikan's moves in calculated dashes, in chapters which, even including experiments and problem sets, close off in half a dozen or perhaps a dozen pages. Anthony and Brackett achieved a well-ordered presentation from which students might learn; Millikan intended his material to be understood. Anthony and Brackett wrote like excellent pedagogues; Millikan wrote like a physicist.

The text for the second part of the course did not appear until five years later, in 1908, even though Millikan took on a coauthor to get it written. It was *Electricity, Sound and Light*, and it was laid out on exactly the same plan, providing an experiment to make the point of every chapter. It is rather less concise than *Mechanics, Molecular Physics, and Heat*, some applications have crept in. The choice of John Mills to be its junior author seems odd since his contact with the course was remarkably limited. He went through the University of Chicago as an undergraduate and after a single graduate year there transferred to the University of Nebraska in 1902. After eight years of teaching, he joined the American Telephone and Telegraph Company as an engineer, winding up in administrative posts in the Bell Laboratories, and when he retired, re-entered Millikan's orbit as an administrative assistant at Cal Tech.[11]

Part of the delay in the appearance of *Electricity, Sound and Light* must be blamed on a distraction, the production of a high school text. The Millikan-Stratton-Michelson program at Chicago was designed for students who came prepared in physics, but many arrived from high schools which never offered the subject. Out of his Oberlin background, Millikan had been offering a summer-quarter course in the pedagogy of physics, and through it he became acquainted with ambitious teachers and their problems. There was a need for a high school text, and it was one which he understood how to fill. For a coauthor he recruited Henry G. Gale, a colleague eight years his junior, who had begun his graduate study at Chicago the year in which Millikan entered the department. Their collaboration produced two volumes, both published in 1906, *A First Course in Physics* and *A Laboratory Course in Physics for Secondary Schools*.

These two sets of books had remarkably different histories. Year in and year out, the college texts were reprinted without change, and when the 28 years of their copyrights were ending, they were still valuable enough to have those copyrights renewed. The high school books, on the other hand, were repeatedly revised, retitled, and rewritten. When the time for modernizing the college books finally arrived in the later 1930s, the first volume was drastically rewritten by its two revisers, Duane Roller, then at the University of Oklahoma, and Earnest Charles Watson at Cal Tech. The *Mechanics, Molecular Physics, Heat, and Sound* which they produced in 1937 was a remarkable job, still laid out on the old plan, but with its more rigorous text and its greatly expanded historical references quite as original as Millikan's earlier version had been. Unfortunately the second volume was cut off by the war and never completed.

In spite of their long life, the college texts had remarkably little influence on the textbooks which followed and competed with them. Millikan made less of mechanics than others have cared to do, and more of molecular physics. Although we all pay tribute to atoms and molecules, few of us ground our ideas as Millikan did on the Laws of Definite Proportions and Multiple Proportions. Most of us, even including Millikan's revisers, consider sound a first-semester topic, an offspring of the vibratory motion of mechanics rather than a useful introduction to wave theory. Few of us will begin optics with diffraction, and fewer still will handle images by the deformation of wave fronts rather than the bending of rays. None of us incorporates a laboratory guide in the text.

In these respects, even Millikan and Gale remained independent of Millikan. There were real differences between the college and the high school texts. In both, that intimidating bugaboo, the mechanics of particles, constituted the beginning as logic requires, but in the high school book, the *First Course in Physics*, it occupied a mere 26 pages. At that point, although work and mechanical energy should logically follow, the *First Course* switched to liquids and gases and to the thermal physics of thermometers and expansions. Only then did mechanics reappear to clarify the concept of heat and to introduce energy under circumstances in which it is really conserved. As compared with the college text, the *First Course* abbreviated electricity, and although it still treated light as a wave phenomenon, its emphasis was on images rather than diffraction. In sum, Millikan and Gale brought out a new book, suited to a different audience of students and intended to be taught by a differently qualified group of teachers.

The laboratory

Millikan's texts deserve to be remembered. They held their markets for 30 and 40 years. Yet they contributed to American physics teaching far less than the laboratory experiments which were devised to support them.

Michelson had wanted precise measurement, Millikan had wanted an experiment in every chapter. Out of those requirements came a collection of apparatus which was universally adopted in American colleges. For those who began teaching in the 1930s, it is a nostalgic exercise to turn the pages of *Mechanics, Molecular Physics, and Heat* and of *Electricity, Sound and Light*. The pieces described there are the pieces we found on our shelves as we explored our departmental inheritances, and whether we taught from Stewart or Spinney or Saunders, these were the pieces around which we built our laboratory courses.

It is the mechanics experiments which stand out particularly in memory: the tuning forks to record accelerations on smoked surfaces, whether for a freely falling body or for a heavy disc pulled into rotation by a descending weight; the optical lever we used for Young's modulus; the torsion lathe for shear modulus with its rim-mounted vernier; whirling masses for centripetal force in a piece which we abandoned in spite of its elegant simplicity. We all stroked metal rods with resined cloth to set up standing waves in a Kundt's tube, vibrated strings with battery-driven tuning forks, and produced resonance by adjusting the water level in glass tubes. In electricity we had the earth-inductor, whose coil flipped over on the release of a spring, and the model generator whose coil jerked by 10° increments between the poles of horseshoe magnets to produce a sine curve of emf.

Not all of them of course were Millikan's inventions. At the introductory level everyone borrows. Nevertheless they are the pieces which served in the laboratory for which he was responsible, the laboratory which he felt to be supremely important. Whether he invented them, improved them, or simply took them over, they became his apparatus – and in fact still are. His textbooks vanished in the 1940s, but it would be possible today to re-equip Millikan's Chicago laboratory of 60 years ago with stock items from current catalogs and leave out no more than a handful of experiments.

Millikan wrote his college textbooks for the University of Chicago; the high school books were written for general adoption. For that reason, Millikan and Gale adopted the conventional separation of text and laboratory guide. Once again, good as the text was, the laboratory course was even better. Here the governing conditions were Millikan's insistent commitment to the laboratory and the pervasive poverty of the schools. Out of those conditions came a highly original collection of invented, adapted, and borrowed experiments, 51 all told, for which a single set of apparatus would cost no more than 50 dollars.

It is worth looking at a few to see how ingeniously they cut the cost. The mechanical equivalent of heat was determined by tumbling lead shot from end to end of a stoppered cardboard tube, calculating the work from the accumulated distance of fall and measuring the change of temperature with a thermometer inserted in the mass of shot at the beginning and end. The specimen of air for Boyle's law was trapped in a fine-bore glass tube between a sealed-off end and a long thread of mercury, whose hydrostatic pressure on the air would increase or diminish as the tube was tilted to different angles.

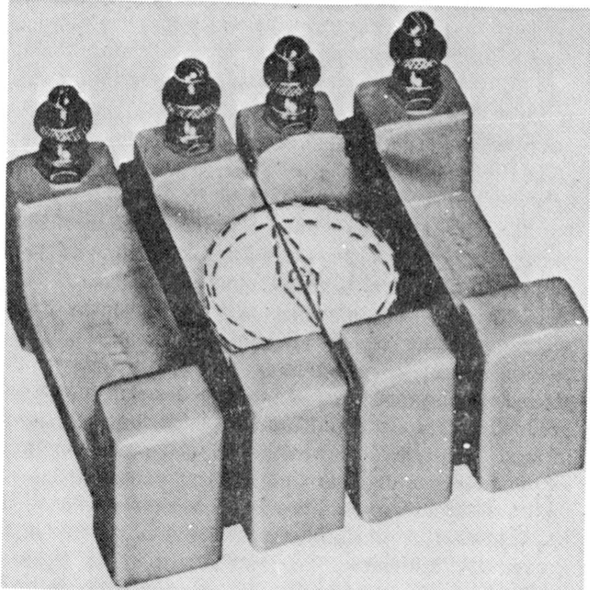

Fig. 6. The galvanoscope, Millikan and Gale's all-purpose instrument for measuring electric currents by the deflection of a compass needle lying within one of the windings on the frame. This is the present-day model with a porcelain base as manufactured by the Sargent-Welch Scientific Company, and shows the longevity of Millikan's laboratory apparatus. As originally designed, the frame was made of wood.

Their electrical experiments required neither ammeters nor voltmeters, but a galvanoscope which did duty for both. As the picture shows, this was a frame with three windings of wire into which a magnetic compass could be slipped. As the sensitivity required, the student might send a current through the single-turn coil, the 25-turn coil, or the 100-turn coil. If necessary, the sensitivity could be fine-tuned by partly withdrawing the compass. For more delicate currents, there was a galvanometer made by hanging an open coil by one of its leads between the poles of a horseshoe magnet of the kind used in the magneto ringers of rural telephone lines. (Either of these instruments could be homemade by a teacher with more time than money.)

Let me quote also the directions for Experiment 38, *The Speed of Sound in Air*. It is interesting because it comes in both an expensive and an inexpensive version. It is interesting also because it gives the flavor of a different era, when high schools were small and elite and a physics teacher could take a class into the open air without disrupting either schedules or discipline.

> *A*. Let the class be divided into two sections and placed exactly a kilometer apart, the distance being measured by laying off fifty times the length of a cord 20 m long. Each group should be provided with a pistol, blank cartridges, and at least one stop watch. Let a member of one group raise and lower a handkerchief three times as a ready signal, and simultaneously with the last lowering let him fire a pistol. Let a member of the other group take with a stop watch the time which elapses between the flash and the report of the pistol. Then let the operations of the two stations be interchanged, in order to elimi-

nate the effect of any wind which may be blowing. In this way take six or more observations, different members of the class timing the interval in turn. Observations which differ badly from the general average and which are evidently the result of awkward handling of the stop watch need not be included in the final mean. From this mean compute the velocity of sound at the temperature of the air.

B. If stop watches are not available, set up a heavy pendulum which beats seconds; attach some white object to it; set up a screen so that the pendulum can be seen only when it is passing the middle point of its swing; let one student stationed near the pendulum pound loudly on some sonorous object at each instant that the pendulum crosses the middle point, and let the class move away until the beats of the hammer appear again to coincide with the passages of the pendulum. The distance from the class to the pendulum is obviously numerically equal to the velocity of sound.

The transition to research

All told the devising of experiments and the writing of textbooks occupied a dozen years of Millikan's life, from his arrival at the University of Chicago in the fall of 1896 to the publication in 1908 of *Electricity, Sound and Light*. That was the year in which he celebrated his fortieth birthday and it might be supposed that his pattern of life was fairly well fixed. His textbooks were becoming successful, the personal research he had been doing, although undistinguished, was adequate in supplying theses for graduate students.

One of those doctoral projects which was in train in 1908 was the repetition and improvement of H.A. Wilson's experiment for determining the charge carried by the ions of a gas. This was in the hands of Louis Begeman, Professor of Physics at Iowa State Teachers College in Cedar Rapids.[12] He used a cloud chamber to surround the ions with water drops and derived the value of the charge from two rates of settling of the clouds, slowly under gravity or more rapidly under the action of an electric field. Preliminary work in the summer of 1907 had shown that conditions could be stabilized so as to make successive clouds reasonably comparable.[13] By the spring of 1910 Begeman had refined his techniques to give reproducible and presumably accurate values which averaged to 4.668×10^{-10} electrostatic units for the unit of ionic charge.[14] (In those units, the modern value would be 4.8029×10^{-10}.) This was far higher than any of the early experiments had given and agreed well with the value of 4.65×10^{-10} which Rutherford and Geiger had obtained by counting alpha particles.[15]

Meanwhile, in the summer of 1909, Millikan had tried reversing the electric field so that it pulled the drops upward rather than down. He found it difficult to hold and study clouds by this method but remarkably easy to hold and follow single drops, and the value for the ionic charge he got from the first few dozen drops he followed agreed well with those which Begeman's careful observations on clouds were giving.[16]

Millikan had been doing research since his Columbia days, but with these drops which rose and fell across the field of his observing microscope, he came alive. His ingenuity, his originality, and his energy (all of which he possessed in large quantity) poured into the problem. It was not simply a question of doing the experiment, but of improving it, polishing it, and bringing it to precision. It was a matter of organizing his resources, of planning all the subsidiary work which must support the oil-drop measurements: the determination of the viscosity of air, the proper law of fall for the drops, the study of the Brownian motion they executed.[17] There were graduate students to do these things, but not without his active supervision since the accuracy of his work depended completely on the accuracy of theirs. Millikan never lost his devotion to teaching, but this is where The Chief as I knew him began.

Acknowledgments

I am greatly indebted to W.E. Bigglestone, Archivist of Oberlin College, for locating the photographs of John F. Peck and of the Oberlin gymnasium, and for much helpful information about Oberlin College. I also thank the Office of Public Information of Columbia University for the pictures of the Madison Avenue building and of Michael Pupin, the Center for the History of Physics of the American Institute of Physics for the picture of the Chicago physics faculty from the W.F. Meggers Collection of the Niels Bohr Library, and the Sargent-Welch Scientific Company for the picture of their current model of the galvanoscope.

References

1. This article is based primarily on Millikan's own reminiscences. Material not otherwise credited is taken from the *Autobiography of Robert A. Millikan* (Prentice-Hall, New York, 1950).
2. R.S. Fletcher, A History of *Oberlin College* (Oberlin College, 1943) pp. 102-107, 665-667.
3. First published as *Elementary Text-book on Physics* (New York, Wiley and Sons, 1884), covering only mechanics and heat. It was reprinted as *Elementary Text-book of Physics* in 1885. The 3rd edition, which added magnetism and electricity, sound, and light, appeared in 1887 and was reprinted in 1888, 1890, and 1891. It was revised in 1897 by William Francis Magie and this edition was reprinted in 1900 and 1908.
4. Fletcher, reference 2, p. 692.
5. *Oberlin College Catalogue For the Year 1888-89*, p. 82.
6. E. M. Avery, *Elementary Physics* (Cleveland, 1876). This was reissued by the American Book Co. and by Sheldon and Co. in 1897.
7. S. P. Thompson, *Dynamo-electric Machinery, a Manual for Students of Electrotechnics*, 3rd ed. (E. and F.N. Spon, London, 1888).
8. G. F. Wright, "Professor Charles Henry Churchill," *The Oberlin Review* June 23, 1897, pp. 449-451.
9. *American Men of Science*, ed. by J. M. Catell (Science Press, New York, 1906), p. 277.
10. *Who Was Who in America*, Vol. 1 (Marquis, Chicago, 1942), p. 1196.
11. *Who Was Who in America*, Vol. 2 (Marquis, Chicago, 1950), p. 375.
12. *Who Was Who in America*, Vol. 3 (Marquis, Chicago, 1960), pp. 62-63.
13. R. A. Millikan and L. Begeman, Phys. Rev. **26**, 197 (1908).
14. L. Begeman, Phys. Rev. **30**, 131 (1910); **31**, 41 (1910).
15. E. Rutherford and H. Geiger, Proc. Roy. Soc. A **81**, 141 (1908).
16. R. A. Millikan, Phys. Rev. **29**, 560 (1909).
17. R. A. Millikan, *The Electron* (Univ. of Chicago Press, 1917).

The two maps
Oersted Medal Response at the joint American Physical Society–American Association of Physics Teachers meeting, Chicago, 22 January 1980

Gerald Holton
Jefferson Physical Laboratory, Harvard University, Cambridge, Massachusetts 02138
(Received 28 August 1980; accepted 26 September 1980)

Mr. President, ladies and gentlemen. I thank you warmly for the award of the Oersted Medal. The names of the previous medalists, from Edwin Hall, Robert A. Millikan, and Arnold Sommerfeld on to our day, form a kind of roll call of our particular Mount Olympus. Each additional entry must necessarily feel more overwhelmed. But there is at least one thing in common: I looked at some of the Responses presented in the past and found that precedent encourages personal reflection. Allow me to adopt this mode today, first to share a serious concern, and then to acknowledge some pleasant debts.

I. THE OERSTED INVISIBILITY

The various efforts mentioned in your gracious citation stemmed from the idea that science should treasure its own history, that historical scholarship should treasure science, and that the full understanding of each is deficient without the other. If I were asked to indicate the chief motivation behind this once unusual, not to say perverse, idea, I would have to do so in the form of a stark statement that can be supported, on this occasion, only sketchily. It is this: At a time when passionate unreason around the globe challenges the fate of Western culture itself, the sciences and the history of their development remains perhaps the best testimony to the potential of mankind's effective reasoning. Therefore, if we do not trouble ourselves to understand and proudly claim our own history, we shall not have done full justice to our responsibility as scientists and as teachers.

A good way to illustrate this point is to glance at the work of the man whose name graces the award. Even among physicists, Hans Christian Oersted (1777–1851) is little known, although an argument can be made that he was a modern kind of scientist. He announced his great discovery, that a magnet needle can be deflected by what was then called a galvanic current, on July 21, 1820 in a broadside which he had privately printed and distributed to the foremost scientists of Europe.[1] The publication led to a veritable explosion of scientific work, for example, the great discoveries in electromagnetism by Ampére and Faraday. Technical applications followed quickly also, starting with an electric telegraph.

Oersted, then 43 years of age, was professor of physics at Copenhagen. An autodidact, he had first studied pharmacy and languages, and also contributed to chemistry and other sciences. He was an early evolutionist in biology, a man with wide interests in science and outside. Among his works are some on the relation of science to poetry and to religion, in which he shows himself to be a gentle but persuasive rationalist.

Oersted saw his chief task to be the discovery of the unities in nature. Far from having stumbled accidentally on the fact that a magnetic field surrounds currents, as the popular myth still has it, he had sought for years for the effect which practically everyone else, during the two decades of widespread experimentations with voltaic cells from 1800 on, had missed. At least eight years before his discovery of 1820, he had declared his faith that light and heat and chemical affinity, as well as electricity and magnetism, are all "different forms of one primordial power," and he had announced that attempts must be launched "to see if electricity has any action on the magnet." Oersted had been deeply impressed by the philosophical works of Immanuel Kant where he found the argument that all physical experiences are due to one force (*Grundkraft*). He also accepted many views of his friend Friedrich W. J. Schelling, a leading exponent of *Naturphilosophie*, who provided an enthusiastic program to find the unity of all natural phenomena, and specifically the unity of physical forces, in such proposals as: "For a long time it has been said that magnetic, electrical, chemical, and finally even organic phenomena would be interwoven in one great association .... This great association, which a scientific physics must set forth, extends over the whole of nature."

R. C. Stauffer, in whose paper the quotation from Schelling is given, also cites the response of Mme. de Staël, on concluding her discussion with him on *Naturphilosophie*, that "systems which aspire to the explanation of the universe cannot be analyzed at all clearly by any discourse: words are not appropriate to ideas of this kind, and the result is that, in order to make them serve, one spreads over all things the darkness which preceded creation, but not the light which followed." Be that as it may, the lack of Cartesian clarity did not impede the influence of *Naturphilosophie* on Oersted (or for that matter, on Ampére, Faraday, Julius Robert Mayer, and others).

The thematic presupposition of the unity of forces led Oersted to look for the connection between electricity and magnetism through a convincing experiment. His success identifies him as the modern initiator of the grand unification program—that great source of motivating energies of modern physics, with a direct genetic influence on Faraday, Maxwell, Einstein, and on to this year's Nobel Prize winners. In the physics meetings we are having here, Oersted, at least in spirit, would surely have felt very much at home.

Moreover, Oersted's apparatus for demonstrating the unity of nature's phenomena was eminently sensible, at least in terms of his presuppositions. He sent progressively larger electric current through high-resistance platinum wires. First the wire got hot, then the wire began to give off light, and then he saw the effect on the magnet needle. Some day I hope to look at Oersted's laboratory books in Denmark to see if he also looked for a gravitational effect as Faraday did later; it would have been a likely extension in the thinking of a unifier.

In his later years, Oersted dedicated himself to many social causes, such as the freedom of the press, and to science education. One of his biographers notes that when Oersted died in 1851, the students arranged for a torch-light cortege in a huge funeral procession in which 20 000 people are said to have participated. (I count on the American Association of Physics Teachers, in due course, to make corresponding arrangements for its Oersted Medalists.)

Now there is no doubt that Oersted's advance in physics changed history in two ways. It opened up physics itself to a succession of unifying theories and discoveries without which the modern state of our science would be unthinkable. And his key discovery, embodied in every electric motor, also triggered the engineering advances that have produced the modern technological landscape. One would expect such a person to be visible in works of history. Yet, despite his role in the initiation of vast changes in science, engineering, and through it our very society, Oersted is virtually absent from both science books and history texts, not least those which young persons typically encounter in school. If one looks into historical encyclopaedias for what happened in 1820, one finds a plethora of events of a quite different sort. For example, Karl Ploetz's compendium of chief dates of history notes for 1820 that "Austrian troops reestablish order in Italy"; and the big *Encyclopaedia of World History*, edited by William Langer, records for 1820 that King George III, having earlier been declared insane, dies.

Let me dwell a little on this phenomenon, which we may call the *Oersted Invisibility*, for we are getting to a significant truth about the influences that shape the intellectual formation of young people throughout the world. William Langer's widely used *Encyclopaedia* was not meant to be a synthetic work and does not represent the more sophisticated work of scholars in history today. But the two main concerns that animate the book are still alive in the teaching of history as most young people encounter it in their formative years: the chronological presentation of historically important "facts" and the "periodization" of the sequence of facts into labeled categories. As to the first, Langer says in his introduction it is his function to provide "a handbook of historical fact"—as it turns out, chiefly political, military and diplomatic history. He explains that for him to have gone also into other achievements such as science would have taken him too "far afield." In my copy (2nd edition, 1948) there *is* a section entitled "Scientific Thought and Progress," precisely 2-1/4 pages out of the total 1270—or less than 0.2% (which happens to be better than the proportion of fundamental science support in our GNP). As an attempt to put some order into the chaos of facts, periodization of history into segments entitled the "Age of ," the "House of ," the "Revolutionary Period," etc., provides a seemingly well-bounded shape to the various fragments of time. But it encourages only occasionally a comment that parts the dark curtain behind which historians are debating. (Was Thucydides right in considering the war of 431–421 B.C. and the war of 414–404 B.C. to be in reality one period, that of the Peloponnesian War? How far back do the causes of the revolutionary outbreaks of A.D. 1848 go?)

The net effect, at least on most young minds, must be that the purpose of history appears to be the provision of a well-labeled place for every miscreant if only his factual mischief was on a big enough scale, and to give some space to every ruler, whether effective and beneficient or not. In such books, genealogies of the mighty abound, from the Ch'in Dynasty and the succession of the Merovingian kings to the sequence of Czars and Presidents. One is reminded of Voltaire's complaint that "for the last 1400 years the only persons in Gaul apparently have been kings, ministers, and generals."

Scientists and their works are virtually taboo. If you look for Newton in my copy of Langer, you will find him on p. 431, under the heading "Third Parliament of William III": "Isaac Newton, master of the mint." The four-word description is not what you or I might have chosen; but at least he is mentioned, which is more than can be said for Galileo or Kepler or even Copernicus. (Others have done worse: Arnold Toynbee's list of "creative individuals," from Xenophon to Lenin and Hindenburg, included not a single scientist.)

On the other hand, Langer's book does give a detailed survey of, say, the vicissitudes and successions of the Ottoman emperors, from Bayazid II, remembered by Langer for being "the least significant of the first ten sultans," through the exploits of Selim I ("The Grim") and Selim II ("The Sot," described as an "indolent ruler, much given to drink"), and so on through the last of them, Mohammed IV, who is described as a boy of ten, followed by a period of anarchy. It might almost sound amusing, until one asks what the costs were for the unmentioned mass of humanity doomed to be born, to live, and to die in the dark back alleys of history.

Every holy war gets its place in Langer's and so many books like it. There is a tiny admixture of humane figures, a Marcus Aurelius, a Jefferson, or a Ghandi; but they are lost in the succession of genocidal maniacs, from Emperor Tiglath III of Babylon, who innovated the idea of consolidating his conquests by "deporting entire populations," to Josef Stalin. As John Locke asked: "What were those conquerors but the great butchers of mankind?" Moreover, from one development to the next, any rational conclusion or extrapolation is seemingly hopeless. No wonder that even many of today's historians, in the words of J. H. Plumb, "have taken refuge in the meaninglessness of history."[2] Thus the Ploetz compendium—which was the model for Langer, and, as it happened, also the book I used in high school in Vienna—started on a promising note, from a rational point of view, with the entry for 19 July 4241 B.C. as the precise date the good Dr. Ploetz proposed for the introduction of the calendar in Egypt. But it, too, rapidly went downhill through the whole sorry list of massacres and delusions, and ended with an entirely unprophetic but appropriate entry, for 27 September 1934, which ran as follows: "Declaration by England, France, and Italy, guaranteeing the integrity and independence of Austria." I can well believe the famous story about another boy, some 15 years earlier but in a school not far from mine. "You are a clever chap," the history teacher said to Viki Weisskopf, "but you don't know any dates!" "Oh, I do," he replied. "I know all the dates; I just don't remember what happened on them."

II. THE BIFURCATED MAPS

I have of course been a little hard on William Langer, a distinguished diplomatic historian, who later became a valued colleague of mine, and who, perhaps partly because of my teasing, did put a good deal more science into the last edition he edited. But the main point I wish to make still stands: wherever their schooling may take place, young people encounter, through the historically oriented set of courses as usually taught, a view of the accomplishments and destiny of mankind that almost celebrates the role of passionate unreason. And if the student is one of the relatively few who also takes substantial science courses, he or she will encounter from that direction a very different picture of mankind's interests and attainments. Indeed, the opposition between these stories is so great that it must seem to many students that there are really two different species involved.

Let me put it in terms of an image from my own schoolroom in Vienna, an image surely not qualitatively different from the one you carry in your mind, or which, *mutatis mutandis*, is being found even today in your town's school. The curriculum at our Gymnasium emphasized history, literature, and ancient languages, a triad that merged into one message. Latin and Caesar's wars; Greek and the Iliad; medieval German and the Nibelungenlied; the chief tragedies of German theater, the Bible, the Edda sagas, and the painstaking probing of the ever-unfinished sequence of historic battles: I must confess it was a powerful and blood-stirring brew—but with only occasional traces of the rational processes.

On the other hand, in our science classes we encountered an entirely different universe. Here was the finished and apparently unchanging product of distant and largely anonymous personages, unchallengeable monuments to their inexorable rationality—but with only occasional traces of historic development. Just as the historian neglects science—Richard Hofstadter said of the historian "he may not disparage science, but he despairs of it"—the scientist is silent about history—the record, a scientist put it to me once, of the errors of the forgotten dead.

It was a cultural schizophrenia which I could not formulate, but which I also could not dismiss. I can capture it best by recalling two very different maps that were hanging in front of my class, to stare at and wonder about, year after year. You probably saw them, too. On the left side was always a geopolitical map of Europe and Asia. (America and most of Africa were evidently on some other planet.) In some storage closet there must have been a whole set of such maps because the left map was regularly changed, a new one for each period, and with each change we students could see the violent, spasmodic, unpredictable pulsation of shapes and colors in the wake of the thrilling story of conquests. On the right side was a very different map—the Periodic Table of the elements: the very embodiment of empirical, testable, reliable, and ordered sets of truths. That map was never changed, although there was a rumor in the benches that some of the blank spaces were being filled in, and even that a previous student of that very Gymnasium, a man named Wolfgang Pauli, had shown that the different features of the table were the consequence of some underlying great idea, rather than some accident on which nature had settled.

To the mind of the child exposed to these two maps, the utter differences between them and thus between the subjects they represented were so profound as to seem unbridgable. On the left side, the forces shaping history were the four horsemen of the Apocalypse. On the right side the forces shaping science were, to use modern terms, the four forces of physics. To the young mind, it seemed like a division that demanded some sort of decision. We know from the autobiographical notes of scientists that this dichotomy can help to focus a career choice. In Einstein's famous remarks, it is the choice, in early youth, between a world "dominated by wishes, hopes, and primitive feelings," and on the other hand, "this huge world which exists independently of us human beings," the contemplation of which "beckoned like a liberation." It is only much later in life that it dawns on such a student at what great costs this separation was being nurtured, that these two kinds of destiny are in fact intertwined, that these two developments stem from two potentials within the same person.

In the meantime, however, students in our classes were left to wonder what the moral point of this bifurcated education would be. Concerning our personal destinies there was in fact a sharp division of opinion between our parents, on one side, and our history books as well as those few teachers who had any interest in discussing such matters, on the other. The parents' theory was that the purpose of the curriculum was chiefly to prepare for the school's final examination without which one could not go on to university. However, our physical education teacher spoke explicitly for the view that presented a different scenario for us: We would be there to fight one day, to revenge the loss of our territories in World War I, and of our honor in the treaties of Versailles, Trianon, and St. Germain. We were being readied to change that map on the left once more.[3]

The great historians we studied seemed, from that second point of view, useful preparations. For example, the father of historical writing, Herodotus of Athens in the Age of Pericles, declares the aim of his great book on the history of Greece to preserve, as he puts it, "the great and wonderful actions of the Greeks and the barbarians [from] losing their due meed of glory." What he means is set forth right at the start. A certain Gyges, upon being made King of Sardis, "made inroads on Miletus and Smyrna .... Afterwards, however, although he reigned for 38 years, he did not perform a single [further] noble exploit. I shall therefore make no further mention of him, but pass on to his son and successor . . . Ardys. This Ardys took Priene, and he made war upon Miletus." Our class quickly got the idea that here was an altogether more glorious type. Even better was his son Alyattes who, Herodotus says, had inherited from his father that war with Miletus, and who "performed other actions very worthy of note." Herodotus tells us one of these: In warfare, "he cut down and utterly destroyed all the trees and all the corn throughout the land and then returned to his own dominion .... The reason that he did not demolish also the buildings was that the inhabitants might be tempted to use them as homesteads from which to go forth to sew and till their land; and so each time that Alyattes invaded the country he might find something to plunder." Indeed, worthy of note.

Turning to a more recent work, we found that Georg Wilhelm Friedrich Hegel, in his great *Philosophy of History*, struggled with the meaning of history and concluded that "the final cause of the world at large, we allege to be the consciousness of its own freedom on the part of spirit, and *ipso facto* the reality of that freedom." I must confess that this formula, when presented in history class, was not easy to unpuzzle; but it had a nice ring to it. History as the evolution of freedom seemed an appealing idea. But just at that point in our studies, this train of thought was deprived of a good deal of its credibility when one Friday evening in

March, the portion of the map of Europe showing Austria turned suddenly brown, and our history teacher, like many teachers in the other subjects, turned up in Nazi regalia on the following Monday.

Hegel also had said, quite correctly, that the most profoundly shaping experience on the mind of the West was Greece of the Homeric Period. The reading of the *Iliad* was meant to be our most permanent memory; I must confess I like it best of all our classes. Just for this reason let me use it further to crystallize my point. You recall the early scene on the beach before Troy. Agamemnon, leader of the Greeks who are laying siege, has lost his own mistress, robs Achilles of his, and thereby launches 1000 pages of dactylic hexameter. Agamemnon explains himself in a way which Homeric Greeks accepted: "Not *I* was the cause of this act, but Zeus." It is the habit and prerogative of the Gods to put *atē* (temptation, infatuation, a clouding of consciousness) into a person's understanding. "So what could I do? The Gods will always have their way." Achilles, the outraged victim of Agamemnon's action, takes the same view. Without Zeus, he agrees, Agamemnon "would never have persisted in rousing the *thumos* (passionate, violent response) in my own chest."

In fact, throughout the *Iliad*, temptations and infatuations, *atē*, and passionate and violent response, *thumos*, are the chief forces motivating human action. As E. R. Dodds explains in the fine book *The Greeks and the Irrational*, *atē* is, in fact, a partial, temporary insanity, "and, like all insanity, it is ascribed not to physiological or psychological causes, but to the external, 'daemonic' agency." Only very rarely does one encounter in the *Iliad* a glimpse of rationality—dispassionate reason sensitive to long-range consequences—as a motivating force. This role falls to Athena, the goddess of good counsel, who occasionally intervenes in the brawl and carnage (even at the risk of conflict of interest, since on the side she is also a deity of war). There is the famous scene in Book I when Athena catches Achilles by the hair and warns him not to strike Agamemnon. She is visible to Achilles alone and persuades him to put back his sword. For a moment, there is sanity. But on the whole, the glorious poem takes place against the background of a dark, archaic world dominated by will and force, dreams and oracles, blood lust and atonement, portents and magical healing, orgiastic cults, superstitious terror, and the obbligato of reckless massacres—the world governed by the irrational self, the *thumos*, the strong force that surfaces today as tribalism, racism, and the longing for combat.

After the Homeric Age—rather precipitously, as history goes—the landscape does change. It is the period identified as the Greek Elightenment, during which there was, in a phrase Dodds quotes approvingly, a "progressive replacement of mythological by rational thinking among the Greeks." The rise of this first enlightenment, in the 6th century B.C., coincides with the first stirrings of Western science as we understand it.

There were bound to be enthusiastic mistakes; for example, Protagoras, an early Sophist, has gone down in history as perhaps the first optimist who thought that virtue could be taught, that history could be cured, that intellectual critique alone could rid us of "barbarian silliness" and lift us to a new level of human life. But at least his intention was right. Science, and rational thought which produced science, are beachheads in the soul that otherwise would be largely given over to *atē* and *thumos*. The existence of such beachheads allows one to hope for a change in the balance of potentials in the individual and, therefore, in the balance of forces that have raged over the geopolitical map since prehistoric times.

I submit that this fact defines, today more than ever, an essential part of the task of all who claim to be teachers. To neglect it is to invite peril. The subsequent history of Greece itself reinforces this point. Around 430 B.C., at the end of the reign of Pericles and the beginning of the Peloponnesian War which finished with Athens' surrender and Sparta's triumph, that first Age of Enlightenment gave way. Teaching astronomy, or expessing disbelief in the supernatural, now could have grave consequences. There followed some thirty years of trials for heresy, with victims such as Anaxagoras, Socrates, Protagoras, perhaps also Euripides, and of course an unknown number of less prominent ones. In the long series of conflicts between reason and passion, Athena, the Weak Force, had lost.

One of the causes for this turn of events appears to be that, from the late period of Plato onward, intellectuals situated themselves not *in* but *beside* society. Dodds writes: "As the intellectuals withdrew further into a world of their own, the popular mind was left increasingly defenseless . . . ; and, left without guidance, a growing number relapsed with a sigh of relief into the pleasures and comforts of the primitive." Greece saw again a great rise in cults, in magical healing practices, in astrology, and other familiar symptoms. It ushered in the long decline or, to give its proper name, the "Return of the Irrational." It was as if the bicameral mind had become aware of its rational strength—and been frightened by the possibility of freedom from the death dance of history, and freedom from the external gods to whom one could spin off the responsibility for the excesses of *atē* and *thumos*. (I am quite aware that a group of philosophers, from Nietzsche to Spengler, from Husserl to our day, puts the blame on the contrary on science itself, on what they call "excesses" of the rational. Their arguments are saved from dismissal as utter absurdity by the unhappy fact that science, too, has frequently lent itself to be a weapon in the service of our Dionysian and antihuman drives.)

The decline of Greece shows parallels with our present predicament. We are at the tail end of the second experiment with rationalism, the fruit of the scientific revolution of the 17th century and the era of enlightenment that followed. That schoolroom I described, and all the others across the globe, may really have functioned as a trap designed to keep us from escaping a destiny that is archaic except for its modern, much larger scale. The most persuasive evidence that the human mind has the power to progress, individually and cumulatively, from ignorance and confusion to sensible, testable, shareable world conceptions—the kind of triumph of man's rational potential of which his science is eloquent testimony—was all but sabotaged by the method of presentation, itself an institutionalization of our reluctance to honor the imperatives of sound reason. Indeed, the very facts of science which we had to memorize seemed the work of some deity that plants them in its passive victim, even as Zeus planted the *thumos* in Agamemnon. Never once was the liberating idea presented to us that the findings and very methods of science are the results of an historical process by which mere humans seek sense and expose nonsense, and that the potential for this process is in each of us.

III. HELPING ATHENA

All that was long ago. Many have worked hard on improving the effectiveness of educators. But the admonition of Max Weber, in his magisterial "Science as a Vocation," has become no less urgent. Rational thought, he reminded his audience, has the moral function of leading to self-clarification, of helping the individual "to give himself an account of the ultimate meaning of his own conduct," and so to be able better to make decisive choices. Without it, life "knows only of an unceasing struggle of these various gods with one another . . . , the ultimately possible attitudes toward life [remain] irreconcilable, and hence their struggle can never be brought to a final conclusion." Today, as we watch the reign of the irrational in world affairs, I find it hard to believe that we have succeeded in making a qualitative change for the better in the products coming from our classrooms. Moreover, in our century the would-be conquerors who are writing themselves into the pages of the new history books have learned how to hire and use for their bloody work students coming from our science classes. It is an ominous conjunction of science and history, making it that much more difficult—and essential—to hold on to the vision that the trajectory of history *can* bring us to a time when, at long last, the goddess Athena in our very soul wins out consistently over the dark passions.

This ancient aim (which Werner Jaeger, in his great work *Paideia: The Ideal of Greek Culture,* identified as the chief hope of education in antiquity) should now be a special concern of those who, through their own life's work, have learned how one distinguishes between fact and delusion, between the demands of eternal law and of internal longing; of those who care most to find out how things work and cohere. The immense authority of, say, an Oersted came of course from the painstaking and repeatable demonstration of a beautiful and useful discovery. The history of science can show that might comes from being right, rather than, as in the rest of history, more often than not the converse. Bringing science and history together in that kind of conjunction—in scholarly research and in the classroom, for scientists and for nonscientists—is one effective way to enlarge the beachhead of reason.

There are others, there have to be others; but it has been my motivating (and perhaps now no longer quite so perverse) view that this way helps focus a young mind exactly on the point where the confrontation should take place, between the habits responsible for the kaleidoscopic sequence of follies on one side, and, on the other, the kind of passionate yet sane thinking that shaped the development and testing of, say, the Table of the elements or of elementary particles.

I have no illusion that more chemistry, more physics, more mathematics will, by themselves and soon, produce wise leaders and wise followers. Protagoras *was* too simple. There is no quick cure for the barbarian silliness within us which shows up so grotesquely in the acts of the present-day Agamemnons, generalissimos, premiers, shahs, and ayatollahs. We must also not be misunderstood to be defending some inevitable benignancy, purity, and progressiveness of science. Paradise will not come upon invoking the name and deeds of Mendeleyev or Wolfgang Pauli. And yet, enough of us must act nevertheless as if something of this sort can happen eventually; for otherwise it will not change. As we, and the future teachers who pass through our hands, face those young students, there is the opportunity to assert and demonstrate the rational powers of Athena as a complementary and balancing element in the productive life of the human spirit.

The risk of failure is high in all educational efforts, and there are always other, more immediately rewarding things one might do instead. But it is a risk worth taking. For those young students in the schoolroom, wondering about the forces that grip the world map and soon caught up themselves in its convulsions—those were you and I; those are now our children; and those should not have to be our children's children, forever.

ACKNOWLEDGMENTS

In retrospect, I am impressed how greatly most of the things I have tried to do depended at crucial points on the willingness of other persons to take a risk on my behalf. This has certainly been true in my work in education, which is the chief subject of this response. Because we are in a time of retrenchment with respect to the support of education, attention should be drawn to the continuing need, especially for those who can afford to do it, to take risks on behalf of people and ideas. On this appropriate occasion, I must mention a few to whom I am especially indebted in this regard.

At Wesleyan University, where I went for my senior year, Professor Walter G. Cady let me be his research assistant in crystal physics, and Professor Vernet Eaton his teaching assistant, thereby giving me my first jobs at a point when my qualifications for these must have been far less obvious than my need and ignorance. Not long afterwards, at Harvard, the same gamble was taken first by Professor Edwin C. Kemble, whose Oersted Medal citation some years ago might well have expanded on his role of providing a splendid example for those of us who taught under and with him while at the same time giving full freedom to develop our own ideas. Professor P. W. Bridgman then agreed to let me do my thesis in his laboratory even though, or perhaps because, the problem I proposed (molecular relaxation at high pressures) was really outside his current work, and even though it threatened to introduce electronic equipment into what had been essentially a dc laboratory.

The next example of risk taking was by a company called Addison-Wesley, then a job printer in Cambridge, which had decided to become a science book publisher. Their new editor, Warren Blaisdell, took a chance when he signed me up for a physics text based on the storyline provided by the history of physics at a time when this was regarded as a very unusual, not to say perverse, idea. And sure enough, when the book was published in 1952, there was for a time only one adoption (mine). Luckily for the company's solvency, the other two authors on their first list of physics texts were Francis Sears and Herbert Goldstein.

In the early 1960s, I had the good luck to meet James Rutherford, then working on his doctorate, and together with Fletcher Watson I encouraged him in his plan to do a physics book that would bring to schools the same point of view about science that had motivated my college text. A little money was needed to free Jim to work on this. His inquiry about funding received a brusk but at least quick refusal from the National Science Foundation—as it happens, from the very same division that Jim has been heading since 1977. Happily, we quickly found persons in the Carnegie Corporation who were willing to take the risk. Not long afterwards, and now in good part at the urging of the NSF, this pilot effort was greatly expanded in 1964 and became the Project Physics Course development, in which literally hundreds of scientists and teachers helped us over an agonizing and exhausting five-year period. Even then, however, it would not have survived the repeated threats to its funding without the personal intervention of a few friends of the Project, above all I. I. Rabi.

The person who next took a risk upon himself was Vincent Alexander who, as head of the School Science Department at Holt, Rinehart and Winston Publishing Company, recommended commercial publication of the course materials. In consequence, since December 1970, a wide range of printed and other components of the course have been in use in U.S. high schools, as well as colleges, and also abroad in local adaptations from Japan and Australia to Brazil and Italy. It is a real pleasure to watch the flowering of these results, though as every gardener knows, it brings with it continuing labor. The third revised U.S. edition is being printed now, and on returning from this meeting I shall probably find a stack of page proofs waiting on my desk.

Let me close my acknowledgment of the risks others have taken on behalf of ideas for which you have kindly honored me by mentioning here the role of Elmer Hutchisson. When he was the director of the American Institute of Physics, a predecessor of William Koch, he encouraged the bold step of starting, in the AIP building itself, an archive, library, and research effort dedicated to the history of recent and current physics. This became the Niels Bohr Library and Center for the History of Physics, now in the able hands of Spencer Weart and Joan Warnow. It was the model for similar facilities set up since by other scientific societies. But there were long periods when the continued existence of the Center was in doubt from one meeting of the AIP Governing Board to the next, and the decision seemed to depend on whether my Eastern Airline shuttle would make it through the snow, or whether a few early friends of the idea, such as Fred Seitz, would stay to the end of the Board meeting. But all that is now passed. And thanks to an endowment fund drive initiated by Emmanuel Piore and continued by Fred Seitz—to whom I urge you to send your contributions!—we shall be able to assure the continued life of a valuable resource for documenting the rise and achievement of our profession.

[1]Shortly thereafter, he also added that the action between magnet needle and current loop is reciprocal, a discovery usually associated with Ampére, who independently published it later. For good accounts of Oersted's work and motivation, see L. Pearce Williams' account in the *Dictionary of Scientific Biography;* Kristine Meyer, "The Scientific Life and Works of H. C. Ørsted," in *H. C. Ørsted, Naturvidenskabelige Skrifter,* edited by K. Meyer (Host, Copenhagen, 1920), Vol. 1, pp. XIII–CLXVI; Robert C. Stauffer, "Speculation and Experiment in the Background of Oersted's Discovery of Electromagnetism," Isis **48,** 33–50 (1957); Bern Dibner, *Oersted and the Discovery of Electromagnetism* (Burndy, Norwalk, 1961); and Barry Gower, "Speculation in Physics: The History and Practice of Naturphilosophie," *Studies in History and Philosophy of Science* (Pergamon, New York, 1973), Vol. 3, pp. 301–356.

[2]For a brilliant analysis of the changing tasks of historians, see Judith N. Shklar, "Learning without Knowing," Daedalus **109**(2), 53–72 (1980), from which the last two quotations are taken.

[3]For a faithful description of the system of education in a Gymnasium of Vienna by an exact contemporary, see Egon Schwarz, *Keine Zeit für Eichendorff: Chronik unfreiwilliger Wanderjahre* (Athenäum Verlag, Königstein, 1979), pp. 11–18.

The Authors

Carl D. Anderson discovered the positron in 1932 and later participated in the discovery of the muon. His Nobel Prize dates from 1936. He has spent his academic life, both as student and as professor, at the California Institute of Technology.

Lawrence Badash is a professor of history at the University of California in Santa Barbara.

Walter H. Brattain won the 1956 Nobel Prize, with John Bardeen and William B. Shockley, for the invention and development of the transistor. He was one of several prominent physicists to have studied at Whitman College, and he returned there as adjunct professor on his retirement from Bell Telephone Laboratories.

Sandford C. Brown (1915–1981) was professor and associate dean at MIT, and was treasurer of AAPT 1958–1967. His definitive biography of Benjamin Thompson, Count Rumford, appeared in 1979.

S. Chandrasekhar is Distinguished Professor both in Astronomy and Astrophysics and in Physics at the University of Chicago. His Nobel Prize was awarded in 1983.

Arthur H. Compton (1892–1962) was awarded the Nobel Prize (1927) for discovering the effect which bears his name. His professional career was divided between Washington University in St. Louis and the University of Chicago.

Edward U. Condon (1902–1974) was associated with several universities, including Princeton, Washington University and the University of Colorado. His career also included directing industrial research laboratories at Westinghouse and at Corning Glass Works. From 1945 to 1951 he was the Director of the Bureau of Standards in Washington, D.C. He was president of AAPT 1964–65.

David M. Dennison (1900–1976) was professor of physics at the University of Michigan in Ann Arbor. His best known research was in the field of molecular spectra.

Samuel Devons has recently retired as professor of physics at Columbia University. His interest in the history of physics has extended to the production of historical films and the development of a historical laboratory for students.

Werner Heisenberg (1901–1976) was one of the chief architects of the quantum theory, best known for his uncertainty principle. He was successively professor at the University of Leipzig, director of the Kaiser Wilhelm Institute in Berlin, and, after the war, director of the Max Planck Institute in Gottingen.

Gerald Holton is Malinckrodt Professor of Physics and Professor of the History of Science at Harvard University. He was recently president of the History of Science Society but is probably best known to physics teachers for his leadership in developing Project Physics, a course for secondary schools.

Marjorie Malley took her PhD in history (history of science) at UC Berkeley, after a BS in physics and philosophy at MIT and an MAT at Harvard.

John J. McCarthy taught at St. John's University.

Albert E. Moyer took his doctorate at the University of Wisconsin and is now in the history department of the Virginia Polytechnic Institute and State University in Blacksburg.

J. Rud Nielsen (1894–1979) was born in Denmark but took his PhD at the California Institute of Technology and spent his professional life at the University of Oklahoma.

Thomas H. Osgood, English by birth, spent the greater part of his professional career at Michigan State University. He was editor of the *American Journal of Physics* from 1948 to 1958.

Melba Phillips, emeritus professor of the University of Chicago, was president of AAPT 1966–67.

Alfred Romer is emeritus professor of St. Lawrence University. He is the author of a history of radioactivity as well as *The Restless Atom*. From 1966 to 1971 he was secretary of AAPT.

Sidney Rosen is Professor Emeritus of Astronomy at the University of Illinois, Urbana-Champaign.

R.S. Shankland (1908–1982) was Ambrose Professor of Physics at Case Western Reserve University. In addition to his repeated analyses of the Michelson-Morley experiments he wrote extensively on architectural acoustics.

Esther B. Sparberg is professor of chemistry at Hofstra University.

Barbara Swartz is a historian who has been in a position to help keep history straight in *The Physics Teacher* through her husband, its long time editor.

George Paget Thomson (1892–1976) was professor at the University of Aberdeen and at the Imperial College of Science, and Master of Corpus Christi College, Cambridge. He was the son of J.J. Thomson, and shared the Nobel Prize with C.J. Davisson for demonstrating the wave nature of the electron.

E.C. Watson (1892–1970) was on the faculty at Cal Tech from 1919 until his retirement in 1962. Over the years he contributed to AJP more than eighty "reproductions of prints, drawings and paintings of interest in the history of physics," in addition to historical articles.